D1295445

A First Course in Probability and Markov Chains

A First Course in Probability and Markov Chains

Giuseppe Modica

and

Laura Poggiolini

University of Firenze, Italy

A John Wiley & Sons, Ltd., Publication

This edition first published 2013

Registered office
John Wiley & Sons Ltd, The Atrium, Southern Gate, Chichester, West Sussex, PO19 8SQ, United Kingdom

For details of our global editorial offices, for customer services and for information about how to apply for permission to reuse the copyright material in this book please see our website at www.wiley.com.

Library of Congress Cataloging-in-Publication Data

Modica, Giuseppe.
A first course in probability and Markov chains / Giuseppe Modica and Laura Poggiolini.
pages cm
Summary: "A first course in Probability and Markov Chains presents an introduction to the basic elements in statistics and focuses in two main areas" – Provided by publisher.
Includes bibliographical references and index.
ISBN 978-1-119-94487-4 (hardback)
1. Markov processes. I. Poggiolini, Laura. II. Title.
QA274.7.M63 2013
519.2′33 – dc23

2012033463

A catalogue record for this book is available from the British Library.

ISBN: 978-1-119-94487-4

Set in 9.5/12pt Times by Laserwords Private Limited, Chennai, India.

Printed and bound in Singapore by Markono Print Media Pte Ltd

Contents

Preface

This book collects topics covered in introductory courses in probability delivered by the authors at the University of Florence. It aims to introduce the reader to typical structures of probability with a language appropriate for further advanced reading. The attention is mainly focused on basic structures.

There is a well established tradition of studies in probability due to the wide range of possible applications of related concepts and structures in science and technology. Therefore, an enormous amount of literature on the subject is available, including treatises, lecture notes, reports, journal papers and web pages. The list of references at the end of this book is obviously incomplete and includes only references used directly in writing the following pages. Throughout this book we adopt the language of measure theory (relevant notions are recalled in the appendices).

The first part of the book deals with basic notions of combinatorics and probability calculus: counting problems and uniform probability, probability measures, probability distributions, conditional probability, inclusion–exclusion principle, random variables, dispersion indexes, independence, and the law of large numbers are also discussed. Central limit theorem is presented without proof. Only a basic knowledge of linear algebra and mathematical analysis is required.

In the second part we discuss, as a first example of stochastic processes, Markov chains with discrete time and discrete states, including the Markov chain Monte Carlo method, and we introduce Poisson process and continuous time Markov chains with finite states. For this part, further notions in mathematical analysis (summarized in the appendices) are required: the Banach fixed point theorem, systems of linear ordinary differential equations, powers and power series of matrices.

We wish to thank all the students who have attended our courses. We also wish to thank our colleagues Matteo Focardi, Mariano Giaquinta, Paolo Maria Mariano, Andrea Mennucci, Marco Romito and Enrico Vicario who helped us with suggestions and comments. Special gratitude goes to Enrico Vicario for many helpful discussions on the applications.

Our special thanks also go to the editorial staff of Wiley for the excellent quality of their work.

We have tried to avoid misprints and errors. However, we would be very grateful to be notified of any errors or misprints and would be glad to receive any criticism or comments. Our e-mail addresses are:

giuseppe.modica@unifi.it laura.poggiolini@unifi.it

We will try to keep up an errata corrige at the following web pages:

```
http://www.dma.unifi.it/~modica
http://www.dma.unifi.it/~poggiolini
http://www.dmi.unifi.it/~modica
http://www.dmi.unifi.it/~poggiolini
```

1

Combinatorics

Combinatorics deals with the cardinality of classes of objects. The first example that jumps to our minds is the computation of how many triplets can be drawn from 90 different balls. In this chapter and the next we are going to compute the cardinality of several classes of objects.

1.1 Binomial coefficients

1.1.1 Pascal triangle

Binomial coeffcients are defined as

$$\binom{n}{k} := \begin{cases} 1 & \text{if } k = 0, \\ 0 & k > n, \\ \dfrac{n(n-1)\dots(n-k+1)}{k!} & \text{if } n \geq 1 \text{ and } 1 \leq k \leq n. \end{cases}$$

Binomial coefficients are usually grouped in an infinite matrix

$$\mathbf{C} := (\mathbf{C}^n_k), \ \ n, k \geq 0, \qquad \mathbf{C}^n_k := \binom{n}{k}$$

called a *Pascal triangle* given the triangular arrangement of the nonzero entries, see Figure 1.1. Here and throughout the book we denote the entries of a matrix (finite or infinite) $\mathbf{A} = (a^i_j)$ where the superscript i and the subscript j mean the ith row and the jth column, respectively. Notice that the entries of each row of $\mathbf{C}$ are zero if the column index is large enough, $\mathbf{C}^i_j = 0\ \forall i, j$ with $j > i \geq 0$. We also recall the *Newton binomial formula*,

$$(1+z)^n = \binom{n}{0} + \binom{n}{1}z + \dots + \binom{n}{n}z^n = \sum_{k=0}^{n} \binom{n}{k} z^k = \sum_{k=0}^{\infty} \binom{n}{k} z^k.$$

A First Course in Probability and Markov Chains, First Edition. Giuseppe Modica and Laura Poggiolini.

$$\begin{pmatrix}
1 & 0 & 0 & 0 & 0 & 0 & 0 & 0 & 0 & 0 & \dots \\
1 & 1 & 0 & 0 & 0 & 0 & 0 & 0 & 0 & 0 & \dots \\
1 & 2 & 1 & 0 & 0 & 0 & 0 & 0 & 0 & 0 & \dots \\
1 & 3 & 3 & 1 & 0 & 0 & 0 & 0 & 0 & 0 & \dots \\
1 & 4 & 6 & 4 & 1 & 0 & 0 & 0 & 0 & 0 & \dots \\
1 & 5 & 10 & 10 & 5 & 1 & 0 & 0 & 0 & 0 & \dots \\
1 & 6 & 15 & 20 & 15 & 6 & 1 & 0 & 0 & 0 & \dots \\
1 & 7 & 21 & 35 & 35 & 21 & 7 & 1 & 0 & 0 & \dots \\
1 & 8 & 28 & 56 & 70 & 56 & 28 & 8 & 1 & 0 & \dots \\
1 & 9 & 36 & 84 & 126 & 126 & 84 & 36 & 9 & 1 & \dots \\
\dots & \dots & \dots & \dots & \dots & \dots & \dots & \dots & \dots & \dots & \dots
\end{pmatrix}$$

Figure 1.1 Pascal matrix of binomial coefficients $(\mathbf{C}^n_k)$, $k, n \geq 0$.

Thus formula can be proven with an induction argument on n or by means of Taylor formula.

1.1.2 Some properties of binomial coefficients

Many formulas are known on binomial coefficients. In the following proposition we collect some of the simplest and most useful ones.

Proposition 1.1 *The following hold.*

(i) $\binom{n}{k} = \dfrac{n!}{k!(n-k)!} \quad \forall k, n,\ 0 \leq k \leq n.$

(ii) $\binom{n}{k} = \dfrac{n}{k}\binom{n-1}{k-1} \quad \forall k, n,\ 1 \leq k \leq n.$

(iii) Stifel formula $\binom{n}{k} = \binom{n-1}{k} + \binom{n-1}{k-1} \quad \forall k, n,\ 1 \leq k \leq n.$

(iv) $\binom{n}{j}\binom{j}{k} = \binom{n}{k}\binom{n-k}{j-k} \quad \forall k, j, n,\ 0 \leq k \leq j \leq n.$

(v) $\binom{n}{k} = \binom{n}{n-k} \quad \forall k,\ 0 \leq k \leq n.$

(vi) the map $k \mapsto \binom{n}{k}$ *achieves its maximum at* $k = \lfloor \frac{n}{2} \rfloor$.

(vii) $\displaystyle\sum_{k=0}^{n} \binom{n}{k} = 2^n \ \forall n \geq 0.$

(viii) $\displaystyle\sum_{k=0}^{n} (-1)^k \binom{n}{k} = \delta_{0,n} = \begin{cases} 1 & \textit{if } n = 0, \\ 0 & \textit{if } n \neq 0. \end{cases}$

(ix) $\binom{n}{k} \leq 2^n \ \forall k, n, 0 \leq k \leq n.$

Proof. Formulas (i), (ii), (iii), (iv) and (v) directly follow from the definition. (vi) is a direct consequence of (v). (vii) and (viii) follow from the Newton binomial formula $\sum_{k=0}^{n}\binom{n}{k}z^k = (1+z)^n$ choosing $z = 1$ and $z = -1$. Finally, (ix) is a direct consequence of (vii).

Estimate (ix) in Proposition 1.1 can be made more precise. For instance, from the Stirling asymptotical estimate of the factorial,

$$\frac{n!}{n^n e^{-n}\sqrt{2\pi n}} \to 1 \qquad \text{as } n \to \infty$$

one gets

$$(2n)! = 4^n n^{2n} e^{-2n}\sqrt{4\pi n}\,(1+o(1)),$$

$$(n!)^2 = n^{2n} e^{-2n} 2\pi n\,(1+o(1)),$$

so that

$$\binom{2n}{n} = \frac{4^n}{\sqrt{\pi n}}\frac{1+o(1)}{1+o(1)}$$

or, equivalently,

$$\frac{\binom{2n}{n}}{\frac{4^n}{\sqrt{\pi n}}} \to 1 \qquad \text{as } n \to \infty. \tag{1.1}$$

Estimate (1.1) is 'accurate' also for small values of n. For instance, for $n = 4$, one has $\binom{8}{4} = 70$ and $4^4\frac{1}{\sqrt{\pi 4}} \simeq 72.2$.

1.1.3 Generalized binomial coefficients and binomial series

For $\alpha \in \mathbb{R}$ we define the sequence $\binom{\alpha}{n}$ of *generalized binomial coefficients* as

$$\binom{\alpha}{n} := \begin{cases} 1 & \text{if } n = 0, \\ \dfrac{\alpha(\alpha-1)(\alpha-2)\cdots(\alpha-n+1)}{n!} & \text{if } n \geq 1. \end{cases}$$

Notice that $\binom{\alpha}{k} \neq 0\ \forall k$ if $\alpha \notin \mathbb{N}$ and $\binom{\alpha}{k} = 0\ \forall k \geq \alpha+1$ if $\alpha \in \mathbb{N}$. The power series

$$\sum_{n=0}^{\infty}\binom{\alpha}{n}z^n \tag{1.2}$$

is called the *binomial series*.

Proposition 1.2 (Binomial series) *The binomial series converges if* $|z| < 1$ *and*

$$\sum_{n=0}^{\infty}\binom{\alpha}{n}z^n = (1+z)^\alpha \qquad \textit{if } |z| < 1.$$

Proof. Since

$$\frac{\left|\binom{\alpha}{n+1}\right|}{\left|\binom{\alpha}{n}\right|} = \frac{|\alpha - n|}{|n+1|} \to 1 \qquad \text{as } n \to \infty,$$

it is well known that $\sqrt[n]{|a_n|} \to 1$ as well; thus, the radius of the power series in (1.2) is 1.

Differentiating n times the map $z \mapsto (1+z)^\alpha$, one gets $D^n(1+z)^\alpha = \alpha(\alpha - 1)\cdots(\alpha - n + 1)(1+z)^{\alpha-n}$, so that the series on the left-hand side of (1.2) is the McLaurin expansion of $(1+z)^\alpha$.

Another proof is the following. Let $S(z) := \sum_{n=0}^\infty \binom{\alpha}{n} z^n$, $|z| < 1$, be the sum of the binomial series. Differentiating one gets

$$(1+z)S'(z) = \alpha S(z), \qquad |z| < 1,$$

hence

$$\left(\frac{S(z)}{(1+z)^\alpha}\right)' = \frac{(1+z)S'(z) - \alpha S(z)}{(1+z)^{\alpha+1}} = 0.$$

Thus there exists $c \in \mathbb{R}$ such that $S(z) = c\,(1+z)^\alpha$ if $|z| < 1$. Finally, $c = 1$ since $S(0) = 1$.

Proposition 1.3 *Let $\alpha \in \mathbb{R}$. The following hold.*

(i) $\displaystyle\binom{\alpha}{k} = \frac{\alpha}{k}\binom{\alpha-1}{k-1} \qquad \forall k \geq 1.$

(ii) $\displaystyle\binom{\alpha}{k} = \binom{\alpha-1}{k} + \binom{\alpha-1}{k-1} \qquad \forall k \geq 1.$

(iii) $\displaystyle\binom{-\alpha}{k} = (-1)^k\binom{\alpha+k-1}{k} \qquad \forall k \geq 0.$

Proof. The proofs of (i) and (ii) are left to the reader. Proving (iii) is a matter of computation:

$$\binom{-\alpha}{k} = \frac{-\alpha(-\alpha-1)\cdots(-\alpha-k+1)}{k!} = (-1)^k\frac{\alpha(\alpha+1)\cdots(\alpha+k-1)}{k!}$$
$$= (-1)^k\binom{\alpha+k-1}{k}.$$

A few negative binomial coefficients are quoted in Figure 1.2.

1.1.4 Inversion formulas

For any N, the matrix $\mathbf{C}_N := (\mathbf{C}^n_k)$, $n, k = 0, \ldots, N$, is lower triangular and all its diagonal entries are equal to 1. Hence 1 is the only eigenvalue of $\mathbf{C}_N$

$$\begin{pmatrix}
1 & 0 & 0 & 0 & 0 & 0 & 0 & 0 & 0 & 0 & \dots \\
1 & -1 & 1 & -1 & 1 & -1 & 1 & -1 & 1 & -1 & \dots \\
1 & -2 & 3 & -4 & 5 & -6 & 7 & -8 & 9 & -10 & \dots \\
1 & -3 & 6 & -10 & 15 & -21 & 28 & -36 & 45 & -55 & \dots \\
1 & -4 & 10 & -20 & 35 & -56 & 84 & -120 & 165 & -220 & \dots \\
1 & -5 & 15 & -35 & 70 & -126 & 210 & -330 & 495 & -715 & \dots \\
1 & -6 & 21 & -56 & 126 & -252 & 462 & -792 & 1\,287 & -2\,002 & \dots \\
1 & -7 & 28 & -84 & 210 & -462 & 924 & -1\,716 & 3\,003 & -5\,005 & \dots \\
1 & -8 & 36 & -120 & 330 & -792 & 1\,716 & -3\,432 & 6\,435 & -11\,440 & \dots \\
1 & -9 & 45 & -165 & 495 & -1\,287 & 3\,003 & -6\,435 & 12\,870 & -24\,310 & \dots \\
\dots & \dots & \dots & \dots & \dots & \dots & \dots & \dots & \dots & \dots & \dots
\end{pmatrix}$$

Figure 1.2 The coefficients $\binom{-n}{k}$.

with algebraic multiplicity N. In particular $\mathbf{C}_N$ is invertible, its inverse is lower triangular, all its entries are integers and its diagonal entries are equal to 1.

Theorem 1.4 *For any* $n, k = 0, \dots, N$, $(\mathbf{C}_N^{-1})^n_k = (-1)^{n+k}\binom{n}{k}$.

Proof. Let $\mathbf{B} := (\mathbf{B}^k_n)$, $\mathbf{B}^k_n := (-1)^{n+k}\binom{n}{k}$ so that both $\mathbf{B}$ and $\mathbf{C}_N\mathbf{B}$ are lower triangular, i.e. $(\mathbf{C}_N\mathbf{B})^n_k = 0$ if $0 \le n < k$. Moreover, (iv) and (viii) of Proposition 1.1 yield for any $n \ge k$

$$\begin{aligned}
(\mathbf{C}_N\mathbf{B})^n_k &= \sum_{j=1}^{N}\binom{n}{j}(-1)^{j+k}\binom{j}{k} = \sum_{j=k}^{n}(-1)^{j+k}\binom{n}{j}\binom{j}{k} \\
&= \binom{n}{k}\sum_{j=k}^{n}(-1)^{j+k}\binom{n-k}{j-k} = \binom{n}{k}\sum_{i=0}^{n-k}(-1)^i\binom{n-k}{i} \\
&= \binom{n}{k}\delta_{0,n-k} = \delta_{n,k}.
\end{aligned}$$

A few entries of the inverse of the matrix of binomial coefficients are shown in Figure 1.3. As a consequence of Theorem 1.4 the following *inversion formulas* hold.

Corollary 1.5 *Two sequences* $\{x_n\}$, $\{y_n\}$ *satisfy*

$$y_n = \sum_{k=0}^{n}\binom{n}{k}x_k, \qquad \forall n \ge 0$$

if and only if

$$x_n = \sum_{k=0}^{n}(-1)^{n+k}\binom{n}{k}y_k, \qquad \forall n \ge 0.$$

$$
\begin{pmatrix}
1 & 0 & 0 & 0 & 0 & 0 & 0 & 0 & 0 & 0 & \dots \\
-1 & 1 & 0 & 0 & 0 & 0 & 0 & 0 & 0 & 0 & \dots \\
1 & -2 & 1 & 0 & 0 & 0 & 0 & 0 & 0 & 0 & \dots \\
-1 & 3 & -3 & 1 & 0 & 0 & 0 & 0 & 0 & 0 & \dots \\
1 & -4 & 6 & -4 & 1 & 0 & 0 & 0 & 0 & 0 & \dots \\
-1 & 5 & -10 & 10 & -5 & 1 & 0 & 0 & 0 & 0 & \dots \\
1 & -6 & 15 & -20 & 15 & -6 & 1 & 0 & 0 & 0 & \dots \\
-1 & 7 & -21 & 35 & -35 & 21 & -7 & 1 & 0 & 0 & \dots \\
1 & -8 & 28 & -56 & 70 & -56 & 28 & -8 & 1 & 0 & \dots \\
-1 & 9 & -36 & 84 & -126 & 126 & -84 & 36 & -9 & 1 & \dots \\
\dots & \dots & \dots & \dots & \dots & \dots & \dots & \dots & \dots & \dots & \dots
\end{pmatrix}
$$

Figure 1.3 The inverse of the matrix of binomial coefficients $\mathbf{C}_N^{-1}$.

Similarly,

Corollary 1.6 *Two N-tuples or real numbers* $\{x_n\}$ *and* $\{y_n\}$ *satisfy*

$$y_n = \sum_{k=n}^{N} \binom{k}{n} x_k, \qquad \forall n,\ 0 \le n \le N,$$

if and only if

$$x_n = \sum_{k=n}^{N} (-1)^{n+k} \binom{k}{n} y_k, \qquad \forall n,\ 0 \le n \le N.$$

1.1.5 Exercises

Exercise 1.7 *Prove Newton binomial formula:*

(i) *directly, with an induction argument on n;*

(ii) *applying Taylor expansion formula;*

(iii) *starting from the formula* $D((1+z)^n) = n(1+z)^{n-1}$.

Exercise 1.8 *Differentiating the power series, see Appendix A, prove the formulas in Figure 1.4.*

Solution. Differentiating the identity $\sum_{k=0}^{\infty} z^k = \frac{1}{1-z}$, $|z| < 1$, we get for $|z|<1$

$$\sum_{k=0}^{\infty} kz^k = z \sum_{k=0}^{\infty} D(z^k) = zD\Big(\sum_{k=0}^{\infty} z^k\Big) = zD\Big(\frac{1}{1-z}\Big) = \frac{z}{(1-z)^2};$$

Let $z \in \mathbb{C}$, $|z| < 1$, and $n \in \mathbb{Z}$. We have the followings.

(i) $$\sum_{k=0}^{\infty} kz^k = \frac{z}{(1-z)^2},$$

(ii) $$\sum_{k=0}^{\infty} k^2 z^{k-1} = D\Big(\frac{z}{(1-z)^2}\Big) = \frac{1+z}{(1-z)^3},$$

(iii) $$\sum_{k=0}^{\infty} \binom{n}{k} z^k = (1+z)^n,$$

(iv) $$\sum_{k=0}^{\infty} k\binom{n}{k} z^k = nz(1+z)^{n-1},$$

(v) $$\sum_{k=0}^{\infty} k^2\binom{n}{k} z^k = nz(1+nz)(1+z)^{n-2}.$$

Figure 1.4 The sum of a few series related to the geometric series.

$$\begin{aligned}\sum_{k=0}^{\infty} k^2 z^{k-1} &= \sum_{k=0}^{\infty} D(kz^k)\\ &= D\Big(\sum_{k=0}^{\infty} kz^k\Big) = D\Big(\frac{z}{(1-z)^2}\Big) = \frac{1+z}{(1-z)^3}.\end{aligned}$$

Moreover, for any non-negative integer n, differentiating the identities

$$\sum_{k=0}^{\infty} \binom{n}{k} z^k = (1+z)^n \qquad \text{and} \qquad \sum_{k=0}^{\infty} \binom{-n}{k} z^k = (1+z)^{-n}$$

for any $|z| < 1$, we get

$$\begin{aligned}\sum_{k=0}^{\infty} k\binom{n}{k} z^k &= z\sum_{k=0}^{n} D\Big(\binom{n}{k} z^k\Big)\\ &= zD\Big(\sum_{k=0}^{n} \binom{n}{k} z^k\Big) = zD((1+z)^n) = nz(1+z)^{n-1};\end{aligned}$$

$$\begin{aligned}\sum_{k=0}^{\infty} k^2\binom{n}{k} z^k &= z\sum_{k=0}^{n} D\Big(k\binom{n}{k} z^k\Big) = zD\Big(\sum_{k=0}^{n} \binom{n}{k} z^k\Big)\\ &= zD(nz(1+z)^{n-1}) = nz(1+nz)(1+z)^{n-2}.\end{aligned}$$

1.2 Sets, permutations and functions

1.2.1 Sets

We recall that a *finite set* A is an *unordered* collection of *pairwise different* objects. For example, the collection A hose objects are 1,2,3 is a finite set which we denote as $A = \{1, 2, 3\}$; the collection 1,2,2,3 is not a finite set, and $\{1, 2, 3\}$ and $\{2, 1, 3\}$ are the same set.

If A is a finite set with n objects (or elements), we may enumerate the elements of A so that $A = \{x_1, \dots, x_n\}$. Therefore, for counting purposes, we can assume without loss of generality that $A = \{1, \dots, n\}$. The number n is the *cardinality* of A, and we write $|A| = n$.

Proposition 1.9 *Let A be a finite set with n elements, $n \geq 1$. There are $\mathbf{C}^n_k = \binom{n}{k}$ subsets of A with k elements.*

Proof. Different proofs can be done. We propose one of them. Let $d_{n,k}$ be the number of subsets of A with k elements. Obviously, $d_{n,1} = n$ and $d_{n,n} = 1$. For $2 \leq k \leq n - 1$, assume we have n football players and we want to select a team of k of them, including the captain of the team. We may proceed in the following way: first we choose the team of k-players: $d_{n,k}$ different teams can be selected. Then, among the team, we select the captain: k different choices are possible: so there are $kd_{n,k}$ ways to select the team and the captain. However, we can proceed in another way: first we choose the captain among the n players: there are n different possible choices. Then we choose $k - 1$ players among the remaining $n - 1$ players: there are $d_{n-1,k-1}$ possible choices. Thus

$$d_{n,k} = \frac{n}{k} d_{n-1,k-1}$$

which by induction, gives

$$\begin{aligned} d_{n,k} &= \frac{n}{k}\frac{n-1}{k-1}\cdots\frac{n-k+2}{2}d_{n-k+1,1} \\ &= \frac{n}{k}\frac{n-1}{k-1}\cdots\frac{n-k+2}{2}\frac{n-k+1}{1} = \binom{n}{k}. \end{aligned}$$

1.2.2 Permutations

Let N be a finite set and let n be its cardinality. Without loss of generality, we can assume $N = \{1, 2, \dots, n\}$. A *permutation* of N is an injective (and thus one-to-one) mapping $\pi : N \to N$. Since composing bijective maps yields another bijective map, the family of permutations of a set N is a group with respect to the composition of maps; the unit element is the identity map; this group is called the *group of permutations of* N. It is denoted as S_n or $\mathcal{P}_n$. Notice that $\mathcal{P}_n$ is a not a commutative group if $n \geq 3$.

Each permutation is characterized by its *image-word* or *image-list*, i.e. by the n-tuple $(\pi(1), \dots, \pi(n))$. For instance, the permutation $\pi \in \mathcal{P}_6$ defined by

$\pi(1) = 2$, $\pi(2) = 3$, $\pi(3) = 1$, $\pi(4) = 4$, $\pi(5) = 6$ and $\pi(6) = 5$ is denoted as

$$\begin{pmatrix} 1 & 2 & 3 & 4 & 5 & 6 \\ 2 & 3 & 1 & 4 & 6 & 5 \end{pmatrix}.$$

or, in brief, with its image-word 231465.

The set of permutations of $N = \{1 \ldots, n\}$ has $n!$ elements,

$$|\mathcal{P}_n| = n!$$

In fact, the image $\pi(1)$ of 1 can be chosen among n possible values, then the image $\pi(2)$ of 2 can be chosen among $n-1$ possible values and so on. Hence

$$|\mathcal{P}_n| = n(n-1)(n-2)\cdots 3 \cdot 2 \cdots 1 = n!$$

1.2.2.1 Derangements

Let $\pi \in \mathcal{P}_n$ be a permutation of $N = \{1, \ldots, n\}$. A point $x \in N$ is a *fixed point* of π if $\pi(x) = x$.

We now compute the cardinality d_n of the set $\mathcal{D}_n$ of *permutations without fixed points*, also called *derangements*.

$$\mathcal{D}_n := \Big\{\pi \in \mathcal{P}_n \,\Big|\, \pi(i) \neq i \ \forall i \in N\Big\}.$$

Proposition 1.10 *The cardinality of $\mathcal{D}_n$ is*

$$d_n = n! \sum_{j=0}^{n} (-1)^j \frac{1}{j!} \qquad \forall n \geq 1.$$

Proof. If a permutation of N has j fixed points, $0 \leq j \leq n$, then it is a derangement of the other $n-j$ points of N. Thus, a permutation with j fixed points splits as a couple: the set of its fixed points and a derangement of $n-j$ points. There are $\binom{n}{j}$ different choices for the j fixed points and d_{n-j} derangements of the remaining $n-j$ points, so that, the possible permutations of N with exactly j fixed points are $\binom{n}{j} d_{n-j}$ (where $d_0 = 1$). Thus $|\mathcal{P}_n| = \sum_{j=0}^{n} \binom{n}{j} d_{n-j}$ $\forall n \geq 1$, i.e.

$$n! = \sum_{j=0}^{n} \binom{n}{j} d_{n-j} \qquad \forall n \geq 0. \tag{1.3}$$

The inversion formula of binomial coefficients, see Corollary 1.5, reads

$$d_n = \sum_{j=0}^{n} (-1)^{(n+j)} \binom{n}{j} j! = n! \sum_{j=0}^{n} \frac{(-1)^j}{j!} \qquad \forall n \geq 0.$$

0, 0, 1, 2, 9, 44, 265, 1 854, 14 833, 133 496, 1 334 961, 14 684 570, ...

Figure 1.5 *From the left, the numbers* $d_0, d_1, d_2, d_3, \dots$ *of derangements of* $0, 1, 2, 3, \dots$ *points.*

Corollary 1.11 *The number* d_n *of derangements of n points is the nearest integer to* $n!/e$.

Proof. The elementary estimate between the exponential and its McLaurin expansion gives

$$\left| e^x - \sum_{j=0}^{n} \frac{x^j}{j!} \right| \leq \frac{|x|^{n+1}}{(n+1)!}, \qquad \forall x \leq 0;$$

hence for $x = -1$ we get

$$\left| \frac{1}{e} - \sum_{j=0}^{n} \frac{(-1)^j}{j!} \right| \leq \frac{1}{(n+1)!},$$

so that, from Proposition 1.10 one gets

$$\left| d_n - \frac{n!}{e} \right| = n! \left| \sum_{j=0}^{n} \frac{(-1)^j}{j!} - \frac{1}{e} \right| \leq \frac{n!}{(n+1)!} = \frac{1}{n+1} \leq \frac{1}{3}$$

for each $n \geq 2$.

Figure 1.5 contains the first elements of the sequence $\{d_n\}$.

1.2.3 Multisets

Another interesting structure is an unordered list of elements taken from a given set A. This structure is called a *multiset* on A. More formally, a *multiset* on a set A is a couple (A, a) where A is a given set and $a : A \to \mathbb{N} \cup \{+\infty\}$ is the *multiplicity function* which counts 'how many times' an element $x \in A$ appears in the multiset. Clearly, each set is a multiset where each object has multiplicity 1. We denote as $\{a^2, b^2, c^5\}$ or $a^2b^2c^5$ the multiset on $A := \{a, b, c\}$ where a and b have multiplicity 2 and c has multiplicity 5. The cardinality of a multiset (A, a) is $\sum_{x \in A} a(x)$ and is denoted by $|(A, a)|$ or $\#(A, a)$. For instance, the cardinality of $a^2b^2c^5$ is 9.

If B is a subset of A, then B is also the multiset (A, a) on A where

$$a(x) = \begin{cases} 1 & \text{if } x \in B, \\ 0 & \text{if } x \notin B. \end{cases}$$

Given two multisets (B, b) and (A, a), we say that (B, b) is *included* in (A, a) if $B \subset A$ and $b(x) \leq a(x)$ $\forall x \in B$. In this case, $(B, b) = (A, \widehat{b})$ where

$$\widehat{b}(x) = \begin{cases} b(x) & \text{if } x \in B, \\ 0 & \text{if } x \notin B. \end{cases}$$

Proposition 1.12 *Let A be a finite set, $|A| = n$. Let (A, a) be a multiset on A and let k be a non-negative integer such that $k \leq a(x)$ $\forall x \in A$. The multisets included in (A, a) with k elements are*

$$\binom{n+k-1}{k}.$$

Proof. Let $A = \{1, \dots, n\}$. A multiset S of cardinality k included in (A, a) contains the element 1 x_1 times, the element 2 x_2 times, and so on, with $x_1 + x_2 + \cdots + x_n = k$. Moreover, the n-tuple $(x_1, \dots, x_n)$ characterizes S. We can associate to a n-tuple $(x_1, \dots, x_n)$ the binary sequence

$$\underbrace{00\dots0}_{x_1}1\underbrace{00\dots0}_{x_2}1\dots1\underbrace{00\dots0}_{x_{n-1}}1\underbrace{00\dots0}_{x_n} \tag{1.4}$$

where the symbol 1 denotes the fact that we are changing the element of A. This is a binary word of length $n + k - 1$ with k zeroes.

The correspondence described above is a one-to-one correspondence between the set of multisets of cardinality k included in (A, a) and the set of binary words of length $n + k - 1$ with k zeroes. There are exactly

$$\binom{n+k-1}{k}$$

different words of this kind, so that the claim is proven.

1.2.4 Lists and functions

Given a set A, a *list* of k objects from the set A or a *k-word* with symbols in A is an ordered k-tuple of objects. For instance, if $A = \{1, 2, 3\}$, then the 6-tuples (1,2,3,3,2,1) and (3,2,1,3,2,1) are two different 6-words of objects in A. In these lists, or words, repetitions are allowed and the order of the objects is taken into account. Since each element of the list can be chosen independently of the others, there are n possible choices for each object in the list. Hence, the following holds.

Proposition 1.13 *The number of k-lists of objects from a set A of cardinality n is n^k.*

A *function* $f : X \to A$ is defined by the value it assumes on each element of X: if $f : \{1, \dots, k\} \to A$, then f is defined by the k-list $(f(1), f(2), \dots, f(k))$, which we refer to as the *image-list* or *image-word* of f. Conversely, each k-list $(a_1, a_2, \dots, a_k)$ with symbols from A defines the function $f : \{1, \dots, k\} \to A$ given by $f(i) := a_i\ \forall i$. If $|A|$ is finite, $|A| = n$, we have a one-to-one correspondence between the set $\mathcal{F}_n^k$ of maps $f : \{1, \dots, k\} \to A$, and the set of the k-lists with symbols in A. Therefore, we have the following.

Proposition 1.14 *The number of functions in $\mathcal{F}_n^k$ is $F_n^k := |\mathcal{F}_n^k| = n^k$.*

1.2.5 Injective functions

We use the symbol $\mathcal{I}_n^k$ to denote the set of injective functions $f : \{1, \dots, k\} \to A$, $|A| = n$, $k \le n$. Let $I_n^k = |\mathcal{I}_n^k|$. Obviously, $I_n^k = 0$ if $k > n$. The image-list of an injective function $f \in \mathcal{I}_n^k$ is a k-word of pairwise different symbols taken from A To form any such image-list, one can choose the first entry among n elements, the second entry can be chosen among $n - 1$ elements, ..., the kth entry can be chosen among the remaining $n - k + 1$ elements of A, so that we have the following.

Proposition 1.15 *The cardinality I_n^k of $\mathcal{I}_n^k$ is*

$$I_n^k = |\mathcal{I}_n^k| = n(n-1) \cdot \dots \cdot (n-k+1) = k!\binom{n}{k} = \frac{n!}{(n-k)!}.$$

Some of the I_n^k's are in Figure 1.6.

1.2.6 Monotone increasing functions

Let $\mathcal{C}_n^k$, $k \le n$, be the set of strictly monotone increasing functions $f : \{1, \dots, k\} \to \{1, \dots, n\}$. The image-list of any such function is an ordered k-tuple of strictly increasing–hence pairwise disjoint–elements of $\{1, \dots, n\}$. The k-tuple is thus identified by the subset of the elements of $\{1, \dots, n\}$ appearing in it, so that we have the following.

$$\left(\begin{array}{rrrrrrrrrrr}
1 & 1 & 1 & 1 & 1 & 1 & 1 & 1 & 1 & 1 & \dots \\
0 & 1 & 2 & 3 & 4 & 5 & 6 & 7 & 8 & 9 & \dots \\
0 & 0 & 2 & 6 & 12 & 20 & 30 & 42 & 56 & 72 & \dots \\
0 & 0 & 0 & 6 & 24 & 60 & 120 & 210 & 336 & 504 & \dots \\
0 & 0 & 0 & 0 & 24 & 120 & 360 & 840 & 1\,680 & 3\,024 & \dots \\
0 & 0 & 0 & 0 & 0 & 120 & 720 & 2\,520 & 6\,720 & 15\,120 & \dots \\
0 & 0 & 0 & 0 & 0 & 0 & 720 & 5\,040 & 20\,160 & 60\,480 & \dots \\
0 & 0 & 0 & 0 & 0 & 0 & 0 & 5\,040 & 40\,320 & 181\,440 & \dots \\
0 & 0 & 0 & 0 & 0 & 0 & 0 & 0 & 40\,320 & 362\,880 & \dots \\
0 & 0 & 0 & 0 & 0 & 0 & 0 & 0 & 0 & 362\,880 & \dots \\
\dots & \dots & \dots & \dots & \dots & \dots & \dots & \dots & \dots & \dots & \dots
\end{array}\right.$$

Figure 1.6 *The cardinality I_n^k of the set of injective maps $\mathcal{I}_n^k$ for $n, k \ge 0$.*

Proposition 1.16 *The cardinality* C_n^k *of* $\mathcal{C}_n^k$ *is*

$$C_n^k := |\mathcal{C}_n^k| = \binom{n}{k} = \mathbf{C}_k^n = (\mathbf{C}^T)_n^k.$$

1.2.7 Monotone nondecreasing functions

Let $\mathcal{D}_n^k$ be the class of monotone nondecreasing functions $f : \{1, \dots, k\} \to \{1, \dots, n\}$. The image-list of any such function is a nondecreasing ordered k-tuple of elements of $\{1, \dots, n\}$, so that elements can be repeated. The functions in $\mathcal{D}_n^k$ are as many as the multisets with cardinality k included in a multiset (A, a), where $A = \{1, \dots, n\}$ and $a(x) \geq k\ \forall x \in A$. Thus, see Proposition 1.12, we have the following.

Proposition 1.17 *The cardinality* D_n^k *of* $\mathcal{D}_n^k$ *is*

$$D_n^k := |\mathcal{D}_n^k| = |\mathcal{C}_{n+k-1}^k| = \binom{n+k-1}{k}.$$

Another proof of Proposition 1.17. Consider the map $\phi : \mathcal{D}_n^k \to \mathcal{F}_{n+k-1}^k$ defined by $\phi(f)(i) := f(i) + i - 1\ \forall i \in \{1, \dots, k\}$, $\forall f \in \mathcal{D}_n^k$. Obviously, if $f \in \mathcal{D}_n^k$, then $\phi(f)$ is strictly monotone increasing, $\phi(f) \in \mathcal{C}_{n+k-1}^k$. Moreover, the correspondence $\phi : \mathcal{D}_n^k \to \mathcal{C}_{n+k-1}^k$ is one-to-one, thus

$$D_n^k = |\mathcal{D}_n^k| = |\mathcal{C}_{n+k-1}^k| = \binom{n+k-1}{k}.$$

Yet another proof of Proposition 1.17. We are now going to define a one-to-one correspondence between a family of multisets and $\mathcal{D}_n^k$. Let (A, a) be a multiset on $A = \{1, \dots, n\}$ with $a(x) \geq k\ \forall k$. For any multiset (S, n_S) of cardinality k included in (A, a), let $f_S : A \to \{0, \dots, k\}$ be the function defined by

$$f_S(x) := \sum_{y \leq x} n_S(y),$$

i.e. for each $x \in A$, $f_S(x)$ is the sum of the multiplicities $n_S(y)$ of all elements $y \in A$, $y \leq x$. f_S is obviously a nondecreasing function and $f_S(n) = k$. Moreover, it is easy to show that the map

$$S \mapsto f_S$$

is a one-to-one correspondence between the family of the multisets included in (A, a) of cardinality k and the family of monotone nondecreasing functions from $\{1, \dots, n\}$ to $\{0, \dots, k\}$ such that $f(k) = 1$. In turn, there is an obvious one-to-one correspondence between this class of functions and the class of monotone nondecreasing functions from $\{1, \dots, n-1\}$ to $\{0, \dots, k\}$. Thus, applying Proposition 1.17 we get

$$|\mathcal{D}_{k+1}^{n-1}| = \binom{k+1+(n-1)-1}{n-1} = \binom{n+k-1}{k}.$$

1.2.8 Surjective functions

The computation of the number of surjective functions is more delicate. Let $\mathcal{S}_n^k$ denote the family of surjective functions from $\{1, \dots, k\}$ onto $\{1, \dots, n\}$ and let

$$S_n^k = \begin{cases} 1 & \text{if } n = k = 0, \\ 0 & \text{if } n = 0,\ k > 0 \\ |\mathcal{S}_n^k| & \text{if } n \geq 1. \end{cases}$$

Obviously, $S_n^k = |\mathcal{S}_n^k| = 0$ if $k < n$. Moreover, if $k = n \geq 1$, then a function $f : \{1, \dots, n\} \to \{1, \dots, n\}$ is surjective if and only if f is injective, so that $S_n^n = |\mathcal{S}_n^k| = I_n^n = n!$

If $k > n \geq 1$, then $\mathcal{S}_n^k \neq \emptyset$. Observe that any function is onto on its range. Thus, for each $j = 1, \dots, n$, consider the set A_j of functions $f : \{1, \dots, k\} \to \{1, \dots, n\}$ whose range has cardinality j. We must have

$$n^k = |\mathcal{F}_n^k| = \sum_{j=1}^{n} |A_j|.$$

There are exactly $\binom{n}{j}$ subsets of $\{1, \dots, n\}$ with cardinality j and there are S_j^k different surjective functions onto each of these sets. Thus, $|A_j| = \binom{n}{j} S_j^k$ and

$$n^k = \sum_{j=1}^{n} \binom{n}{j} S_j^k \qquad \forall n \geq 1.$$

Since we defined $S_0^k = 0$, we get

$$n^k = \sum_{j=0}^{n} \binom{n}{j} S_j^k \qquad \forall n \geq 0. \tag{1.5}$$

Therefore, applying the inversion formula in Corollary 1.5 we conclude the following.

Proposition 1.18 *The cardinality S_n^k of the set $\mathcal{S}_n^k$ of surjective functions from $\{1, \dots, k\}$ onto $\{1, \dots, n\}$ is*

$$S_n^k = \sum_{j=0}^{n} (-1)^{n+j} \binom{n}{j} j^k = \sum_{j=0}^{n} (-1)^j \binom{n}{j} (n-j)^k \qquad \forall n, k \geq 1.$$

We point out that the equality holds also if $k \leq n$ so that

$$\frac{1}{n!}\sum_{j=0}^{n}(-1)^j\binom{n}{j}(n-j)^k = \frac{1}{n!}S_n^k = \begin{cases} 1 & \text{if } k = n, \\ 0 & \text{if } k < n. \end{cases}$$

Another useful formula for S_n^k is an inductive one, obtained starting from $S_n^n = n!$ $\forall n \geq 0$ and $S_n^k = 0$ for any k and n with $k < n$.

Proposition 1.19 *We have*

$$\begin{cases} S_n^k = n(S_n^{k-1} + S_{n-1}^{k-1}) & \textit{if } k \geq 1, n \geq 0, \\ S_n^n = n!, & \\ S_0^k = 0 & \textit{if } k \geq 1. \end{cases} \tag{1.6}$$

Proof. Assume $n \geq 1$ and $k \geq 1$ and let $f : \{1, \ldots, k\} \to \{1, \ldots, n\}$ be a surjective function. Let $A \subset \mathcal{S}_n^k$ be the class of functions such that the restriction $f : \{1, \ldots, k-1\} \to \{1, \ldots, n\}$ of f is surjective and let $B := \mathcal{S}_n^k \setminus A$. The cardinality of A is nS_n^{k-1} because there are S_n^{k-1} surjective maps from $\{1, \ldots, k-1\}$ onto $\{1, \ldots, n\}$ and there are n possible choices for $f(k)$. Since the maps on B have a range of $(n-1)$ elements, we infer that there are nS_{n-1}^{k-1} maps of this kind. In fact, there are $\binom{n}{n-1} = n$ subsets E of $\{1, \ldots, n\}$ of cardinality $n-1$ and there are S_{n-1}^{k-1} surjective functions from $\{1, \ldots, k-1\}$ onto E. Therefore,

$$S_n^k = |A| + |B| = nS_n^{k-1} + nS_{n-1}^{k-1}.$$

i.e. (1.6).

Some of the S_n^k's are in Figure 1.7.

$$\begin{pmatrix}
0 & 0 & 0 & 0 & 0 & 0 & 0 & 0 & 0 & \ldots \\
0 & 1 & 0 & 0 & 0 & 0 & 0 & 0 & 0 & \ldots \\
0 & 1 & 2 & 0 & 0 & 0 & 0 & 0 & 0 & \ldots \\
0 & 1 & 6 & 6 & 0 & 0 & 0 & 0 & 0 & \ldots \\
0 & 1 & 14 & 36 & 24 & 0 & 0 & 0 & 0 & \ldots \\
0 & 1 & 30 & 150 & 240 & 120 & 0 & 0 & 0 & \ldots \\
0 & 1 & 62 & 540 & 1\,560 & 1\,800 & 720 & 0 & 0 & \ldots \\
0 & 1 & 126 & 1\,806 & 8\,400 & 16\,800 & 15\,120 & 5\,040 & 0 & \ldots \\
0 & 1 & 254 & 5\,796 & 40\,824 & 126\,000 & 191\,520 & 141\,120 & 40\,320 & \ldots \\
0 & 1 & 510 & 18\,150 & 186\,480 & 834\,120 & 1\,905\,120 & 2\,328\,480 & 1\,451\,520 & \ldots \\
\ldots & \ldots & \ldots & \ldots & \ldots & \ldots & \ldots & \ldots & \ldots & \ldots
\end{pmatrix}$$

Figure 1.7 *The cardinality S_n^k of the set of surjective maps $\mathcal{S}_n^k$ for $n, k \geq 0$.*

1.2.9 Exercises

Exercise 1.20 *How many diagonals are there in a polygon having n edges?*

1.3 Drawings

A *drawing* or *selection* of k objects from a population of n is the choice of k elements among the n available ones. We want to compute how many of such selections are possible. In order to make this computation, it is necessary to be more precise, both on the composition of the population and on the rules of the selection as, for instance, if the order of selection is relevant or not. We consider a few cases:

- The population is made by pairwise different elements, as in a lottery: in other words, the population is a set.
- The population is a multiset (A, a). In this case, we say that we are dealing with a *drawing from A with repetitions*.
- The selected objects may be given an order. In this case we say that we consider an *ordered selection*. Unordered selections are also called *simple selections*.

Some drawing policies simply boil down to the previous cases:

- In the lottery game, numbers are drawn one after another, but the order of drawings is not taken into account: it is a simple selection of objects from a set.
- In ordered selections the k-elements are selected one after another and the order is taken into account.
- A drawing with replacement, i.e. a drawing from a set where each selected object is put back into the population before the next drawing is equivalent to a drawing with repetitions, i.e. to drawing from a multiset where each element has multiplicity larger than the total number of selected objects.

1.3.1 Ordered drawings

Ordered drawings of k objects from a multiset (A, a) are k-words with symbols taken from A.

1.3.1.1 Ordered drawings from a set

Each ordered drawing of k objects from a set A is a k-list with symbols in A that are pairwise different. Thus the number of possible ordered drawings of k elements from A is the number of k-lists with pairwise different symbols in A.

If $|A| = n$, there are n possible choices for the first symbol, $n - 1$ for the second and so on, so that there are

$$n(n-1)\ldots(n-k+1)$$

different k-words with pairwise different symbols.

1.3.1.2 Ordered drawings from a multiset

Let (A, a) be a multiset where $|A| = n$ and let $k \in \mathbb{N}$ be less than or equal to $\min\{a(x) \mid x \in A\}$. Each ordered drawing of k elements from (A, a) is a k-list with symbols in A, where the same symbol may appear more than once. We have already proven that there are n^k possible k-lists of this kind, so that the following holds.

Proposition 1.21 *The number of ordered drawings of k elements from a multiset (A, a) where $k \leq \min\{a(x) \mid x \in A\}$ is n^k.*

In particular, the number of ordered drawings with replacement of k elements from A is n^k.

1.3.2 Simple drawings

1.3.2.1 Drawings from a set

The population from which we make the selection is a set A. To draw k objects from A is equivalent to selecting a subset of k elements of A: we do not distinguish selections that contain the same objects with a different ordering.

Proposition 1.22 *The number of possible drawings of k elements from a set of cardinality n is $\binom{n}{k}$.*

1.3.2.2 Drawings from a multiset

Let (A, a) be a multiset, $|A| = n$, and let $k \leq \min\{a(x) \mid x \in A\}$. Each sequence S drawn from (A, a) is a sequence of symbols in A where repetitions may occur and the order of the symbols is not taken into accout, e.g.

$$FABADABDF = FBFDDAABA$$

i.e. S is a multiset of k elements included in (A, a) (cf. Proposition 1.17).

Proposition 1.23 *The number of simple drawings of k elements from a multiset (A, a), is $\binom{n+k-1}{k}$ provided $k \leq \min\{a(x) \mid x \in A\}$.*

1.3.3 Multiplicative property of drawings

The previous results on drawings can also be obtained from the following *combinatorics properties* of drawings.

Theorem 1.24 *For each non-negative integer k let a_k and b_k be the numbers of drawings of k objects from the multisets (A, a) and (B, b) made according to policies P_1 and P_2, respectively. If A and B are disjoint, then the number of drawings of k elements from the population obtained by the union of (A, a) and (B, b) made according to policies P_1 and P_2 for the drawings from (A, a) and (B, b), respectively, is*

$$c_k = \sum_{j=0}^{k} a_j b_{k-j}.$$

Proof. A drawing of k objects from the union of the two populations contains, say, j elements from (A, a) and $k - j$ elements from (B, b), where j is an integer, $0 \le j \le k$. The j elements drawn from (A, a) can be chosen in a_j different ways, while the $n - j$ elements drawn from (B, b) can be chosen in b_{k-j} different ways and the two choices are independent. Thus,

$$c_k = \sum_{j=0}^{k} a_j b_{k-j}.$$

A similar result holds for ordered drawings

Theorem 1.25 *For each non-negative integer k let a_k and b_k be the number of ordered drawings from the multisets (A, a) and (B, b) made according to policies P_1 and P_2, respectively. If A and B are disjoint, then the number of ordered drawings from the population union of (A, a) and (B, b) made according to policy P_1 for the elements of (A, a) and according to policy P_2 for elements of (B, b) are*

$$c_k = \sum_{j=0}^{k} \binom{k}{j} a_j b_{k-j}.$$

Proof. A drawing of k elements from the union of the two populations contains j elements from (A, a) and $n - j$ elements from (B, b) for some integer $j \in \{0, \dots, k\}$. The j elements from (A, a) can be chosen in a_j different ways, the $k - j$ elements drawn from (B, b) can be chosen in b_{k-j} different ways and the two chosen groups are independent. Finally, there are $\binom{k}{j}$ ways to order such selections. Thus,

$$c_k = \sum_{j=0}^{k} \binom{k}{j} a_j b_{k-j}.$$

1.3.4 Exercises

Exercise 1.26 *A committee of 7 people has to be chosen among 11 women and 8 men. In each of the following cases compute how many different committees can be chosen:*

- *No constraint is imposed.*
- *At least two women and at least one man must be present.*
- *There must be more women than men.*
- *At least two women and no more than three men must be present.*

1.4 Grouping

Many classical counting problems amount to a *collocation* or *grouping problem*: how many different arrangements of k objects in n boxes are there? Putting it another way, how many different ways of grouping k objects into n groups are there? Also in this case a definite answer cannot be given: we must be more precise both on the population to be arranged, on the rules (or *policy*) of the procedure, and on the way the groups are evaluated. For example, one must say whether the objects to be arranged are pairwise different or not, whether the order of the objects in each box must be taken into account or not, whether the boxes are pairwise distinct or not, and if further constraints are imposed. Here we deal with a few cases, all referring to *collocation* or *grouping* in *pairwise different boxes*. We consider the formed groups as a list instead of as a set: for instance, if we start with the objects $\{1, 2, 3\}$ then the two arrangements in two boxes $(\{1\}, \{2, 3\})$ and $(\{2, 3\}, \{1\})$ are considered to be different.

1.4.1 Collocations of pairwise different objects

Arranging k distinct objects in n pairwise different boxes is the act of deciding the box in which each object is going to be located. Since both the objects and the boxes are pairwise distinct, we may identify the objects and the boxes with the sets $\{1, \dots, k\}$ and $\{1, \dots, n\}$, respectively. Each arrangement corresponds to a *grouping map* $f : \{1, \dots, k\} \to \{1, \dots, n\}$ that puts the object j into the box $f(j)$.

1.4.1.1 No further constraint

In this case the set of possible locations is in a one-to-one correspondence with the set $\mathcal{F}_n^k$ of all maps $f : \{1, \dots, k\} \to \{1, \dots, n\}$. Therefore, there are n^k different ways to locate k-different objects in n boxes.

A different way to do the computation is the following. Assume $i_1, \dots, i_n$ objects are placed in the boxes $1, \dots, n$, respectively, so that $i_1 + \dots + i_n = k$. There are $\binom{k}{i_1}$ different choices for the elements located in the first box, $\binom{k-i_1}{i_2}$ different choices for the elements in the second box, and so on, so that there are

$$\binom{k - i_1 - \dots - i_{n-1}}{i_n}$$

different choices for the elements in the nth box. Thus the different possible arrangements are

$$\binom{k}{i_1}\binom{k-i_1}{i_2}\cdots\binom{k-i_1-\cdots-i_{n-1}}{i_n} =$$
$$= \frac{k!}{i_1!(k-i_1)!}\frac{(k-i_1)!}{i_2!(k-i_1-i_2)!}\cdots = \frac{k!}{i_1!\,i_2!\cdots i_n!}; \tag{1.7}$$

the ratio in (1.7) is called the *multinomial coefficient* and is denoted as

$$\binom{k}{i_1\ i_2\ \cdots\ i_n}.$$

From (1.7) we infer that the possible collocations of k pairwise different objects in n pairwise different boxes are

$$\sum_{i_1+\cdots+i_n=k}\binom{k}{i_1\ i_2\ \cdots\ i_n}$$

where the sum is performed over all the n-tuples $i_1, \ldots, i_n$ of non-negative integers such that $i_1 + \cdots + i_n = k$. Thus, from the two different ways of computing collocations, we get the equality

$$n^k = \sum_{i_1+\cdots+i_n=k}\binom{k}{i_1\ i_2\ \cdots\ i_n}.$$

1.4.1.2 At least one object in each box

We now want to compute the number of different arrangements with at least one object in each box. Assuming we have k objects and n boxes, collocations of this type are in a one-to-one correspondence with the class of *surjective maps* $\mathcal{S}_n^k$ from $\{1, \ldots, k\}$ onto $\{1, \ldots, n\}$, thus there are

$$S_n^k = \sum_{j=0}^{n}(-1)^j\binom{n}{j}(n-j)^k$$

collocations of k pairwise different into n pairwise different boxes that place at least one object in each box.

Another way to compute the previous number is the following. Assume $i_1, \ldots, i_n$ objects are located in the boxes $1, \ldots, n$, respectively, with at least one object in each box, i.e. $i_1 + \cdots + i_n = k$ and $i_1, \ldots, i_n \geq 1$. As in (1.7), there are

$$\binom{k}{i_1\ i_2\ \cdots\ i_n} \tag{1.8}$$

ways to arrange k different objects in n boxes with i_j objects in the box j. Thus the number of arrangements with no empty box is

$$\sum_{\substack{i_1+\cdots+i_n=k \\ i_1,\ldots,i_n\geq 1}} \binom{k}{i_1\ i_2\ \cdots\ i_n};$$

here, the sum is performed over all the n-tuples $i_1, \ldots, i_n$ with *positive* components with $i_1 + \cdots + i_n = k$. The above two ways of computing the number of such collocations yield the identity

$$S_n^k = \sum_{\substack{i_1+\cdots+i_n=k \\ i_1,\ldots,i_n\geq 1}} \binom{k}{i_1\ i_2\ \cdots\ i_n}. \tag{1.9}$$

1.4.1.3 At most one object in each box

We now impose a different constraint: each box may contain at most one object. Assuming we have k objects and n boxes, collocations of this type are in a one-to-one correspondence with the class of *injective grouping maps* $\mathcal{I}_n^k$ from $\{1, \ldots, k\}$ onto $\{1, \ldots, n\}$, thus there are

$$I_n^k = k!\binom{n}{k}$$

collocations of this type.

1.4.1.4 Grouping into lists

Here, we want to compute the number of ways of grouping k pairwise different objects in n pairwise different boxes and pretend that the order of the objects in each box matters. In other words we want to compute how many different ways exist to group k objects in a list of n lists of objects. We proceed as follows.

The first object can be collocated in one of the n boxes, that is in n different ways. The second object can be collocated in $n + 1$ different ways: in fact, it can be either collocated in each of the $n - 1$ empty boxes, or it can be collocated in the same box as the first object. In the latter case it can be collocated either as the first or as the second object in that box. So the possible arrangements of the second object are $(n - 1) + 2 = n + 1$. The third object can be collocated in $n + 2$ ways. In fact, if the first two objects are collocated in two different boxes, then the third object can either be collocated in one of the $n - 2$ empty boxes or in two different ways in each of the two nonempty boxes. Thus, there are $(n - 2) + 2 + 2 = n + 2$ possible arragements. If the first two objects are in the same box, then the third object can either be collocated in one of the $n - 1$ empty boxes or in the nonempty one. In the latter case, it can be collocated in three different ways: either as the first, or between the two objects already present, or as the last one. Again, the third object can be collocated in $(n-1)+3=n+2$ ways. By an induction argument, we infer that there are $n+k-1$ different arrangements for the kth object. Thus, *the number of different ordered locations of k objects in n boxes is*

$$n(n+1)(n+2)\dots(n+k-1)=k!\binom{n+k-1}{k}.$$

1.4.2 Collocations of identical objects

We want to compute the number of ways to arrange k identical objects in n pairwise different boxes. In this case each arrangement is characterized by the number of elements in each box, that is by the map $x:\{1,\dots,n\}\to\{0,\dots,k\}$ which counts how many objects are in each box. Obviously, $\sum_{s=1}^{n}x(s)=k$. If the k objects are copies of the number '0', then each arrangement is identified by the binary sequence

$$\underbrace{00\dots0}_{x(1)}1\underbrace{00\dots0}_{x(2)}1\dots1\underbrace{00\dots0}_{x(n-1)}1\underbrace{00\dots0}_{x(n)} \tag{1.10}$$

where the number '0' denotes the fact that we are changing box.

1.4.2.1 No further constraint

Let us compute the number of such arrangements with no further constraint. There is a one-to-one correspondence between such arrangements and the set of all binary sequences of the type (1.10). Therefore, see Proposition 1.12, *the different collocations of k identical objects in n pairwise different boxes is*

$$\binom{n+k-1}{k}. \tag{1.11}$$

1.4.2.2 At least one in each box

We add now the constraint that each box must contain at least one object. If $k<n$ no such arrangement is possible. If $k\geq n$, we then place one object in each box so that the constraint is satisfied. The remaining $k-n$ objects can be now collocated without constraints. Therefore, cf. (1.11), there are

$$\binom{n+(k-n)-1}{k-n}=\binom{k-1}{k-n}=\binom{k-1}{n-1}$$

ways to arrange k identical objects in n boxes, so that no box remains empty.

1.4.2.3 At most one in each box

We consider arrangements of k identical objects in n pairwise different boxes that place at most one object into each box. In this case, each arrangement is completely characterized by the subset of filled boxes. Since we can choose them in $\binom{n}{k}$ different ways, we conclude that *the collocations of k identical objects in n pairwise different boxes with at most one object per box* is

$$\binom{n}{k}.$$

1.4.3 Multiplicative property

Combinatorial properties hold for collocations as well as for drawings.

Theorem 1.27 *For each non-negative integer k, let a_k and b_k be the number of collocations of k pairwise different objects in two sets S_1 and S_2 of pairwise different boxes with policies P_1 and P_2, respectively. If $S_1 \cap S_2 = \emptyset$, then the different collocations of the k objects in $S_1 \cup S_2$ following policy P_1 for collocations in boxes of S_1 and policy P_2 for collocations in boxes of S_2 is*

$$c_k = \sum_{j=0}^{k} \binom{k}{j} a_j b_{k-j}.$$

Proof. Let j objects, $0 \le j \le k$ be collocated in the boxes of the set S_1 and let the other $k - j$ objects be collocated in the boxes of S_2. There are a_j different ways of placing j objects in the boxes of S_1 and b_{k-j} different ways of placing $(k - j)$ objects in the boxes of S_2. Moreover, there are $\binom{k}{j}$ different ways to choose which objects are collocated in the boxes of S_1. Hence,

$$c_k = \sum_{j=0}^{k} \binom{k}{j} a_j b_{k-j} \qquad \forall k \ge 0.$$

A similar result holds for the collocations of identical objects.

Theorem 1.28 *For each non-negative integer k, let a_k and b_k be the number of collocations of k identical objects in two sets S_1 and S_2 of pairwise different boxes with policies P_1 and P_2, respectively. If $S_1 \cap S_2 = \emptyset$, then the collocations of the k objects in the boxes of $S_1 \cup S_2$ made according to policy P_1 for the collocations in the boxes of S_1 and according to policy P_2 for the collocations in the boxes of S_2 is*

$$c_k = \sum_{j=0}^{k} a_j b_{k-j}.$$

Proof. Let j objects, $0 \le j \le k$ be collocated in the boxes of the set S_1 and let the other $k - j$ objects be collocated in the boxes of S_2. There are a_j ways of placing j objects in the boxes of S_1 and b_{k-j} different ways of placing $(k - j)$

objects in the boxes of S_2. Since the objects are identical, there is no way to select which the j objects to be placed in the boxes of S_1 are. Then the possible different collocations are

$$c_k = \sum_{j=0}^{k} a_j b_{k-j} \qquad \forall k \geq 0.$$

1.4.4 Collocations in statistical physics

In statistical physics, each 'particle' is allowed to be in a certain 'state'; an 'energy level' is associated with each state. The total energy of a system of particles depends on how many particles are in each of the possible states; the mean value of the energy depends on the probabilities that particles stay in a certain state. Thus, the number of possible collocations of the particles in the available states must be evaluated.

1.4.4.1 Maxwell-Boltzmann statistics

This is the case of classical statistical physics: the particles are distinct and no constraint is imposed on their distribution in different states. The number of possible collocations of k particles in n states is thus the number of collocations of k pairwise different objects in n pairwise different boxes, i.e. n^k.

1.4.4.2 Bose–Einstein statistics

The particles are indistinguishable and no constraint is imposed on their distribution in different states. Particles with this behaviour are called *bosons*. The number of collocations of k particles in n states is then the number of collocations of k identical objects in n pairwise different boxes, i.e.

$$\binom{n+k-1}{k} = \binom{n+k-1}{n-1}.$$

1.4.4.3 Fermi–Dirac statistics

The particles are indistinguishable and each state can be occupied by at most one particle (Pauli exclusion principle). Particles following this behaviour are called *fermions*. Then the collocations of k particles in n states is the number of possible choices for the states to be occupied, i.e. $\binom{n}{k}$. Obviously, the Pauli exclusion principle implies $n \geq k$.

1.4.5 Exercises

Exercise 1.29 *A group of eight people sits around a table with eight seats. How many different ways of sitting are there?*

Exercise 1.30 *Compute the number $g_{n,k}$ of subsets of $\{1 \ldots, n\}$ having cardinality k and that do not contain two consecutive integers.*

Solution. There is a one-to-one correspondence between the family of the subsets of cardinality k and the set of binary n-words given by mapping a subset $A \subset \{1, \ldots, n\}$ to its characteristic function $\mathbb{1}_A$. Namely, to the subset $A \subset \{1, \ldots, n\}$ we associate the binary n-word $(a_1, a_2, \ldots, a_n)$ where $a_i = 1$ if $i \in A$ and $a_i = 0$ otherwise. Consequently, the family we are considering is in a one-to-one correspondence with the binary n-words in which there cannot be two consecutive 1's, in

$$0001000101000101$$

Considering the 0's as the sides of a box that contains at most one 1, we have k 1's and $n - k + 1$ boxes with at most one 1 per box. Thus, each collocation is uniquely detected by the choice of the k nonempty boxes. Thus, see Section 1.4.2, $g_{n,k} = \binom{n-k+1}{k}$.

Exercise 1.31 *A physical system is made by identical particles. The total energy of the system is $4E_0$ where E_0 is a given positive constant. The possible energy levels of each particle are $k\ E_0$, $k = 0, 1, 2, 3, 4$. How many different configurations are possible if;*

(i) the kth energy level is made of $k^2 + 1$ states;

(ii) the kth energy level is made of $2(k^2 + 1)$ states;

(iii) the kth energy level is made of $k^2 + 1$ states and two particles cannot occupy the same state.

Exercise 1.32 *For any non-negative integer $n \geq 0$, define*

$$x^{\overline{n}} := x(x+1)(x+2)\cdots(x+n-1),$$

$$x^{\underline{n}} := x(x-1)(x-2)\cdots(x-n+1).$$

Prove that

$$x^{\overline{n}} = (-1)^n(-x)^{\underline{n}}. \tag{1.12}$$

[*Hint.* Take advantage of (iii) of Proposition 1.3.]

Exercise 1.33 *n players participate in a single-elimination tennis tournament. How many matches shall be played?*

Exercise 1.34 *You are going to place 4 mathematics books, 3 chemistry books and 2 physics books on a shelf. Compute how many different arrangements are possible. What if you want to keep books of the same subject together?*

Exercise 1.35 *A comittee of* 4 *is to be chosen from a group of 20 people. How many different choices are there for the roles of president, chairman, secretary, and treasurer?*

Exercise 1.36 *How many bytes of N digits exist with more zeroes than ones?*

Exercise 1.37 *There are* 40 *cabinet ministers sitting around a circular table. One seat is reserved for the Prime Minister. In how many ways can the other ministers seat themselves?*

Exercise 1.38 *Thirty people meet after a long time and they all shake hands. How many handshakes are there?*

Exercise 1.39 *Find the number of solutions to equation* $x + y + z + w = 15$*:*

(i) *in the set of non-negative integers;*

(ii) *in the set of positive integers;*

(iii) *in the set of integers such that* $x > 2$, $y > -2$, $z > 0$, $w > 3$.

Exercise 1.40 *Find the number of non-negative integer solutions* (x, y, z, t) *of* $x + y + z + t \leq 6$.

2

Probability measures

The correspondence between Blaise Pascal (1623–1662) and Pierre de Fermat (1601–1665) on some matters related to card games suggested by the nobleman and gambler Chevalier de Méré is considered to be the birth of probability. The first treatise, *De Ratiociniis in ludo aleae*, by Christiaan Huygens (1629–1695) was published in 1647. Other publications followed in the eighteenth century: at the beginning of the century, the *Ars conjectandi* by Jacob Bernoulli (1654–1705) and the *Doctrine of Chances* by Abraham de Moivre (1667–1754) were published in 1713 and 1718, respectively. Then the *Théorie analytique des probabilités* by Pierre-Simon Laplace (1749–1827), which appeared in 1812, summarized the knowledge on probability of the whole century.

The definition of probability given by Pascal for games of chance is the so-called *classical interpretation*: the probability of an event E is the ratio between the number of successes (the cases in which the event E happens) and the number of all possible cases:

$$\mathbb{P}(E) := \frac{\text{favourable cases}}{\text{possible cases}}. \tag{2.1}$$

Clearly, the definition can be criticized: it cannot define the probability of those events for which we do not know a priori the number of successes and it does not make sense when there are an infinite number of possible cases. Generally speaking, the definition does not apply to a multitude of cases which we would like to speak of in probability terms.

Another definition, given by Richard von Mises (1883–1953), is the *frequentist interpretation*: the probability of an event is the *limit of the frequencies* of successes as the number of trials goes to infinity:

$$\mathbb{P}(E) = \lim_{n\to\infty} \frac{\text{favourable cases among } n \text{ trials}}{n} \tag{2.2}$$

A First Course in Probability and Markov Chains, First Edition. Giuseppe Modica and Laura Poggiolini.

This definition can also be easily criticized: how can the limit of a sequence be computed if the sequence itself is not known? Moreover, the limit may not exist and in many situations, such as in economics, experiments cannot be repeated.

A further definition, first given by Jacob Bernoulli (1654–1705), is the *subjectivist interpretation*: the probability of an event E is a number $p = \mathbb{P}(E)$ that measures the *degree of belief* that E may happen, expressed by a coherent individual. The 'coherence' is expressed by the fact that if the individual measures as $p \in [0, 1]$ the degree of belief that E happens, then $1 - p$ is the degree of belief that E does not happen.

Another way to describe the subjectivist interpretation is the bet that an individual, according to the information in his possession, regards as a fair cost to pay (to get paid) in order to get (pay) one if the event E happens and in order to get (pay) nothing if E does not happen. The *coherence* is in a *fair* evaluation of the probability that E happens: the individual may be either the gambler or the bookmaker but the evaluation remains the same. The following example is from [1].

Example 2.1 Betting on horse racing helps to illustrate the subjectivist point of view. In *fixed-odds betting* the bookmaker decides a different bet for each horse. The gambler, if he or she wins, receives the amount $a(1+q)$ where a is the money bet and q is the bet. In other words, the gambler pays a to win $a(1+q)$ if the horse bet on wins the race and pays a to get no money if the horse bet on does not win the race.

Let p be the probability that the horse wins the race. Then the ratio between a and $a(1+q)$ must be equal to the ratio between p and 1, i.e. $p = 1/(1+q)$ or $q = \frac{1}{p} - 1$. The game is fair if the sum of the probabilities of all horses in the race is 1. UNIRE, the public corporation that regulates Italian horse racing, has fixed an upper bound on the bookmakers bets in order to protect gamblers: the betting is fair if the sum of the probabilities of the horses is not greater than 1.6, $\sum_i p_i \le 1.6$.

2.1 Elementary probability

In the classic definition by Pascal, the *probability* of an event is the ratio between favourables cases (successes) and possible ones. Let Ω be the set of all possible cases and let $A \subset \Omega$ be the subset of favourable cases, then

$$P(A) = \frac{|A|}{|\Omega|} \qquad \forall A \subset \Omega.$$

Calculus of probabilities thus boils down to a counting problem. In mathematics nowadays, this probability is known as the *uniform distribution on a finite set*.

Example 2.2 In flipping a coin, the possible results are 'heads' or 'tails'. The set of all possible cases is $\Omega = \{H, T\}$. For a fair coin, one has $\mathbb{P}(\{T\}) = \mathbb{P}(\{H\}) = 0.5$, i.e. the uniform probability on a set of two elements.

Example 2.3 When throwing a dice, the possible results are the six faces of the dice. The set of possible cases is $\Omega = \{1, 2, 3, 4, 5, 6\}$. If the dice is fair, then all the possible results have the same probability $1/6$, i.e. the uniform probability on a set of 6 elements.

Example 2.4 When throwing two dice, the possible results are the couples (i, j), $i, j = 1, \dots, 6$, where the first and the second components are the result of the first and second dice, respectively. In other words, the set of possible cases is $\Omega := \{1, 2, 3, 4, 5, 6\}^2$. If the dice are fair and do not interfere with each other, then each couple has the same probability, i.e. we have the uniform probability on a set of 36 elements. For instance, the probability to obtain two sixes or to obtain $(4, 6)$, that is, 4 on the first dice and 6 on the second dice or, to obtain $(6, 4)$, i.e. 6 on the first dice and 4 on the second dice are equal to $1/(36)$. Notice that in dice games the outcomes are not ordered, thus the probability to obtain six on both dices is $1/(36)$ while the probability of obtaining one 4 and one 6, i.e. either $(4, 6)$ or $(6, 4)$, is $1/(18)$.

Example 2.5 Draw a number among ninety available ones. The set of possible cases is $\Omega = \{1, 2, \dots, 90\}$. In a lottery game five different numbers are drawn; the order of the drawing is not taken into account. If the drawing is fair, each subset of 5 elements of Ω has the same probability. The set of possible cases is the family of all the subsets of Ω with 5 elements, and the probability of each element of the family is $1/\binom{90}{5}$.

Example 2.6 We draw 5 numbers among a set of 90 different ones. The order of the drawing is taken into account, so that the possible cases are all the ordered lists of five pairwise different integers in $1, 2, \dots, 90$. Thus the possible cases are $5!\binom{90}{5} = \frac{90!}{85!}$. If the drawing is fair, then any drawn list has the same probability $p = 1/(5!\binom{90}{5}) \simeq 1.9 \cdot 10^{-10}$.

2.1.1 Exercises

Exercise 2.7 *In a group of n people, each person writes his or her name on a card and drops the card in a common urn. Later each person gets one card from the urn. Compute the probability that one (and only one) person in the group gets the card with his or her own name. Compute the probability that no one in the group gets the card with his or her own name.*

Exercise 2.8 *Five balls are drawn from an urn containing 90 balls labelled from 1 to 90. A player can play each of the following gambles:*

- *He or she tries to guess three of the five drawn balls.*
- *He or she tries to guess four of the five drawn balls.*
- *He or she tries to guess all five drawn balls.*

Compute the probability of winning in each of the cases above.

Exercise 2.9 *k items of a stock of N objects are subject to testing. If only one item of the stock is defective, compute the probability that the test detects that item. If two items are defective, compute the probability that the test identifies at least one of them and both of them.*

Exercise 2.10 (Birthday paradox) *A group of n people meets at a party, $n \leq 365$. Assuming that all the years are 365 days long (i.e. 29th February does not exist) and that childbirth is equiprobable on each day of the year, compute the probability that the n people at the party are born on n different days of the year.*

Solution. We label the people of the party from 1 to n and compute the number of favourable cases. Person '1' can be born on any day of the year, person '2' can be born on any of the other 364 days of the year, person '3' can be born on any of the other 363 days of the year and so on. Thus the favourable cases are as many as the *collocations of n pairwise distinct objects in* 365 *pairwise different boxes with at most one object per box*, i.e. $\dfrac{365!}{(365-n)!}$.

Let us now compute the number of possible cases. Each person can be born on any of the 365 days of the year, i.e. the possible cases are as many as the *collocations of n pairwise objects* in 365 pairwise different boxes with no further constraint, i.e. there are 365^n possible cases. Thus the probability that the n people at the party are born on n different days of the year is $\dfrac{365!}{365^n\,(365-n)!}$.

We leave it to the reader to prove that the map

$$p : n \in \{1, \ldots, 365\} \mapsto \frac{365!}{365^n\,(365-n)!} \in \mathbb{R}$$

is strictly monotone decreasing, that $p(n) \geq 0.5$ if and only if $n \leq 22$ and $p(n) \geq 0.1$ if and only if $n \leq 40$. In particular, $1 - p(n) \geq 0.5$ if $n \geq 23$ and $1 - p(n) \geq 0.9$ if $n \geq 41$: in words, if at least 23 people are at the party, then the probability that at least two of them are born on the same day of the year is greater than 0.5, while if there are at least 41 people, then the probability that at least two of them are born on the same day of the year is greater than 0.9.

Exercise 2.11 *An urn contains n balls labelled* $1, 2, \ldots, n$*; another urn contains k balls labelled* $1, 2, \ldots, k$*. Assume $k \geq n$ and draw* randomly *a ball from each urn. Compute the following:*

- *The probability that the two balls are labelled with the same number.*
- *The probability that the two balls are labelled with an even number.*

Solution. The set of possible cases is

$$\Omega := \{(a, b) \,|\, a \in \{1, \ldots, n\},\ b \in \{1, \ldots, k\}\}.$$

Clearly the cardinality of Ω is $|\Omega| = kn$. In the first case, the set of successes is

$$A = \{(a, a) \mid a \in \{1, \dots, n\}\}$$

with cardinality $|A| = \min(k, n) = n$. Thus $P(A) = \frac{|A|}{|\Omega|} = \frac{1}{k}$.

In the second case the set of successes is

$$B = \left\{(2i, 2j) \,\middle|\, i \in \left\{1, \dots, \left\lfloor \frac{n}{2} \right\rfloor\right\}, j \in \left\{1, \dots, \left\lfloor \frac{k}{2} \right\rfloor\right\}\right\}$$

with cardinality $|B| = \left\lfloor \frac{n}{2} \right\rfloor \left\lfloor \frac{k}{2} \right\rfloor$ so that $P(B) = \frac{|B|}{|\Omega|} = \frac{\left\lfloor \frac{n}{2} \right\rfloor \left\lfloor \frac{k}{2} \right\rfloor}{kn}$.

Exercise 2.12 *An urn contains n balls labelled* $1, 2, \dots, n$. *We* randomly *draw two balls. Compute the probability that both balls are labelled with even numbers.*

Solution. We are now drawing two balls from the same urn, thus the set Ω of all possible cases is the family of the subset of $\{1, 2, \dots, 20\}$ having exactly two elements. Thus,

$$|\Omega| = \binom{n}{2} = \frac{n(n-1)}{2}.$$

Since the balls are randomly drawn, we are dealing with the uniform probability over a finite set. The set of favourable cases A is the family of all subsets of $\{2j \mid 1 \le j \le \lfloor \frac{n}{2} \rfloor\}$ having two elements. Thus,

$$|A| = \binom{\lfloor \frac{n}{2} \rfloor}{2}$$

and

$$P(A) = \frac{|A|}{|\Omega|} = \frac{\binom{\lfloor \frac{n}{2} \rfloor}{2}}{\binom{n}{2}}.$$

Exercise 2.13 *An urn contains 20 white balls, 30 red balls, 10 green balls and 40 black balls. Ten balls are randomly drawn. In each of the following cases compute the probability of drawing 2 white balls, 3 red balls, 1 green ball and 4 black balls:*

- *The balls are pairwise distinguishable, i.e. they are labelled* $1, 2, \dots, 100$.
- *The balls can be distinguished only by their colour.*

Solution. In the first case we draw 10 balls from an urn containing 100 balls labelled $1, 2, \dots, 100$. Thus the set Ω of all possible events is the family of all the subsets of $\{1, \dots, 100\}$ having cardinality 10, i.e.

$$|\Omega| = \binom{100}{10}.$$

We can assume that the white balls are labelled $1, 2, \ldots, 20$, the red balls $21, \ldots, 50$, the green balls $51, \ldots, 60$ and the black ones $61, \ldots, 100$.

The set A of successes is the Cartesian product of the following events:

- the set of all the subsets of $\{1, \ldots, 20\}$ made by 2 elements;
- the set of all the subsets of $\{21, \ldots, 50\}$ made by 3 elements;
- the singletons in $\{51, \ldots, 60\}$;
- the set of all the subsets of $\{61, \ldots, 100\}$ made by 4 elements.

Thus

$$|A| = \mathbf{C}_2^{20}\mathbf{C}_3^{30}\mathbf{C}_1^{10}\mathbf{C}_4^{40} = \binom{20}{2}\binom{30}{3}\binom{10}{1}\binom{40}{4}$$

and

$$P(A) = \frac{\binom{20}{2}\binom{30}{3}\binom{10}{1}\binom{40}{4}}{\binom{100}{10}} \simeq 0.407$$

When the balls can be distinguished only by their colour, we draw 10 balls from a population of 4 different objects, where each object has multiplicity greater than or equal to 10; thus we are playing a simultaneous drawing from a multiset, so the number of possible drawings is

$$\binom{4+10-1}{10} = \binom{13}{10} = 246.$$

Moreover, there is only one way to draw 2 white balls, 3 red balls, 1 green ball and 4 black ones, i.e. there is only one favourable case. Thus the probability of success is

$$P(A) = \frac{1}{\binom{13}{10}} = \frac{1}{246} \simeq 0.004.$$

Exercise 2.14 *An urn contains 5 white balls, 8 red balls and 7 black balls, labelled $1, 2, \ldots, 20$. Two balls are drawn without replacement. Compute the probability that two balls are painted the same colour.* [$\simeq 0.31$]

Exercise 2.15 *An urn contains 6 white balls, 5 red balls and 9 black balls, labelled $1, 2, \ldots, 20$. Three balls are drawn without replacement. Compute the probability that the three drawn balls are three different colours.* [$9/38 \simeq 0.24$]

Exercise 2.16 *We are given two urns. The first urn contains 6 white balls and 4 red ones. The second urn contains 4 white balls and 6 red ones. The balls are pairwise distinguishable. A ball is* randomly *drawn from each urn. Compute the probability that the two balls are the same colour.* [0.48]

Exercise 2.17 *Let A be a finite set, $|A| = k$. Compute the average cardinality of its subsets.*

Solution. By definition the average cardinality of the subsets of A is

$$\frac{1}{|\mathcal{P}(A)|} \sum_{S \subset \mathcal{P}(A)} |S|.$$

For any $j \in \{0, 1, \dots, k\}$, there are $\binom{k}{j}$ subsets with j elements. Thus

$$\frac{1}{|\mathcal{P}(A)|} \sum_{S \subset \mathcal{P}(A)} |S| = \frac{1}{2^k} \sum_{j=0}^{k} j \binom{k}{j} = \frac{1}{2^k} \sum_{j=1}^{k} k \binom{k-1}{j-1} = \frac{k 2^{k-1}}{2^k} = \frac{k}{2}.$$

Exercise 2.18 *Compute the average number of fixed points in the set of permutations of n elements. The average number of fixed points is the ratio between the total number of fixed points in all the permutations and the number of permutations.*

Solution. For each interger $j = 0, \dots, n$, there are $\binom{n}{j} d_{n-j}$ permutations with j fixed points, see the proof of Proposition 1.10; thus the average number of fixed points is

$$\frac{1}{n!} \sum_{j=0}^{n} j \binom{n}{j} d_{n-j}.$$

This ratio can be computed explicitly. We have

$$\begin{aligned}\frac{1}{n!} \sum_{j=0}^{n} j \binom{n}{j} d_{n-j} &= \frac{1}{n!} n \sum_{j=1}^{n} \binom{n-1}{j-1} d_{n-j} = \frac{1}{n!} n \sum_{j=1}^{n} \binom{n-1}{n-j} d_{n-j} \\ &= \frac{1}{(n-1)!} \sum_{j=0}^{n-1} \binom{n-1}{j} d_j = \frac{(n-1)!}{(n-1)!} = 1,\end{aligned}$$

see (1.3). Thus the average number of fixed points is 1.

2.2 Basic facts

Independently of the interpretation–and the subjective interpretation is the most flexible one–one gets to an *axiomatic definition* of probability, which is due to Andrey Kolmogorov (1903–1987) and is based on the definition of *abstract measure*. In this setting, the choice of the set of possible cases, the set family of possible events and the probability itself are not specified. The choice will be made from time to time on the basis of *statistical* considerations and *model theory*. For our purposes, the choice is basically irrelevant: in this volume we deal with the computation of the probability of events defined by means of other events whose probability is already known.

2.2.1 Events

Intuitively, an *event* is a mathematical proposition to which we assign a number in [0,1] which is called the *probability* of the event. In order to make negations meaningful and to avoid paradoxes, one should consider only *predicates*, i.e. mathematical propositions indexed by the elements of a given set Ω. This way, each predicate defines a subset

$$E = \left\{x \in \Omega \,\middle|\, p(x) \text{ is true}\right\},$$

and two predicates are equivalent if they define the same subset of Ω. For the sake of clarity, it is better to call an *event* a subset of Ω instead of the different equivalent predicates that define it.

Logical connectives are thus 'translated' into operators between sets: unions, intersections, subsets and complementary sets. In fact, if p and q are two predicates on Ω and

$$A = \left\{x \in \Omega \,\middle|\, p(x) \text{ is true}\right\} \qquad B = \left\{x \in \Omega \,\middle|\, q(x) \text{ is true}\right\}$$

then

$$\begin{aligned} A \cup B &= \left\{x \in \Omega \,\middle|\, p(x) \text{ or } q(x) \text{ is true}\right\} \\ A \cap B &= \left\{x \in \Omega \,\middle|\, \text{both } p(x) \text{ and } q(x) \text{ are true}\right\} \\ A^c &= \left\{x \in \Omega \,\middle|\, p(x) \text{ is false}\right\} \\ A \subset B \quad &\text{if and only if} \quad p(x) \Rightarrow q(x) \quad \forall x \in \Omega. \end{aligned}$$

Propositional logic suggests that the *set of events* $\mathcal{E}$ has the following properties:

(i) $\emptyset \in \mathcal{E}$.

(ii) If $A \in \mathcal{E}$, then $A^c := \Omega \setminus A \in \mathcal{E}$.

(iii) If A and $B \in \mathcal{E}$, then $A \cup B \in \mathcal{E}$ and $A \cap B \in \mathcal{E}$.

So that, by induction:

(iv) If $A_1, \ldots, A_n \in \mathcal{E}$, then $\cup_{i=1}^n A_i \in \mathcal{E}$ and $\cap_{i=1}^n A_i \in \mathcal{E}$.

Notice that by (i) and (ii), $\Omega \in \mathcal{E}$ and that, by (iv), if $A, B \in \mathcal{E}$ then $A \setminus B$ and $B \setminus A \in \mathcal{E}$.

The event $\emptyset$ (characterized by $x \in \emptyset$ is always false) is called the *impossible event* while the event Ω ($x \in \Omega$ is always true) is called the *certain event*. The event $A^c := \Omega \setminus A$ is called the *complementary event* of A. Finally, two events $A, B \in \mathcal{E}$ are said to be *incompatible* if $A \cap B = \emptyset$, i.e. if and only if $A \subset B^c$.

Beware, one should resist the temptation to think that every subset of Ω is an event, i.e. that the set of all events $\mathcal{E}$ is the set of all subsets of Ω. Although this is possible, and preferable, if Ω is finite or denumerable, it leads to contradictions if Ω is not denumerable, see Section 2.2.7; for the moment, let us think of $\mathcal{E}$ as a possible proper class of subsets of Ω with the properties (i)–(iv) above.

A family $\mathcal{E}$ of events of Ω satisfying properties (i), (ii) and (iii) above is called an *algebra* of subsets of Ω.

When the family of events $\mathcal{E}$ is infinite, we require a stronger property for the family $\mathcal{E}$: the family of events $\mathcal{E}$ is closed with respect to denumerable unions and intersections, i.e.

(v) For any sequence $\{A_i\} \subset \mathcal{E}$ we have $\cup_{i=1}^{\infty} A_i \in \mathcal{E}$ and $\cap_{i=1}^{\infty} A_i \in \mathcal{E}$.

This property will bring many further properties that will be shown later. Clearly this property boils down to (iii) when $\mathcal{E}$ is a finite family. Moreover, by *De Moivre formulas* it can be also simplified to:

(vi) If (ii) holds, then for any sequence $\{A_i\} \subset \mathcal{E}$ either $\cup_{i=1}^{\infty} A_i \in \mathcal{E}$ or $\cap_{i=1}^{\infty} A_i \in \mathcal{E}$.

We summarize the previous requests in a formal definition.

Definition 2.19 *Let Ω be a nonempty set and let $\mathcal{P}(\Omega)$ be the family of all subsets of Ω.*

- *An* algebra *of subsets of Ω is a family $\mathcal{E} \subset \mathcal{P}(\Omega)$ such that:*

  (i) $\emptyset \in \mathcal{E}$.

  (ii) If $A \in \mathcal{E}$, then $A^c := \Omega \setminus A \in \mathcal{E}$.

  (iii) If $A, B \in \mathcal{E}$, then $A \cup B \in \mathcal{E}$.

- *A* σ-algebra *of subsets of Ω is a family $\mathcal{E} \subset \mathcal{P}(\Omega)$ such that:*

  (i) $\emptyset \in \mathcal{E}$.

  (ii) If $A \in \mathcal{E}$, then $A^c := \Omega \setminus A \in \mathcal{E}$.

  (iii) For any sequence $\{A_i\} \subset \mathcal{E}$ we have $\cup_{i=1}^{\infty} A_i \in \mathcal{E}$.

Notice that if $\mathcal{D} \subset \mathcal{P}(\Omega)$ is a family of subsets of Ω, then the family

$$\mathcal{S} := \bigcap \left\{ \mathcal{E} \,\middle|\, \mathcal{E} \text{ is a } \sigma\text{-algebra, } \mathcal{E} \supset \mathcal{D} \right\}$$

is a well defined family of subsets of Ω. It is easy to show that $\mathcal{S}$ itself is a σ-algebra, and is in fact *the smallest σ-algebra that includes $\mathcal{D}$*. $\mathcal{S}$ is called the σ-algebra *generated* by $\mathcal{D}$.

Example 2.20 Let $\Omega := \{1, 2, 3, 4, 5\}$. Let us describe the algebra $\mathcal{E}$ generated by the subsets $E_1 = \{1, 2\}$ and $E_2 = \{2, 3\}$ of Ω.

Clearly, E_1, E_2, $\emptyset$ and $\Omega \in \mathcal{E}$. Moreover $\mathcal{E}$ contains the sets $E_1^c := \{3, 4, 5\}$ and $E_2^c = \{1, 4, 5\}$ and their intersections $E_1 \cap E_2^c$, $E_2 \cap E_1^c$. Thus

$$\{1\}, \quad \{2\}, \quad \{3\}, \quad \{4, 5\} \in \mathcal{E}.$$

The singletons $\{4\}$ and $\{5\}$ are not in $\mathcal{E}$. The finite unions of the above sets must be in $\mathcal{E}$, so that $|\mathcal{E}| = 2^4 = 16$. More precisely, the subsets of Ω that are in $\mathcal{E}$ are:

$$\begin{aligned}
&\emptyset, \quad \{1, 2, 3, 4, 5\}\\
&\{1\}, \quad \{2\}, \quad \{3\}, \quad \{4, 5\}\\
&\{1, 2\}, \quad \{1, 3\}, \quad \{1, 4, 5\}, \quad \{2, 3\}, \quad \{2, 4, 5\}, \quad \{3, 4, 5\},\\
&\{1, 2, 3\}, \quad \{1, 2, 4, 5\}, \quad \{1, 3, 4, 5\}, \quad \{2, 3, 4, 5\}.
\end{aligned}$$

Definition 2.21 *We denote by $\mathcal{B}(\mathbb{R}^n)$ the smallest σ-algebra of subsets of $\mathbb{R}^n$ that contains all the open sets of $\mathbb{R}^n$. The subsets $E \in \mathcal{B}(\mathbb{R}^n)$ are called* Borel sets *of $\mathbb{R}^n$ and $\mathcal{B}(\mathbb{R}^n)$ is called the* Borel σ-algebra *of $\mathbb{R}^n$.*

In what follows the σ-algebra $\mathcal{B}(\mathbb{R}^n)$ will play an inportant role.

2.2.2 Probability measures

Let Ω be the certain event and let $\mathcal{E} \subset \mathcal{P}(\Omega)$ be a σ-algebra. We 'measure' the probability of an event $E \in \mathcal{E}$ by a real number $\mathbb{P}(E) \in [0, 1]$; in particular, we assign probability 0 to the impossible event and probability 1 to the certain event Ω. The subjective interpretation of probability suggests that the map $\mathbb{P} : \mathcal{E} \to \mathbb{R}$ has the following properties:

- If $A \subset B$, then $\mathbb{P}(A) \leq \mathbb{P}(B)$.
- $\mathbb{P}(A^c) = 1 - \mathbb{P}(A)$.
- If A, B are incompatible, $A \cap B = \emptyset$, then $\mathbb{P}(A \cup B) = \mathbb{P}(A) + \mathbb{P}(B)$.

Definition 2.22 *Let Ω be a nonempty set. A* non-negative measure *on Ω is a couple $(\mathcal{E}, \mathbb{P})$ where $\mathcal{E} \subset \mathcal{P}(\Omega)$ is a σ-algebra of subsets of Ω, called the* family of measurable sets *and $\mathbb{P} : \mathcal{E} \to \overline{\mathbb{R}}_+$ is a map with the following properties:*

(i) $\mathbb{P}(\emptyset) = 0$.

(ii) (Monotonicity) *If $A, B \in \mathcal{E}$ and $A \subset B$, then $\mathbb{P}(A) \leq \mathbb{P}(B)$.*

(iii) (σ-additivity) *For any disjoint sequence $\{A_i\} \subset \mathcal{E}$ we have*

$$\mathbb{P}\Big(\bigcup_{i=1}^{\infty} A_i\Big) = \sum_{i=1}^{\infty} \mathbb{P}(A_i).$$

If, moreover, $\mathbb{P}(\Omega) = 1$, *we say that the non-negative measure* $(\mathcal{E}, \mathbb{P})$ *is a* probability measure *on* Ω. *In this case* $\mathcal{E}$ *is called a* family of events *and the triplet* $(\Omega, \mathcal{E}, \mathbb{P})$ *is called a* probability space.

Remark 2.23 Notice the following:

- If $\mathcal{E}$ is finite, then (iii) boils down to finite additivity for incompatible events.
- If $\mathcal{E}$ is infinite, then the sum on the right-hand side is to be understood as a *series* with non-negative addenda. σ-additivity is crucial and is natural in a large number of situations.
- For a non-negative measure $(\mathcal{E}, \mu)$ we may have $\mu(\Omega) = +\infty$.

Definition 2.24 *Let* $(\Omega, \mathcal{E}, \mathbb{P})$ *be a probability space. An event* $A \in \mathcal{E}$ *such that* $\mathbb{P}(A) = 0$ *is said to be* almost impossible, *an event* $A \in \mathcal{E}$ *such that* $\mathbb{P}(A) = 1$ *is* almost sure. *If A is a singleton, then A is called an* atomic event.

Let $(\mathcal{E}, \mathbb{P})$ be a measure on a set Ω. We say that $N \subset \Omega$ is $\mathbb{P}$-*negligible*, or simply a *null set*, if there exists $E \in \mathcal{E}$ such that $N \subset E$ and $\mathbb{P}(E) = 0$. Let $\widetilde{\mathcal{E}}$ be the collection of all the subsets of Ω of the form $F = E \cup N$ where $E \in \mathcal{E}$ and N is $\mathbb{P}$-negligible. It is easy to check that $\widetilde{\mathcal{E}}$ is a σ-algebra which is called the $\mathbb{P}$-completion of $\mathcal{E}$. Moreover, setting $\mathbb{P}(F) := \mathbb{P}(E)$ if $F = E \cup N \in \widetilde{\mathcal{E}}$, then $(\widetilde{\mathcal{E}}, \mathbb{P})$ is again a measure on Ω called the $\mathbb{P}$-completion of $(\mathcal{E}, \mathbb{P})$. It is customary to consider measures as $\mathbb{P}$-complete measures.

2.2.3 Continuity of measures

The following proposition collects some properties that easily follow from Definition 2.22.

Proposition 2.25 *Let* $(\Omega, \mathcal{E}, \mathbb{P})$ *be a probability space.*

(i) If $A \in \mathcal{E}$, *then* $0 \le \mathbb{P}(A) \le \mathbb{P}(\Omega) = 1$.

(ii) If $A, B \in \mathcal{E}$ *and* $A \subset B$, *then* $\mathbb{P}(B \setminus A) = \mathbb{P}(B) - \mathbb{P}(A)$.

(iii) If $A, B \in \mathcal{E}$, *then* $\mathbb{P}(A \cup B) + \mathbb{P}(A \cap B) = \mathbb{P}(A) + \mathbb{P}(B)$.

(iv) If $A, B \in \mathcal{E}$, *then* $\mathbb{P}(A \cup B) \le \mathbb{P}(A) + \mathbb{P}(B)$.

(v) (σ-subadditivity) *For any sequence* $\{A_i\} \subset \mathcal{E}$ *we have*

$$\mathbb{P}\Big(\bigcup_{i=1}^{\infty} A_i\Big) \le \sum_{i=1}^{\infty} \mathbb{P}(A_i). \tag{2.3}$$

Another easy consequence of Definition 2.22 is the *law of total probability*.

Proposition 2.26 (Law of total probability) *Let* $(\Omega, \mathcal{E}, \mathbb{P})$ *be a probability space and let* $\{D_i\}_i \subset \mathcal{E}$ *be a finite or denumerable partition of* Ω, *i.e.* $\Omega = \cup_{i=1}^{\infty} D_i$ *and* $D_i \cap D_j = \emptyset \ \forall i, j, i \neq j$. *Then*

$$\mathbb{P}(A) = \sum_{i=1}^{\infty} \mathbb{P}(A \cap D_i) \qquad \forall A \in \mathcal{E}. \tag{2.4}$$

We finally point out the following important continuity property of probability measures.

Proposition 2.27 (Continuity) *Let* $(\Omega, \mathcal{E}, \mathbb{P})$ *be a probability space and let* $\{E_i\} \subset \mathcal{E}$ *be a denumerable family of events.*

(i) *If* $E_i \subset E_{i+1} \ \forall i$, *then* $\cup_{i=1}^{\infty} E_i \in \mathcal{E}$ *and*

$$\mathbb{P}\Big(\bigcup_{i=1}^{\infty} E_i\Big) = \lim_{i \to +\infty} \mathbb{P}(E_i). \tag{2.5}$$

(ii) *If* $E_i \supset E_{i+1} \ \forall i$, *then* $\cap_{i=1}^{\infty} E_i \in \mathcal{E}$ *and*

$$\mathbb{P}\Big(\bigcap_{i=1}^{\infty} E_i\Big) = \lim_{i \to +\infty} \mathbb{P}(E_i). \tag{2.6}$$

Proof. We prove (i). Write $E := \cup_{k=1}^{\infty} E_k$ as

$$E = E_1 \bigcup \Big(\bigcup_{k=2}^{\infty} (E_k \setminus E_{k-1})\Big).$$

The sets E_1 and $E_k \setminus E_{k-1}$, $k \geq 2$, are pairwise disjoint events of $\mathcal{E}$. By the σ-additivity property of measures, we get

$$\begin{aligned} \mathbb{P}(E) &= \mathbb{P}(E_1) + \sum_{k=2}^{\infty} \mathbb{P}(E_k \setminus E_{k-1}) \\ &= \mathbb{P}(E_1) + \sum_{k=2}^{\infty} (\mathbb{P}(E_k) - \mathbb{P}(E_{k-1})) = \lim_{k \to \infty} \mathbb{P}(E_k). \end{aligned}$$

In order to prove (ii) we point out that from $\mathbb{P}(E_1) < +\infty$ and $E_k \subset E_1 \ \forall k \geq 2$, we get $\mathbb{P}(E_1) - \mathbb{P}(E_k) = \mathbb{P}(E_1 \setminus E_k)$ for any $k \geq 2$. Moreover, the sets $E_1 \setminus E_k$ are an increasing sequence of events of Ω. Thus, applying (i) to the family of events $\{E_1 \setminus E_k\}$ one gets

$$\begin{aligned} \mathbb{P}(E_1) - \lim_{k \to \infty} \mathbb{P}(E_k) &= \lim_{k \to \infty} \mathbb{P}(E_1 \setminus E_k) \\ &= \mathbb{P}\Big(\bigcup_k (E_1 \setminus E_k)\Big) = \mathbb{P}(E_1) - \mathbb{P}\Big(\bigcap_{k=2}^{\infty} E_k\Big). \end{aligned}$$

2.2.4 Integral with respect to a measure

Given a measure $(\mathcal{E}, \mathbb{P})$ on a nonempty set Ω, one defines an integral with respect to such measure,

$$\int_\Omega f(x)\,\mathbb{P}(dx)$$

whenever f is in an appropriate class of functions, called *random variables*, see Section 3.1. Here we define only the integral of functions $f : \Omega \to \mathbb{R}$ that have a finite range. For the *characteristic function* $\mathbb{1}_E$

$$\mathbb{1}_E(x) = \begin{cases} 1 & \text{if } x \in E, \\ 0 & \text{if } x \notin E \end{cases}$$

of $E \in \mathcal{E}$, set

$$\int_\Omega \mathbb{1}_E(x)\,\mathbb{P}(dx) := \mathbb{P}(E).$$

A *simple function* is a linear combination of characteristic functions of pairwise disjoint events $\{E_i\}$. The integral of a simple function $\varphi(x) = \sum_{i=1}^n c_i \mathbb{1}_{E_i}(x)$, $x \in \Omega$, is then defined by

$$\int_\Omega \varphi(x)\,\mathbb{P}(dx) := \sum_{i=1}^n c_i \mathbb{P}(E_i).$$

Integration is a linear operator on the family of simple functions. In fact, the following holds.

Proposition 2.28 *Let φ and ψ be simple functions, and let $a, b \in \mathbb{R}$. Then $a\varphi(x) + b\psi(x)$ is a simple function and*

$$\int_\Omega (a\varphi(x) + b\psi(x))\,\mathbb{P}(dx) = a\int_\Omega \varphi(x)\,\mathbb{P}(dx) + b\int_\Omega \psi(x)\,\mathbb{P}(dx).$$

Proof. Let $\varphi(x) = \sum_{i=1}^n c_i \mathbb{1}_{E_i}(x)$ and $\psi(x) = \sum_{j=1}^m d_j \mathbb{1}_{F_j}(x)$ where the sets $\{E_i\}$ are pairwise disjoint and so are the sets $\{F_j\}$. Without any loss of generality, we can assume that both the families $\{E_i\}$ and $\{F_j\}$ are finite partitions of Ω. Then $\{E_i \cap F_j\}$ is a finite partition of Ω and

$$a\varphi(x) + b\psi(x) = ac_i + bd_j \qquad \text{if } x \in E_i \cap F_j.$$

Thus $a\varphi + b\psi$ is a simple function and by the definition of integral

$$\int_\Omega (a\varphi(x) + b\psi(x))\,\mathbb{P}(dx) = \sum_{\substack{i=1,n\\ j=1,m}} (ac_i + bd_j)\mathbb{P}(E_i \cap F_j)$$

$$= \sum_{i=1}^{n} ac_i \sum_{j=1}^{m} \mathbb{P}(E_i \cap F_j) + \sum_{j=1}^{m} bd_j \sum_{i=1}^{n} \mathbb{P}(E_i \cap F_j)$$

$$= a \sum_{i=1}^{n} c_i \mathbb{P}(E_i) + b \sum_{j=1}^{m} d_j \mathbb{P}(F_j)$$

$$= a \int_\Omega \varphi(x)\, \mathbb{P}(dx) + b \int_\Omega \psi(x)\, \mathbb{P}(dx).$$

A function φ having a finite range, $\varphi(x) = \sum_{i=1}^{n} a_i \mathbb{1}_{A_i}(x)$, where the events $\{A_i\}$ are not necessarily disjoint, is a linear combination of simple functions. Proposition 2.28 then yields

$$\int_\Omega \varphi(x)\; \mathbb{P}(dx) = \sum_{i=1}^{n} a_i \int_\Omega \mathbb{1}_{A_i}(x)\, \mathbb{P}(dx) = \sum_{i=1}^{n} a_i \mathbb{P}(A_i), \tag{2.7}$$

as for simple functions.

2.2.5 Probabilities on finite and denumerable sets

2.2.5.1 Probabilities on finite sets

In Example 2.20 the certain event is $\Omega = \{1, 2, 3, 4, 5\}$ and the σ-algebra $\mathcal{E}$ generated by the sets $E_1 = \{1, 2\}$, $E_2 = \{2, 3\}$ is given by all the finite unions of

$$\{1\}, \quad \{2\}, \quad \{3\}, \quad \{4, 5\}.$$

In particular, the singletons $\{4\}$ and $\{5\}$ are not events. Thus, if $(\mathcal{E}, \mathbb{P})$ is a probability measure on Ω, then $\mathbb{P}(\{4\})$ and $\mathbb{P}(\{5\})$ are not defined. If one is willing to define the probability on any subset of Ω, then one of the two following procedures may be followed:

(i) Define $\mathbb{P}(\{4\})$ and $\mathbb{P}(\{5\})$ so that $\mathbb{P}(\{4\}) + \mathbb{P}(\{5\}) = \mathbb{P}(\{4, 5\})$. Obviously there are infinitely many ways to do that.

(ii) Make Ω 'smaller' so that the events by which the σ-algebra can be generated by unions only are atomic. In the example we are considering, one identifies 4 and 5 by setting $\{A\} = \{4, 5\}$ and chooses $\Omega = \{1, 2, 3, A\}$.

Thus, if Ω is a finite set, by applying one of the previous procedures, one may always assume that all the subsets of Ω are events, $\mathcal{E} = \mathcal{P}(\Omega)$, so that $\mathbb{P}(A)$ is defined for all $A \subset \Omega$. *In what follows, we always assume that $\mathcal{E} = \mathcal{P}(\Omega)$ if Ω is finite, without explicitly stating it.*

In particular, assuming $\Omega = \{x_1, \dots, x_n\}$, each singleton $\{x_k\}$ is an atomic event, with its own probability $p_k := \mathbb{P}(\{x_k\})$. We call p_k the *density* of $\mathbb{P}$ at x_k

and the vector $p := (p_1, \dots, p_n)$ the *mass density* of $\mathbb{P}$. Clearly,

$$p_k \geq 0 \ \ \forall k, \qquad \sum_{k=1}^{n} p_k = 1$$

and for any subset $A \subset \Omega$,

$$\mathbb{P}(A) = \sum_{x_k \in A} \mathbb{P}(\{x_k\}) = \sum_{x_k \in A} p_k.$$

2.2.5.2 Uniform probability

Let Ω be a finite set. If all the atomic events have the same probability, $\mathbb{P}(\{x\}) = p \ \forall x \in \Omega$, then, by the additivity property, one gets,

$$1 = \mathbb{P}(\Omega) = \sum_{x \in \Omega} p = p\,|\Omega| \qquad \text{i.e.} \qquad p = \frac{1}{|\Omega|}.$$

Thus, for any set $A \subset \Omega$, we may compute

$$\mathbb{P}(A) = p\,|A| = \frac{|A|}{|\Omega|}.$$

This case is called *uniform probability on a finite set* and the probability of the event A coincides with the classical interpretation of probability: the *ratio between the number of favourable events and the number of possible events*. Thus the computation of the probability of an event implies only two counting operations, see Chapter 1.

Given a finite set Ω, the everyday-language phrase 'choose $x \in \Omega$ randomly' will here be interpreted as: consider Ω as equipped with the uniform probability, i.e. $\mathbb{P}(\{x\}) := \frac{1}{|\Omega|} \ \forall x \in \Omega$.

2.2.5.3 Finite Bernoulli process

Assume you perform an experiment which has only two possible results: *success*, with probability p, $0 \leq p \leq 1$, or *failure* with probability q, $q = 1 - p$. This is called a *Bernoulli trial* of parameter p. Thus the certain event has only two elements: $\Omega = \{0, 1\}$, where 0 is for failure and 1 is for success. The mass density vector is (q, p).

We then perform the same experiment N times, and assume that each trial does not interfere with the results of the other trials. This procedure is called a *finite Bernoulli process*. The result of the repeated experiment is a N-tuple of zeroes and ones, where the kth component is 1 if and only if we get a success at the kth trial. Thus the certain event is $\Omega = \{0, 1\}^N$, the set of all binary N-tuples, which is a finite set with 2^N elements. The probability p_v of a given sequence of results $v = (v_1, \dots, v_N) \in \{0, 1\}^N$ depends only on the number of successes (or failures) that are in it: if k components of v are equal to 1 and $N - k$ components of v are equal to zero, then

$$p_v = \mathbb{P}(\{v\}) := p^k q^{N-k}$$

The mass density vector $p = (p_v)_{v \in \{0,1\}^N}$, which has 2^N components, completely describes a probability measure on $\Omega := \{0, 1\}^N$, called a *Bernoulli distribution of parameter* p on $\{0, 1\}^N$ and denoted by $(\mathcal{P}(\{0, 1\}^N), Ber(N, p))$. We then have

$$Ber(N, p)(A) := \sum_{v \in A} p_v \qquad \forall A \subset \{0, 1\}^N.$$

Observe that, in the case $p = q = 1/2$, then $p_v = \frac{1}{2^N}$ $\forall v \in \{0, 1\}^N$, i.e. $Ber(N, 1/2)$ is the uniform probability on the set $\{0, 1\}^N$.

Various examples fit in the finite Bernoulli process: the repeated flipping of a coin, games where further matches are played, such as championships, random walks, the behaviour (up and down) of the stock exchange, medical tests (positive or negative) or even answers (yes or no) of a population of N individuals.

The meaning of the sentence 'the trials of the experiment do not interfere one another' must be made precise, see Section 4.3.1, and is a crucial request which must be accurately checked any time one wishes to apply the finite Bernoulli process model and its distribution.

Example 2.29 Compute the probability of getting k successes in n trials in a finite Bernoulli process.

Let p, $0 \le p \le 1$ be the probability of success in each trial. We want to compute the probability of the event $E_k \subset \{0, 1\}^n$ defined by

$$E_k := \Big\{x \in \{0, 1\}^n \;\Big|\; \text{exactly } k \text{ components of } x \text{ are equal to } 1\Big\}$$

The probability of each atomic event $v \in E_k$ is $\mathbb{P}(\{v\}) = Ber(n, p)(\{v\}) = p^k (1 - p)^{n-k}$ and the cardinality of E_k is $\binom{n}{k}$, hence

$$\mathbb{P}(E_k) = \sum_{v \in E_k} \mathbb{P}(\{v\}) = p^k (1 - p)^{n-k} \sum_{v \in E_k} 1 = \binom{n}{k} p^k (1 - p)^{n-k}.$$

2.2.6 Probabilities on denumerable sets

Let Ω be a denumerable set. As in the finite case, one may always assume that all the subsets of Ω are events, $\mathcal{E} = \mathcal{P}(\Omega)$, so that $\mathbb{P}(A)$ is defined for all $A \subset \Omega$. This will be tacitly assumed if the certain event Ω is denumerable.

Assume $\Omega = \{x_k\}$. As in the finite case, each singleton $\{x_k\}$ is an atomic event, with its own probability $p_k := \mathbb{P}(\{x_k\})$. We call p_k the *density* of $\mathbb{P}$ at x_k

and the sequence $p := \{p_k\}$ the *mass density* of $\mathbb{P}$. Clearly,

$$p_k \geq 0 \;\; \forall k, \qquad \sum_{k=1}^{\infty} p_k = 1$$

and for any subset $A \subset \Omega$,

$$\mathbb{P}(A) = \sum_{x_k \in A} \mathbb{P}(\{x_k\}) = \sum_{x_k \in A} p_k.$$

A noticeable departure from the finite case is the following.

Proposition 2.30 *If Ω is denumerable, then there exists no uniform probability on Ω.*

Proof. Assume, by contradiction, that $(\mathcal{P}(\Omega), \mathbb{P})$ is a uniform probability measure on Ω, $\mathbb{P}(\{x\}) = p \;\forall x \in \Omega$. Then σ-additivity yields

$$1 = \mathbb{P}(\Omega) = \sum_{x \in \Omega} p = p|\Omega| = \begin{cases} +\infty & \text{if } p > 0, \\ 0 & \text{if } p = 0, \end{cases}$$

a contradiction.

If one wants to define some uniform 'probability' $\mathbb{P}$, say, on $\mathbb{N}$, then $\mathbb{P}$ cannot be countably additive. It may be useful, anyhow, to have a uniform and finitely additive 'probability measure' $\mathbb{P} : \mathcal{P}(\mathbb{N}) \to \mathbb{R}$ such that a non-negative integer is even with probability $1/2$, or is divisible by 3 with probability $1/3$ and so on. For any $E \subset N$ we may define $\mathbb{P}(E)$ using the frequentist interpretation

$$\mathbb{P}(E) := \lim_{n \to \infty} \frac{\mathbb{P}(E \cap \{1, 2, \dots, n\})}{n},$$

but the limit on the right-hand side may not exist. For instance, let

$$E := \{1, 10, 11, 12, 100, \dots, 129, 1000, \dots, 1299, \dots\}$$

It can be shown that the limit $\lim_n \mathbb{P}(E \cap \{1, \dots, n\})/n$ does not exist.

Another possible definition of a finitely additive 'probability' on $\mathbb{N}$ is due to Lejeune Dirichlet (1805–1859). For any $\alpha > 1$, consider the probability measure $(\mathcal{P}(\mathbb{N}), \mathbb{P}_\alpha)$ (it is a true probability measure) whose mass density is $p_\alpha(n) := \frac{1}{n^\alpha}$ so that, for any $E \subset N$ one gets

$$\mathbb{P}_\alpha(E) = \sum_{n \in E} \frac{1}{n^\alpha} \Big/ \sum_{n=1}^{\infty} \frac{1}{n^\alpha}.$$

Define $\mathbb{P}(E) := \limsup_{\alpha \to 1^+} \mathbb{P}_\alpha(E)$. One can easily verify that

$$\mathbb{P}(\{\text{even numbers}\}) = 1/2.$$

More generally, we have the following.

Proposition 2.31 *If* $\mathbb{P}(E \cap \{1, \dots, n\})/n \to L$ *as* $n \to \infty$, *then* $\mathbb{P}(E) = L$.

We omit the proof. The interested reader may refer to, e.g. [2].

2.2.7 Probabilities on uncountable sets

2.2.7.1 Uniform probability on an interval

Let $\Omega = [0, 1] \subset \mathbb{R}$. One may think of doing an experiment whose result is a random number $x \in [0, 1]$, where each number has the same chance of being the result to the experiment. In particular, we require any closed interval $[a, b]$ $0 \le a \le b \le 1$ to be an event whose probability equals its length, $\mathbb{P}([a, b]) = b - a$. Since the class of the events $\mathcal{E}$ must be a σ-algebra, choosing $a = b$, we get $\{a\} = \cap_n [a, a + 1/n] \in \mathcal{E}$ and $\mathbb{P}(\{a\}) \le 1/n \; \forall n$, i.e. $\mathbb{P}(\{a\}) = 0$. Thus the family of the events $\mathcal{E}$ contains the closed intervals, the singletons, hence the open and the half-closed intervals; moreover, $\mathbb{P}(]a, b[) = \mathbb{P}(]a, b]) = \mathbb{P}([a, b[) = b - a$. Since any open set in $\mathbb{R}$ can be written as the union of a denumerable family of open intervals, one gets that open and closed sets in [0,1] are events and so is any countable union of countable instersections of closed and open sets and so on. One wishes to choose $\mathcal{E}$ as the σ-algebra of all subsets of [0,1], but a famous example by Giuseppe Vitali (1875–1932) shows that this is not possible: one cannot define a σ-additive function on $\mathcal{P}([0, 1])$ which is consistent with the elementary length for intervals.

It can be shown that one can choose as the family of events the σ-algebra generated by the closed intervals, called the *σ-algebra of Borel sets* of [0,1]. In fact, it can be proven that the map $\mathbb{P}$ defined on the closed intervals of [0,1] as $\mathbb{P}([a, b]) := b - a$ can be extented to a probability measure $(\mathcal{B}([0, 1]), \mathbb{P})$ and that the extension is unique. The procedure above actually defines the *Lebesgue measure* on [0,1] starting from the elementary measure of the intervals, see Appendix B.

Notice that each singleton $\{a\}$ is an event of zero probability. Thus, atomic events do not provide any information on the probability of the uncountably infinite events. In other words, the probability measure cannot be described by its mass density, which is identically zero.

2.2.7.2 Infinite Bernoulli process

Even in problems related to counting, the certain event may not be countable. For instance, we want to compute the time of the first success in the iterated flipping of a coin, i.e. the first time we get a head. We assume that the trials do not interfere with each other. In principle, we may get the first head after an arbitrary number of coin flips or we may even never get it. Thus we must consider an infinite number of coin flips or, more generally, a denumerable number of Bernoulli trials that do not interfere with each other. The process of doing infinitely many Bernoulli trials of parameter p that do not interfere with each other is called an *infinite Bernoulli process*.

Let 1 and 0 denote the success (head) or the failure (tail) in each trial. Then the certain event Ω is the set of all sequences with values in $\{0, 1\}$

$$0, 0, \dots, 0, 1, \dots, 1, 0, 1, 0 \dots .$$

which we denote as $\Omega = \{0, 1\}^\infty$. It is worth recalling that $\{0, 1\}^\infty$ is also the set of all maps $a : \mathbb{N} \to \{0, 1\}$, or the set of all *sequences of binary digits*

$$00\dots01\dots1000\dots,$$

which form an uncountable set (one may prove this fact by means of the Cantor diagonal process). To be more precise, the following holds.

Proposition 2.32 $\{0, 1\}^\infty$ *has the cardinality of the continuum, i.e. the same cardinality of* $\mathbb{R}$.

Proof. Consider the map $T : \{0, 1\}^\infty \to [0, 1]$ defined as

$$T(\{a_n\}) := \sum_{i=1}^{\infty} \frac{a_i}{2^i}. \tag{2.8}$$

Clearly, $T : \Omega \to [0, 1]$ is surjective, since for any $x \in [0, 1[$, the binary sequence of x, $x = 0, a_0 a_1 a_2 \cdots$ is such that $T(\{a_n\}) = x$ and, for $\{a_n\} = \{1, 1, 1, 1, 1, \dots\}$ we have $T(\{a_n\}) = 1$. Thus, the cardinality of $\{0, 1\}^\infty$ is greater than or equal to the cardinality of $[0, 1]$, i.e. of the continuum.

T is not injective since there are binary sequences that give the same real number (e.g. $0, 00001111111 \cdots = 0, 00010000 \dots$). These sequences are constant for large enough n's hence they form a denumerable set. Thus, $\{0, 1\}^\infty$ has the same cardinality of $[0, 1]$, i.e. of the continuum.

We want to define a probability measure on $\{0, 1\}^\infty$ related to the Bernoulli distributions $Ber(n, p)$ constructed by means of the finite Bernoulli process.

Intuitively, n-tuples of trials must be events. This cannot be imposed as it is since n-tuples are not sequences, so we proceed as follows. To any binary n-tuple $a = (a_1, \dots, a_n)$ and any set $J = (j_1, \dots, j_n)$ of index there corresponds a set of sequences of trials, the 'cylinder',

$$E_{J,a} := \left\{ x \in \{0, 1\}^\infty \,\middle|\, x_{j_i} = a_i \ \forall i = 1, \dots, n \right\}$$

that is the set of all sequences of trials that at step j_k have value a_k. We require that the sets $E_{J,a} \subset \{0, 1\}^\infty$ are events, and that the probability of the cylinders $E_{j,a}$ in $\{0, 1\}^\infty$ is the same as the probability of the n-tuple a in the finite Bernoulli process, i.e.

$$\mathbb{P}(E_{J,a}) := Ber(n, p)(\{a\}) = p^k q^{n-k}$$

if a contains k ones and $n - k$ zeroes.

Consider the σ-algebra $\mathcal{E}$ generated by the cylinders. It can be shown that the map $E_{J,a} \mapsto \mathbb{P}(E_{J,a})$ extends from the family of cylinders to $\mathcal{E}$ as a probability measure $(\mathcal{E}, Ber(\infty, p))$ and that this extension is unique, see Appendix B. $(\mathcal{E}, Ber(\infty, p))$ is the *Bernoulli distribution* of parameter p on the set of binary sequences.

Notice that we are close to the construction of the Lebesgue measure on $\mathbb{R}$. In fact, if we 'identify' real numbers in $[0, 1]$ with their binary representation through the map T in (2.8), then $\mathcal{E}$ agrees with the family of Borel sets $\mathcal{B}([0, 1])$ and $Ber(\infty, 1/2)$ agrees with the Lebesgue measure $\mathcal{L}^1$ on the interval $[0, 1]$. For a precise statement, see Theorem 3.43.

Example 2.33 The probability of getting successes only in an infinite Bernoulli process of parameter $p < 1$ is zero.

Let $v = (1, 1, \dots,)$. We want to compute $\mathbb{P}(\{v\}) := Ber(\infty, p)(\{v\})$. Let $A_n = \{x = (x_i) \in \{0, 1\}^\infty \mid x_i = 1 \ \forall i = 1, \dots, n\}$. Then

$$\{v\} = \bigcap_{n=1}^{\infty} A_n \qquad \text{and} \qquad A_n \supset A_{n+1}.$$

Thus $\{v\}$, being a countable intersection of events, is an event and $\mathbb{P}(\{v\}) = \lim_{n\to\infty} \mathbb{P}(A_n)$. Let p be the probability of success in each trial. If $p = 1$, then $\mathbb{P}(A_n) = 1 \ \forall n$ so that $\mathbb{P}(\{v\}) = 1$. If $p < 1$, then $\mathbb{P}(A_n) = p^n$ so that $\mathbb{P}(\{v\}) = 0$.

Proceeding as in Example 2.33, one can prove that any singleton $\{v\} \in \{0, 1\}^\infty$ is an almost impossible event for the Bernoulli distribution $(\mathcal{E}, Ber(\infty, p))$ of parameter p if $0 < p < 1$, i.e. $\{v\} \in \mathcal{E}$ and $Ber(\infty, p)(\{v\}) = 0$.

2.2.8 Exercises

Exercise 2.34 *Show that $\mathcal{B}(\mathbb{R})$ is the smallest σ-algebra generated by one of the following families of sets:*

- *the closed sets;*
- *the open intervals;*
- *the closed intervals;*
- *the intervals $[a, b[$, $a, b \in \mathbb{R}$, $a < b$;*
- *the intervals $]a, b]$, $a, b \in \mathbb{R}$, $a < b$;*
- *the closed half-lines $]-\infty, t]$, $t \in \mathbb{R}$.*

Solution. [*Hint*. Show that any open set can be written as the union of a countable family of intervals.]

Exercise 2.35 *A multiple choice test comprises 10 questions. Each question has four possible answers: one is right, three are wrong. To pass the test the candidate must correctly answer at least eight questions. Compute the probability of passing the test if the candidate answers each question randomly.*

Solution. The test is a finite Bernoulli process of 10 Bernoulli trials, and the probablity of success is $p = 0.25$ at each trial. Let E_k be the event 'the candidate answers k questions correctly'. Then we want to compute the probability of

$$E = \left\{ x \in \{0, 1\}^{10} \,\middle|\, \text{there are at least 8 successes} \right\} = E_8 \cup E_9 \cup E_{10}.$$

Thus,

$$\mathbb{P}(E) = \mathbb{P}(E_8) + \mathbb{P}(E_9) + \mathbb{P}(E_{10}) = \sum_{k=8}^{10} \binom{10}{k} (.25)^k (.75)^{10-k} \simeq 0.0004.$$

Exercise 2.36 *It is more probable to obtain a 6 by throwing a single dice 4 times than obtaining a double 6 by throwing two dice 24 times. This is the remark made by the nobleman and hard gambler Chevalier de Méré to Pascal Blaise (1623–1662) around 1650. The following (wrong) consideration in a letter from Pascal to Pierre de Fermat (1601–1665), dated 1654, seems to reject the empiric remark of de Méré: the probability of obtaining a 6 when throwing a dice is 1/6; the probability of getting a double 6 when throwing two dice is 1/36; since 4/6 = 24/36, the two events have the same probability. Find the mistake.*

Solution. At each throw of a dice, the probability of failure is 5/6, so that the probability of 4 failures in 4 throws is $(5/6)^4$; hence the probability of at least 1 success in 4 throws is

$$1 - \left(\frac{5}{6}\right)^4 \simeq 0.5177.$$

Similarly the probability of never getting a double 6 in 24 throws of two dice is $(35/36)^{24}$, so that the probability of getting at least a double 6 is

$$1 - \left(\frac{35}{36}\right)^{24} \simeq 0.4914,$$

which is less than the previous probability.

In general, the probability of *at least* one success in n trials is

$$1 - (1 - p)^n,$$

not np, as Pascal thought. Notice that if p is small enough, then $1 - (1 - p)^n$ and np are quite close since $1 - (1 - p)^n = np + O(p^2)$ as $p \to 0$.

Exercise 2.37 *Compute the probability of getting the first success at the kth trial of a Bernoulli processs of n trials, $n \geq k \geq 1$.*

Solution. Let p be the probability of success in a single trial, $0 \le p \le 1$. Clearly, it suffices to play k trials and compute the probability of the k-tuple $\overline{x} = (0, 0, 0, \dots, 0, 1)$, i.e.

$$\mathbb{P}(\{\overline{x}\}) = (1-p)^{k-1} p.$$

Exercise 2.38 *Compute the probability of getting k failures before obtaining the first success in a Bernoulli process of trials.*

Solution. Let p be the probability of success in a single trial, $0 \le p \le 1$. It suffices to play $k+1$ trials and compute the probability of the $(k+1)$-tuple $\overline{x} = (0, 0, 0, \dots, 0, 1)$. Therefore,

$$\mathbb{P}(\{\overline{x}\}) = (1-p)^k p.$$

Exercise 2.39 *Compute the probability of getting k failures before obtaining the nth success in a Bernoulli process of trials.*

Solution. Let p be the probability of success in a single trial, $0 \le p \le 1$. It suffices to play $n+k$ trials to get the answer. We want to compute the probability of the event E containing all the sequences of $\{0, 1\}^{n+k}$ with k zeroes and n ones, whose $(n+k)$ component is one. Thus $|E| = \binom{n+k-1}{k}$, cf. Section 1.4.2, so that

$$\mathbb{P}(E) = \binom{n+k-1}{k}(1-p)^k p^n = \binom{-n}{k} p^n (p-1)^k.$$

Exercise 2.40 *Two players A and B play a 7 rubber game. In each match A wins with probability p and loses with probability* $1-p$. *Compute the following:*

- *The probability that the game ends after the fourth, the fifth, the sixth or the seventh match.*
- *The probability that A wins the game.*

Exercise 2.41 (Rubber games) *Two players A and B play a series of fair matches until one of them wins s matches. After some time A has won a matches and B has won b matches and* $a, b < s$. *Compute the probability that A wins the game.*

Solution. A must obtain a further $s-a$ successes before obtaining a further $s-b$ failures. Let E be this event. For any $k = 0, 1, \dots, s-b-1$, let E_k be the event 'A obtains k failures before obtaining $s-a$ successes'. The events $\{E_k\}$ are pairwise disjoint and $E = E_0 \cup E_1 \cup E_2 \dots E_{s-b-1}$. Thus,

$$\mathbb{P}(E) = \sum_{k=0}^{s-b-1} \mathbb{P}(E_k) = \sum_{k=0}^{s-b-1} \binom{s-a+k-1}{k} (0.5)^k (0.5)^{s-a}.$$

Exercise 2.42 (Production lines) *In a production line, each produced item has probability 0.03 of being faulty, independently of all the other produced items. Compute the following:*

- *The probability that* 3 *items out of* 100 *are faulty.*
- *The probability that the quality control inspector finds the first faulty item when inspecting the* 15*th item.*
- *The probability that the quality control inspector finds* 100 *nonfaulty items before finding the first faulty one.*
- *The probability that the quality control inspector finds at least* 5 *faulty items when checking a stock of* 200 *produced items.*
- *The probability that the quality control inspector finds exactly* 5 *faulty items when checking a stock of* 200 *produced items.*

Exercise 2.43 *We repeatedly throw a fair dice and count how many times we obtain a* 6*. What is the probability of having to throw the dice more than* 8 *times before getting a* 6*?*

Solution. We need to consider the infinite Bernoulli process and the Bernoulli distribution $\mathbb{P}$ of parameter $p = 1/6$. For any k let E_k be the event 'The first success is at the kth throw'. The events E_k, $k \geq 0$, are pairwise disjoint and we need to compute the probability of $E := \cup_{k=9}^{\infty} E_k$. Thus, by the σ-additivity property, we get

$$\begin{aligned}\mathbb{P}(E) &= \sum_{k=9}^{\infty} \mathbb{P}(E_k) = \sum_{k=9}^{\infty} p(1-p)^{k-1} = p\sum_{k=8}^{\infty}(1-p)^k \\ &= p(1-p)^8 \sum_{k=0}^{\infty}(1-p)^k = p(1-p)^8 \frac{1}{1-(1-p)} = (1-p)^8.\end{aligned}$$

Another procedure is the computation of the complementary event 'Either there is a success during the first 8 trials or there never is a success'. The probability of never obtaining a success is zero, see Example 2.33. Thus,

$$\begin{aligned}\mathbb{P}(E) &= 1 - \mathbb{P}(E^c) = 1 - \sum_{k=1}^{8} \mathbb{P}(E_k) = 1 - \sum_{k=0}^{7} p(1-p)^k \\ &= 1 - p\frac{1-(1-p)^8}{1-1+p} = (1-p)^8.\end{aligned}$$

Exercise 2.44 *Compute the probability of having to throw a fair dice more than* 50 *times in order to obtain* 6 *eight times.*

Solution. We consider the infinite Bernoulli process and the Bernoulli distribution $\mathbb{P}$ of parameter $p = 1/6$. We must compute the probability of having $n = 8$ successes, at least $k = 51 - n = 43$ failures and that the last trial is a success. For any k let E_k be the event 'We obtain k failures before getting 8 successes'. We want to compute the probability of the event $E = \cup_{k=43}^{\infty} E_k$, hence, see Exercise 2.39,

$$\mathbb{P}(E) = \sum_{k=43}^{\infty} \binom{8+k-1}{k} p^8 (1-p)^k = p^8 \sum_{k=43}^{\infty} \binom{8+k-1}{k} (1-p)^k$$

Since

$$\begin{aligned}\sum_{k=0}^{\infty} \binom{8+k-1}{k} (1-p)^k &= \sum_{k=0}^{\infty} \binom{-8}{k} (-(1-p))^k \\ &= (1-1+p)^{-8} = p^{-8},\end{aligned}$$

i.e.

$$p^8 \sum_{k=0}^{\infty} \binom{8+k-1}{k} (1-p)^k = 1,$$

we finally get

$$\mathbb{P}(E) = 1 - p^8 \sum_{k=0}^{42} \binom{8+k-1}{k} (1-p)^k.$$

Exercise 2.45 *Two players A and B flip in turn a fair coin. Player A is the first to play. The winner is the one who obtains the first head. What is the probability of the event 'A wins the game'?*

Solution. We consider the Bernoulli distribution of parameter $p = 0.5$. It is evident that A has more chances than B of winning the game, in fact, the probability of the event 'A wins the game at the first flipping' is $1/2$. Let E_k be the event 'A wins the game at the kth flipping'. Since A does only the odd flippings, then k must be odd, $k = 2h + 1$. The probability of obtaining the first success at the $(2h + 1)$th trial is $\mathbb{P}(E_{2h+1}) = (1 - p)^{2h} p$. Since the events $\{E_{2h+1}\}$ are pairwise disjoint and the event 'Player A wins the game' is $E = \cup_h E_{2h+1}$, we obtain

$$\begin{aligned}\mathbb{P}(E) &= \sum_{h=0}^{\infty} \mathbb{P}(E_{2h+1}) = p \sum_{h=0}^{\infty} (1-p)^{2h} \\ &= p \sum_{h=0}^{\infty} \left((1-p)^2\right)^h = \frac{p}{1-(1-p)^2} = \frac{1}{2-p}.\end{aligned}$$

Since $p = 0.5$, player A wins the game with probability $2/3$. Notice that $\frac{1}{2-p} > \frac{1}{2}$ for any p, $0 < p < 1$, so that, however unbalanced the coin is, player A has always more chances than player B.

2.3 Conditional probability

In this section we introduce the notion of *conditional probability*, a very useful tool when we want to compute the probability that two or more events are verified.

2.3.1 Definition

Definition 2.46 *Let $(\Omega, \mathcal{E}, \mathbb{P})$ be a probability space and let $A, B \in \mathcal{E}$ such that $\mathbb{P}(B) > 0$. The conditional probability of A given B, denoted by $\mathbb{P}(A \mid B)$ is defined as*

$$\mathbb{P}(A|B) := \frac{\mathbb{P}(A \cap B)}{\mathbb{P}(B)}. \tag{2.9}$$

$\mathbb{P}(A|B)$ is the probability of A when B is taken as the certain event. In fact, let $\mathbb{P}_B : A \in \mathcal{E} \mapsto \mathbb{P}(A|B) \in [0, 1]$. Then $(\mathcal{E}, \mathbb{P}_B)$ is a new measure on Ω such that $\mathbb{P}_B(B) = 1$ and, more generally, $\mathbb{P}_B(C) = 1$ $\forall C \in \mathcal{E}$ such that $C \supset B$.

Example 2.47 Throw a fair dice. The probability of the event A ='we get 6' is $1/6$, the probability of the event B ='we get an even number' is $3/6 = 1/2$ and the probability of getting 6 *knowing that we get an even number* is

$$\mathbb{P}(A|B) = \frac{\mathbb{P}(A \cap B)}{\mathbb{P}(B)} = \frac{\mathbb{P}(A)}{\mathbb{P}(B)} = \frac{1/6}{1/2} = 1/3.$$

By definition $\mathbb{P}(A|B)$ can be used to compute $\mathbb{P}(A \cap B)$. In fact, formula (2.9) rewrites as the *multiplication formula*

$$\mathbb{P}(A \cap B) = \mathbb{P}(A|B)\mathbb{P}(B) \tag{2.10}$$

if $\mathbb{P}(B) > 0$. If $\mathbb{P}(B) = 0$, $\mathbb{P}(A|B)$ is not defined; to make (2.10) meaningful also in the case $\mathbb{P}(B) = 0$, one sets $\mathbb{P}(A|B)\mathbb{P}(B) = 0$ if $\mathbb{P}(B) = 0$.

Let $\{D_i\}$ be a partition of Ω. Then the *total probability formula* $\mathbb{P}(A) = \sum_{i=1}^{\infty} \mathbb{P}(A \cap D_i)$ reads

$$\mathbb{P}(A) = \sum_{i=1}^{\infty} \mathbb{P}(A \mid D_i)\mathbb{P}(D_i). \tag{2.11}$$

Example 2.48 Two cards are picked from a deck of 40. Compute the probability of picking two aces. We already know that the required probability is the ratio between successful and possible cases, i.e. $\binom{4}{2}/\binom{40}{2} = \frac{1}{130}$.

Another procedure is the following. Let A and B be the events 'The first picked card is an ace' and 'The second picked card is an ace', respectively. Then we must compute $\mathbb{P}(A \cap B)$. Clearly, $\mathbb{P}(A) = 4/40 = 1/10$ and the probability of the second card being an ace, given that the first card was an ace also, is $3/39$. Thus, by the multiplication formula,

$$\mathbb{P}(A \cap B) = \mathbb{P}(A|B)\mathbb{P}(B) = \frac{4 \cdot 3}{40 \cdot 39} = \frac{1}{130}.$$

We conclude observing that formula (2.10) can be iterated. For three events A, B and C with $\mathbb{P}(B \cap C) > 0$ we have

$$\begin{aligned}\mathbb{P}(A \cap B \cap C) &= \mathbb{P}(A|B \cap C)\mathbb{P}(B \cap C) \\ &= \mathbb{P}(A|B \cap C)\mathbb{P}(B \mid C)\mathbb{P}(C);\end{aligned} \tag{2.12}$$

for a finite family $\{A_1, A_2, \dots, A_n\}$ of events, such that $\mathbb{P}(A_2 \cap \dots \cap A_n) > 0$ one gets

$$\begin{aligned}\mathbb{P}(A_1 \cap A_2 \cap \dots \cap A_n) =&\mathbb{P}(A_1 \mid A_2 \cap \dots \cap A_n) \\ &\cdot \mathbb{P}(A_2 \mid A_3 \cap \dots \cap A_n) \\ &\cdot \mathbb{P}(A_3 \mid A_4 \cap \dots \cap A_n) \\ &\dots \\ &\cdot \mathbb{P}(A_{n-1} \mid A_n)\mathbb{P}(A_n).\end{aligned} \tag{2.13}$$

2.3.2 Bayes formula

Let $(\Omega, \mathcal{E}, \mathbb{P})$ be a probability space and let $A, B \in \mathcal{E}$. If both $\mathbb{P}(A)$ and $\mathbb{P}(B)$ are positive, then

$$\mathbb{P}(A|B)\mathbb{P}(B) = \mathbb{P}(A \cap B) = \mathbb{P}(B \cap A) = \mathbb{P}(B|A)\mathbb{P}(A);$$

since

$$\mathbb{P}(A) = \mathbb{P}(A \cap B) + \mathbb{P}(A \cap B^c) = \mathbb{P}(A|B)\mathbb{P}(B) + \mathbb{P}(A|B^c)(1 - \mathbb{P}(B))$$

we get the formula due to Thomas Bayes (1702–1761).

Proposition 2.49 (Bayes formula) *Let $(\Omega, \mathcal{E}, \mathbb{P})$ be a probability space and let $A, B \in \mathcal{E}$ such that $\mathbb{P}(A)\mathbb{P}(B)\mathbb{P}(B^c) > 0$. Then*

$$\mathbb{P}(B|A) = \frac{\mathbb{P}(A|B)\mathbb{P}(B)}{\mathbb{P}(A|B)\mathbb{P}(B) + \mathbb{P}(A|B^c)(1 - \mathbb{P}(B))}.$$

In other words, one can compute the probability of B given A in terms of the probability of A given B *and the absolute probability of* B.

Example 2.50 Clinical tests are a classical example. Clinical tests detect individuals ill with a disease. The test gives a result that can be either positive or negative. In order to qualify the ability of detecting ill individuals, an 'unbiased sample' of a population is examined and divided into two groups of healthy and ill individuals, respectively. Individuals of the sample are then subjected to the test. Let α and β be the percentages of positive results for ill and healthy individuals, respectively. Since the sample is assumed to be representative of the whole population, we assume α and β to be the percentages of positive results for all the ill individuals and all the healthy individuals of the population, respectively. Denoting by H and I the sets of healthy and ill individuals, respectively, and denoting by P the set of individuals that would get a positive result in the test, we assume

$$\alpha := \mathbb{P}(P|I), \qquad \beta = \mathbb{P}(P|H);$$

From α, β and the percentage $x := \mathbb{P}(I)$ of ill individuals, Bayes formula allows us to compute the probability that an individual is ill given a positive result was obtained:

$$\mathbb{P}(I|P) = \frac{\mathbb{P}(P|I)\mathbb{P}(I)}{\mathbb{P}(P|I)\mathbb{P}(I) + \mathbb{P}(P|H)(1 - \mathbb{P}(I))} = \alpha \frac{1}{(\alpha - \beta) + \frac{\beta}{x}}. \tag{2.14}$$

If the test is a good one, then α is close to 1 and β is close to zero. Formula (2.14) shows how, in practice, the test loses efficacy in the field with respect to the quality shown in the laboratory. In fact, if $x = \beta$ one gets

$$\frac{1}{(\alpha - \beta) + \frac{\beta}{x}} \simeq \frac{1}{\alpha - \beta + 1} \simeq \frac{1}{2}.$$

If the test is to be helpful in detecting ill individuals, then the ratio $\beta/x = \mathbb{P}(P|H)/\mathbb{P}(I)$ must be small.

Of course, in the previous example, one can replace the population with the items produced in a production process, where healthy individuals are replaced by non-faulty items and ill individuals correspond to faulty items. The clinical test is, of course, substituted by a quality control. The mathematics involved is exactly the same. Knowing the probability that the test has of detecting known faulty and non faulty items and the percentage of faulty items, we can compute the probability that the test is able to detect the faulty items.

Example 2.51 In modelling an email spam filter, one needs to compute the probability that the messages containing a given word are spam messages. Let M be the set of spam messages and let W be the set of the messages that contain the given word. We want to compute $\mathbb{P}(M|W)$. By Bayes formula, it suffices to know:

(i) $\mathbb{P}(W|M)$ and $\mathbb{P}(W|M^c)$, i.e. the probabilities that the word appears in spam messages and in non-spam messages, respectively. These probabilities can be obtained by a statistical analysis on the messages already arrived.

(ii) $\mathbb{P}(M)$, the probability that a message is a spam one. This is an absolute number that can be found in the net or via a statistical analysis of the traffic.

We then get

$$\mathbb{P}(M|W) = \mathbb{P}(W|M) \frac{\mathbb{P}(M)}{\mathbb{P}(W|M)\mathbb{P}(M) + \mathbb{P}(W|M^c)(1 - \mathbb{P}(M))}.$$

2.3.3 Exercises

Exercise 2.52 *An urn contains* 3 *white balls and* 5 *red balls. The balls are pairwise distinguished, for example by numbering them. Two balls are extracted* randomly.

- *Compute the probability of drawing 2 red balls.*
- *Compute the probability of drawing 2 red balls, given that at least one is red.*

Solution. For the first question, the certain event Ω is given by all the subsets having cardinality 2 of a set of 8 elements, thus

$$|\Omega| = \binom{8}{2} = 28.$$

The favourable event A is in a one-to-one correspondence with the family of the subsets of cardinality 2 of a set of 5 elements, i.e.

$$|A| = \binom{5}{2} = 10, \qquad \mathbb{P}(A) = \frac{|A|}{|\Omega|} = \frac{10}{28} = \frac{5}{14}.$$

For the second question, we take advantage of Bayes formula: Let B be the event 'At least one of the extracted ball is red'. One gets

$$|B| = \mathbf{C}_2^5 + \mathbf{C}_1^5\mathbf{C}_1^3 = \binom{5}{2} + \binom{5}{1}\binom{3}{1} = 10 + 15 = 25,$$

$$\mathbb{P}(B) = \frac{|B|}{|\Omega|} = \frac{25}{28}.$$

Thus

$$\mathbb{P}(A|B) = \frac{\mathbb{P}(A \cap B)}{\mathbb{P}(B)} = \frac{\mathbb{P}(A)}{\mathbb{P}(B)} = \frac{|A|}{|B|} = \frac{10}{25} = \frac{2}{5}.$$

Find the mistake. In the second question the certain event is smaller. In fact, since one of the drawn balls is red, the problem reduces to a simpler one: we may think that the urn contains 3 white balls and 4 red ones and that we are to extract only one ball. The successful cases are the ones where the drawn ball is red. Denote by $\widetilde{\Omega}$ the new certain event and let $\widetilde{A}$ be the new event of the favourable cases

$$|\widetilde{\Omega}| = \binom{7}{1} = 7, \quad |\widetilde{A}| = \binom{4}{1} = 4.$$

Thus,

$$\mathbb{P}(\widetilde{A}) = \frac{4}{7}.$$

This wrong answer is in fact the right answer for another question. Which one?

Exercise 2.53 *We throw two fair dice:*

- *Compute the probability that the sum of the two dice is at least* 7.
- *On one of the two dice we get a* 4. *Compute the probability that the sum of the two dice is at least* 7.

Solution. The certain event Ω is the set of all the possible results of the throwing: $\Omega = \{1, \ldots, 6\}^2$, $|\Omega| = 36$. The event of successes is

$$A = \{(i, j) \in \Omega : i + j \geq 7\}$$

whose cardinality is $|A| = 1 + 2 + \ldots + 6 = \frac{6 \cdot 7}{2} = 21$. Since the dice are fair, the probability measure $\mathbb{P}$ on Ω is the uniform one. Thus,

$$\mathbb{P}(A) = \frac{|A|}{|\Omega|} = \frac{21}{36} = \frac{7}{12}.$$

In the second case, it is known that on one of the two dice we get a 4. Consider the event

$$B = \left\{(i, j) \in \Omega \,\middle|\, i = 4 \text{ or } j = 4\right\}.$$

Bayes formula then gives

$$\mathbb{P}(A|B) = \frac{\mathbb{P}(A \cap B)}{\mathbb{P}(B)} = \frac{|A \cap B|}{|B|} = \frac{7}{11}.$$

Find the mistake. In the second question the certain event can be reduced from Ω to a smaller one. In fact, since on a dice we get a 4, we may assume we are throwing only a dice. The successful event is 'We get a 3 or more on the thrown dice'. Thus, denoting by $\widetilde{\Omega}$ the new certain event and by $\widetilde{A}$ the successful event, we get

$$|\widetilde{\Omega}| = 6, \quad |\widetilde{A}| = 4,$$

so that

$$\mathbb{P}(\widetilde{A}) = \frac{4}{6} = \frac{2}{3}.$$

Exercise 2.54 *An urn contains w white balls and r red balls. Compute the probability of drawing one white ball in one drawing only. Assume you do not know the colour of the first drawn ball. Without replacing it, you draw another ball. Compute the probability of drawing a white ball in this second drawing.*

Solution. In this case, the certain event is the fact that two balls are drawn and each of them can be either red or white. So we take $\Omega = \{0, 1\}^2$ where the ith component of $z = (z_1, z_2) \in \Omega$ is zero if the ith drawn is white and 1 if the ith drawn is red. Thus the event 'The first ball is white' is associated with

$$W_1 = \{(0, 0), (0, 1)\}$$

so that $\mathbb{P}(W_1) := w/(w+r)$. In fact, when we draw 'randomly' the first ball we are performing a Bernoulli trial where the probability of success is $w/(w+r)$. Similarly, the event 'The first ball is red' is associated with

$$R_1 = \{(1,0), (1,1)\} = \Omega \setminus W_1$$

so that $\mathbb{P}(R_1) = 1 - \mathbb{P}(W_1) = r/(r+w)$. We are asked to compute the probability of the event 'The second ball is white', which is is associated with

$$W_2 = \{(0,0), (1,0)\}.$$

Once we have drawn the first ball, two cases may occur. If the first ball is white, then we are left with $w-1$ white balls and r red balls, i.e.

$$\mathbb{P}(W_2 \mid W_1) = \frac{w-1}{w+r-1},$$

and, similarly,

$$\mathbb{P}(W_2 \mid R_1) = \frac{w}{w+r-1}.$$

The law of total probability then yields

$$\begin{aligned}
\mathbb{P}(W_2) &= \mathbb{P}(W_2 \cap W_1) + \mathbb{P}(W_2 \cap R_1) \\
&= \mathbb{P}(W_2|W_1)\mathbb{P}(W_1) + \mathbb{P}(W_2|R_1)\mathbb{P}(R_1) \\
&= \frac{w-1}{w+r-1}\frac{w}{w+r} + \frac{w}{w+r-1}\frac{r}{w+r} = \frac{w}{w+r}.
\end{aligned}$$

Notice that, without any knowledge of the colour of the first drawn ball, the probability of drawing a ball of the same colour in the second drawing remains the same. Thus, iterating the result, we get the following: without any knowledge of the colour of the previously drawn balls, the probability remains the same also in the following drawings [as far as the drawn balls are less than $\min(r, w)$].

Exercise 2.55 *Two submarines comunicate with each other by Morse code, a binary code whose two letters are dots and dashes. The operator of the first submarine sends a message where 45% of the letters are dashes.*

The operator of the second submarine receives dashes and dots correctly with probability 96% and 98% , respectively. Compute the following:

- *The probability that the operator of the second submarine receives a dash.*
- *The probability that the first operator has sent a dash, given that the second operator has received a dash.*
- *The probability that the first operator has sent a dot given that the second operator has received a dot.*

Solution. For a single bit of the comunication the following cases may occur:

- When considering its transmission, the letter can be either a dot (0) or a dash (1).
- When considering its reception, the letter can be either a dot (0) or a dash (1).

Thus we may identify the certain event with $\Omega = \{0, 1\}^2$ where the first component of $z = (z_1, z_2) \in \Omega$ is 0 if a dot is transmitted and is 1 if a dash is transmitted, and the second component z_2 is 0 if a dot is received and 1 otherwise. Thus, the event 'A dot is transmitted' is associated with

$$S_0 = \{(0, 0),\ (0, 1)\},$$

the event 'A dash is transmitted' is associated with

$$S_1 = \{(1, 0),\ (1, 1)\} = \Omega \setminus S_0$$

while the events 'A dot is received' and 'A dash is received' are associated with

$$R_0 = \{(0, 0),\ (1, 0)\} \quad \textit{and} \qquad R_1 = \{(0, 1),\ (1, 1)\} = \Omega \setminus R_0,$$

respectively. Assumption tells us that $\mathbb{P}(S_1) := 0.45$ so that $\mathbb{P}(S_0) = 1 - \mathbb{P}(S_1) = 0.55$ and that $\mathbb{P}(R_0 \mid S_0) = 0.98$ and $\mathbb{P}(R_1 \mid S_1) = 0.96$.

We are asked to compute $\mathbb{P}(R_1)$, $\mathbb{P}(S_1|R_1)$ and $\mathbb{P}(S_0|R_0)$. We have

$$\begin{aligned}
\mathbb{P}(R_1) &= \mathbb{P}(R_1|S_0)\mathbb{P}(S_0) + \mathbb{P}(R_1|S_1)\mathbb{P}(S_1) \\
&= (1 - \mathbb{P}(R_0|S_0))\mathbb{P}(S_0) + \mathbb{P}(R_1|S_1)\mathbb{P}(S_1) \\
&= (1 - 0.98)(0.55) + (0.96)(0.45) = 0.443,
\end{aligned}$$

$$\mathbb{P}(S_1|R_1) = \frac{\mathbb{P}(S_1 \cap R_1)}{\mathbb{P}(R_1)} = \frac{\mathbb{P}(R_1|S_1)\mathbb{P}(S_1)}{\mathbb{P}(R_1)} \simeq 0.975,$$

$$\mathbb{P}(S_0|R_0) = \frac{\mathbb{P}(S_0 \cap R_0)}{\mathbb{P}(R_0)} = \frac{\mathbb{P}(R_0|S_0)\mathbb{P}(S_0)}{1 - \mathbb{P}(R_1)} \simeq 0.968.$$

Exercise 2.56 *We have n urns, labelled 1 to n. The ith urn contains i white balls and $n + 1 - i$ red balls. Two balls are randomly drawn from a randomly chosen urn:*

- *Compute the probability of drawing a white ball and a red one.*
- *Compute the probability of having chosen the ith urn given that a white ball and a red ball are drawn. Which urn has the maximum probability of having been chosen?*

Solution. Here the certain event is that fact that one urn is chosen and two balls are drawn from that urn, so we identify the certain event with

$$\Omega := \{1, \dots, n\} \times \big\{z = (z_1, z_2) \mid z_i \in \{0, 1\}\big\}$$

where $z_j = 0$ if the jth drawn ball is red and $z_j = 0$ if the jth drawn ball is white.

Denote $N := \{1, \dots, n\}$ and $Z = \{0, 1\}^2$ so that $\Omega = N \times Z$. We know that the urn is 'randomly chosen', i.e.

$$\mathbb{P}(\{i\} \times Z) = \frac{1}{n} \qquad \forall i = 1, \dots, n.$$

The event 'A white ball and a red ball are drawn' is associated with

$$A := N \times \{(0, 1),\ (1, 0)\}$$

Since the balls are 'randomly' drawn, we know that

$$\mathbb{P}(A \mid U_i) = \frac{\binom{i}{1}\binom{n+1-i}{1}}{\binom{n+1}{2}} = \frac{2i(n+1-i)}{n(n+1)}.$$

Thus, since $A = \bigcup_{i=1}^{n} (A \cap U_i)$, we get

$$\begin{aligned}\mathbb{P}(A) &= \sum_{i=1}^{n} P(A \cap U_i) = \sum_{i=1}^{n} P(A|U_i)\,\mathbb{P}(U_i)\\ &= \sum_{i=1}^{n} \frac{2i(n+1-i)}{n(n+1)}\frac{1}{n} = \frac{n+2}{3n}.\end{aligned}$$

To answer the second question, we apply Bayes formula, hence

$$\mathbb{P}(U_i|A) = \frac{\mathbb{P}(A|U_i)\mathbb{P}(U_i)}{\mathbb{P}(A)} = \frac{2i(n+1-i)}{n(n+1)}\frac{1}{n}\frac{3n}{n+2} = \frac{6i(n+1-i)}{n^2(n+1)(n+2)}.$$

The urn with the maximum probability of having been chosen is U_i where

$$i = \begin{cases} k,\ k+1 & \text{if } n = 2k,\\ k+1 & \text{if } n = 2k+1.\end{cases}$$

Exercise 2.57 *A firm produces a certain instrument. The failure rate in the production process is 8%. A sample of the produced instruments is subjected to a quality check. Eighty per cent of the faulty instruments fail the quality check, while 1% of the nonfaulty instruments fails the check. Compute the probability that an instrument that passed the test is faulty.*

Exercise 2.58 *We have three cards. Both sides of one card are black, both sides of another card are red and the other card has one black side and one red side. We choose a card: one of its side is red. Compute the probability of having chosen the card with two red sides.*

Exercise 2.59 *An urn* a *contains* 2 *red balls and* 1 *black ball. Another urn,* b*, contains* 3 *red balls and* 2 *black ones. An urn is randomly chosen and a ball is randomly drawn from that urn. Compute the probability of having drawn a black ball. Given that you have drawn a black ball, what is the probability of having chosen urn* a*?*

Exercise 2.60 *In a TV quiz show, the host shows three doors to a contestant and states that behind one of the doors there is a prize. The contestant randomly chooses a door and marks it. Then the host, who knows where the prize is, opens a door that was not marked by the contestant and that does not hide the prize. The host now asks the contestant whether he would like to change his choice. What should the contestant do?*

Exercise 2.61 *A firm produces two kinds of bolt. Bolts of type* A *are* 70% *of the production. Only* 95% *of them pass the quality check and can be sold, while only* 80% *of bolts of type* B *pass the check and can be sold. What percentage of the production is fit to be sold?* [90.5%]

Exercise 2.62 *A web server may be either working or broken. When broken, it cannot be accessed. Even when it is working, an access attempt may fail, due to web congestion (which the web server cannot control). Assume the probability that the server is working is* 0.8 *and that each access attempt is successful with probability* 0.9*. Compute the following:*

(i) *The probability that the first access attempt fails.*

(ii) *The probability that the web server is working, given that the first access attempt has failed.*

(iii) *The probability that the second attempt fails, given that the first attempt has failed.*

(iv) *The probability that the server is working, given that two consecutive access attempts have failed.*

Exercise 2.63 *Assume that each vertex of a cube may be coloured in blue with probability* p*, independently of the other vertices. Let* B *be the event 'All the vertices of at least one face of the cube are coloured in blue'. Compute;*

(i) *The probability of* B *given that* 5 *vertices are blue.*

(ii) *The probability of* B*.*

Exercise 2.64 *We are given 3 urns a, b and c. Urn a contains 5 red balls, 5 five white balls and 5 black ones; urn b contains 3 red balls, 5 white balls and 7 black balls; urn c contains 6 red balls, 7 white balls and 2 black balls. Two balls are randomly drawn by one randomly chosen urn (no replacement occurs). Given that the drawn balls are both black, compute the probability that they were drawn from urn b.* $\left[21/32\right]$

Exercise 2.65 *We are given six urns, which we label as U_j $j = 1, \dots, 6$. Each urn U_j contains j red balls and $6 - j$ black balls. An urn is selected. The probability of selecting the urn U_j is proportional to j. A ball is randomly drawn from the selected urn. Compute the probability that the drawn ball is red.* $\left[13/18\right]$

2.4 Inclusion–exclusion principle

A typical problem is the computation of the probability of at least one event among n given ones: In this section we show how to compute;

- the probability of at least one event among $n$given ones;
- the probability of exactly k events among n;
- the probability of at least k events among n.

With the *inclusion–exclusion principle* we can compute these probabilities knowing the probabilities that two or more events happen, something which is usually more easily evaluated. For example, if $A, B \in \mathcal{E}$, then $A \setminus B$, $A \cap B$ and $B \setminus A$ are pairwise disjoint, so that

$$\begin{aligned} \mathbb{P}(A \cup B) &= \mathbb{P}(A \setminus B) + \mathbb{P}(A \cap B) + \mathbb{P}(B \setminus A) + \mathbb{P}(A \cap B) - \mathbb{P}(A \cap B) \\ &= \mathbb{P}(A) + \mathbb{P}(B) - \mathbb{P}(A \cap B) \end{aligned} \tag{2.15}$$

and

$$\mathbb{P}((A \cup B)^c) = 1 - \mathbb{P}(A \cup B) = 1 - \mathbb{P}(A) - \mathbb{P}(B) + \mathbb{P}(A \cap B). \tag{2.16}$$

In order to discuss the case of many events, it is useful to introduce a convenient notation. A *multi-index* is a subset $I \subset \{1, \dots, n\}$. The cardinality $|I|$ of the multi-index I is called the *length* of I. Let Ω be a set and let $A_1, \dots, A_n$ be nonempty subsets of Ω. For any multi-index $I \subset \{1, \dots, n\}$, define

$$A_I := \begin{cases} \Omega & \text{if } I = \emptyset, \\ A_{i_1} \cap A_{i_2} \cap \dots \cap A_{i_k} & \text{if } k \geq 1 \text{ and } I = \left\{i_1, \dots, i_k\right\}. \end{cases} \tag{2.17}$$

Example 2.66 Let $A_1, A_2, A_3 \subset \Omega$ be subsets of Ω. The multi-indexes of $\{1, 2, 3\}$ are

$$\emptyset,\ \{1\},\ \{2\},\ \{3\},\ \{1,2\},\ \{1,3\},\ \{2,3\},\ \{1,2,3\}$$

and the associated subsets A_I, $I \subset \{1, 2, 3\}$, are, respectively,

$$\begin{aligned}
&A_\emptyset = \Omega, && && \\
&A_{\{1\}} = A_1, && A_{\{2\}} = A_2, && A_{\{3\}} = A_3, \\
&A_{\{1,2\}} = A_1 \cap A_2, && A_{\{2,3\}} = A_2 \cap A_3, && A_{\{1,3\}} = A_1 \cap A_3, \\
&A_{\{1,2,3\}} = A_1 \cap A_2 \cap A_3. && &&
\end{aligned}$$

Definition 2.67 *Let Ω be a set and let $A_1, \dots, A_n \subset \Omega$. We call the* multiplicity *of the family $A_1, \dots, A_n$ at $x \in \Omega$ the number* $\mathrm{val}(x)$ *of subsets of the family $A_1, \dots, A_n$ that contain x:*

$$\mathrm{val}(x) := \left|\left\{j \in \{1, \dots, n\} \,\middle|\, x \in A_j\right\}\right|.$$

Moreover, for every $k = 0, \dots, n$, let

$$E_k := \left\{x \in \Omega \,\middle|\, \mathrm{val}(x) = k\right\}$$

be the set of points of Ω that belong to exactly k subsets of the family $A_1, \dots, A_n$.

Clearly,

$$(A_1 \cup A_2 \cup \dots \cup A_n)^c = \left\{x \in \Omega \,\middle|\, \mathrm{val}(x) = 0\right\} = E_0.$$

Proposition 2.68 *Let $A_1, \dots, A_n$ be nonempty subsets of a set Ω, let* $\mathrm{val}(x)$ *be the multiplicity of the family $A_1, \dots, A_n$ at $x \in \Omega$ and let E_k, $k = 0, 1, \dots, n$, be the kth level set of the multiplicity function* $\mathrm{val}(x)$. *Then*

$$\sum_{|J|=j} \mathbb{1}_{A_J}(x) = \binom{\mathrm{val}(x)}{j} = \sum_{k=0}^{n} \binom{k}{j} \mathbb{1}_{E_k}(x) \qquad \forall j = 0, \dots, n. \tag{2.18}$$

and

$$\mathbb{1}_{E_k}(x) = \sum_{j=k}^{n} (-1)^{k+j} \binom{j}{k} \Big(\sum_{|J|=j} \mathbb{1}_{A_J}(x) \Big) \qquad \forall j = 0, \dots, n. \tag{2.19}$$

Proof. The number

$$\sum_{|J|=j} \mathbb{1}_{A_J}(x) = \sum_{\substack{|J|=j \\ x \in A_J}} 1$$

is the number of subfamilies of $\{A_1, \ldots, A_n\}$ of cardinality j whose elements contain x. Since x belongs to $\mathrm{val}(x)$ subsets, there are $\binom{\mathrm{val}(x)}{j}$ ways to choose any such subfamily, hence the equality

$$\sum_{|J|=j} \mathbb{1}_{A_J}(x) = \binom{\mathrm{val}(x)}{j}$$

in (2.18). Moreover, the second equality in (2.18) follows since $\binom{\mathrm{val}(x)}{j}$ if and only if $x \in E_k$,

Formula (2.19) is an equivalent formulation of (2.18) as an application of the inversion formula for the Pascal matrix of binomial coefficients, see Corollary 1.6.

Theorem 2.69 *Let* $(\Omega, \mathcal{E}, \mathbb{P})$ *be a probability space and let* $A_1, \ldots, A_n \in \mathcal{E}$. *Let* $\mathrm{val}(x)$, $x \in \Omega$, *be the multiplicity function of* $A_1, \ldots, A_n$ *and for every* $k = 0, \ldots, n$ *let* E_k *be the k-level set of* $\mathrm{val}(x)$. *Then*

$$\sum_{|J|=j} \mathbb{P}(A_J)(x) = \sum_{k=0}^{n} \binom{k}{j} \mathbb{P}(E_k)(x) \qquad \forall j = 0, \ldots, n \tag{2.20}$$

and

$$\mathbb{P}(E_k) = \sum_{j=k}^{n} (-1)^{k+j} \binom{j}{k} \Big(\sum_{|J|=j} \mathbb{P}(A_J) \Big) \qquad \forall k = 0, \ldots, n. \tag{2.21}$$

Proof. Notice that all the terms in the sums in (2.18) and (2.19) are linear combinations of simple functions. Thus, (2.20) and (2.21) follow integrating with respect to $\mathbb{P}$ both terms of (2.18) and (2.19), respectively, taking into account the linearity of the integral of simple functions, see Proposition 2.28.

Equation (2.21) reduces the computation of the probability that exactly k events among n happen to the computation of the probabilities of the intersections of the events of the family. As a consequence, we get the *inclusion–exclusion principle*, a particularly useful tool for computing the probability that at least one event among n happens.

Corollary 2.70 *With the notation above, we have the* Sylvester formula

$$\mathbb{P}\Big(\Big(\bigcup_{i=1}^{n} A_i \Big)^c \Big) = \sum_{j=0}^{n} (-1)^j \Big(\sum_{|J|=j} \mathbb{P}(A_J) \Big) \tag{2.22}$$

and the inclusion–exclusion principle

$$\mathbb{P}\Big(\bigcup_{i=1}^{n} A_i \Big) = \sum_{j=1}^{n} (-1)^{j+1} \Big(\sum_{|J|=j} \mathbb{P}(A_J) \Big). \tag{2.23}$$

Proof. Since $(A_1 \cup \cdots \cup A_n)^c = E_0$, (2.22) is (2.21) with $k = 0$. Moreover, since $\mathbb{P}(\cup_{i=1}^n A_i) = 1 - \mathbb{P}((\cup_{i=1}^n)A_i)^c)$, and $\sum_{|J|=0} \mathbb{P}(A_J) = \mathbb{P}(\Omega) = 1$, (2.23) follows from, and, actually, is equivalent to, (2.22).

Summing (2.21) for $k = k, k+1, \dots, n$, we get an useful formula for computing the probability that at least k events among n happen. In fact, the following holds.

Corollary 2.71 *Let $(\Omega, \mathcal{E}, \mathbb{P})$ be a probability space, let $A_1, \dots, A_n \in \mathcal{E}$ and, for $k = 0, 1, \dots, n$, let F_k be the set of points of Ω that belong to at least k sets of the family $A_1, \dots, A_n$, $k \geq 1$,*

$$F_k := \left\{x \in \Omega \,\middle|\, \mathrm{val}(x) \geq k\right\}.$$

Then

$$\mathbb{P}(F_k) = \sum_{j=k}^{n} (-1)^{j+k} \binom{j-1}{k-1} \Bigg(\sum_{\substack{J \subset \{1,\dots,n\} \\ |J|=j}} \mathbb{P}(A_J) \Bigg). \tag{2.24}$$

Proof. For $p = 0, 1, \dots, n$, let $E_p := \{x \in \Omega \,|\, \mathrm{val}(x) = p\}$. Obviously, the sets E_p are pairwise disjoint and $F_k = \cup_{p=k}^n E_p$. Thus, by (2.19) for any $x \in \Omega$

$$\begin{aligned}
\mathbb{1}_{F_k}(x) &= \sum_{p=k}^{n} \mathbb{1}_{E_p}(x) = \sum_{p=k}^{n} \sum_{j=p}^{n} (-1)^{p+j} \binom{j}{p} \Bigg(\sum_{\substack{J \subset \{1,\dots,n\} \\ |J|=j}} \mathbb{1}_{A_J}(x) \Bigg) \\
&= \sum_{j=k}^{n} (-1)^j \Bigg(\sum_{p=k}^{j} (-1)^p \binom{j}{p} \Bigg) \Bigg(\sum_{\substack{J \subset \{1,\dots,n\} \\ |J|=j}} \mathbb{1}_{A_J}(x) \Bigg).
\end{aligned}$$

Since

$$\sum_{p=k}^{j} (-1)^p \binom{j}{p} = -\sum_{p=0}^{k-1} (-1)^p \binom{j}{p} = -(-1)^{k-1} \binom{j-1}{k-1},$$

see Exercise A.13, one gets

$$\mathbb{1}_{F_k}(x) = \sum_{j=k}^{n} (-1)^{j+k} \binom{j-1}{k-1} \Bigg(\sum_{\substack{J \subset \{1,\dots,n\} \\ |J|=j}} \mathbb{1}_{A_J}(x) \Bigg).$$

The claim then follows by integrating the previous equality with respect to $\mathbb{P}$.

2.4.1 Exercises

Exercise 2.72 (Derangements) *Taking advantage of the inclusion–exclusion principle (2.23), compute the number of derangements of n objects, see Section 1.2.2,*

$$|\mathcal{D}_n| := n! \sum_{j=0}^{n} (-1)^j \frac{1}{j!}.$$

Solution. Let $\mathcal{P}_n$ and $\mathcal{D}_n$ be the set of permutations and of derangements of n objects, respectively. For any $i = 1, \dots, n$, let $A_i \subset \mathcal{P}_n$ be the set of permutations that do not move i,

$$A_i = \left\{\pi \in \mathcal{P}_n \,\middle|\, \pi(i) = i\right\}.$$

Clearly,

$$\mathcal{D}_n = (A_1 \cup A_2 \cup \dots \cup A_n)^c$$

so that, by the inclusion–exclusion principle,

$$|\mathcal{D}_n| = \sum_{J \subset \{1,\dots,n\}} (-1)^{|J|} |A_J| = \sum_{j=0}^{n} (-1)^j \sum_{\substack{J \subset \{1,\dots,n\} \\ |J|=j}} |A_J|.$$

If $J = \{k_1, \dots, k_j\}$ then A_J is the set of permutations that do not move the points $\{k_1, \dots, k_j\}$. These permutations are as many as the permutations of the $n - j$ remaining elements, i.e. $|A_J| = (n - j)!$ Thus, denoting by c_j the number of multi-indexes J of length j in $\{1, \dots, n\}$, we get

$$|\mathcal{D}_n| = \sum_{j=0}^{n} (-1)^j (n - j)! \Bigg(\sum_{\substack{J \subset \{1,\dots,n\} \\ |J|=j}} 1 \Bigg) = \sum_{j=0}^{n} (-1)^j (n - j)!\, c_j.$$

Since $c_j = \binom{n}{j}$ we finally get

$$|\mathcal{D}_n| = \sum_{j=0}^{n} (-1)^j (n - j)! \binom{n}{j} = n! \sum_{j=0}^{n} (-1)^j \frac{1}{j!}.$$

Exercise 2.73 (Ménage problem) *We ask for the number of different ways in which it is possible to seat a set of n heterosexual couples at a dining round table so that men and women alternate and nobody sits next to his or her partner. The* ménage problem *is a classical problem in combinatorics, first formulated by Édouard Lucas in 1891.*

Solution. Let K_n be the possible arrangements. If $n = 1$ or $n = 2$ then obviously $K_n = 0$. If $n = 3$, there are $3! = 6$ possible arrangements with the given constraints. For larger n, $n \geq 4$, we can proceed as follows: there are $n!$ seating arrangements for women, thus it remains to count the number $U_n \subset \mathcal{P}_n$ of permutations of seating men away from their wives.

Theorem (Touchard) *We have*

$$K_n = n!\, U_n, \qquad U_n := \sum_{k=0}^{n} (-1)^k \frac{2n}{2n - k} \binom{2n - k}{k} (n - k)! \tag{2.25}$$

Proof. Without loss of generality, we may assume that the seats are anticlockwise numbered from 1 to $2n$. For any $h = 1, \dots, n$, let

$$A_{2h-1} = \Big\{\text{permutations in } \mathcal{P}_n \text{ that sit the husband } h \text{ on the left of his wife}\Big\},$$

$$A_{2h} = \Big\{\text{permutations in } \mathcal{P}_n \text{ that sit the husband } h \text{ on the right of his wife}\Big\}.$$

The acceptable arrangements are one $(A_1 \cup \dots \cup A_{2n})^c \subset \mathcal{P}_n$, thus

$$U_n = |(A_1 \cup \dots \cup A_{2n})^c|.$$

By the inclusion–exclusion principle (2.22),

$$U_n = |(A_1 \cup \dots \cup A_{2n})^c| = \sum_{J \subset \{1,\dots,2n\}} (-1)^{|J|} |A_J|$$
$$= \sum_{j=0}^{2n} (-1)^j \Big(\sum_{\substack{J \subset \{1,\dots,2n\} \\ |J|=j}} |A_J| \Big),$$

thus, we have to compute $|A_J|$ for $J \subset \{1, 2, \dots, 2n\}$.

If $|J| = j$ and $J = \{k_1, \dots, k_j\}$, then $A_J = A_{k_1} \cap \dots \cap A_{k_j}$. Notice that $A_{2h} \cap A_{2h+1}$ for $h = 1, \dots, n-1$ are empty since any permutation of one of these intersections puts husbands h and $h+1$ on the same chair. Similarly, $A_{2n} \cap A_1$ is empty since the permutations in it force the husbands 1 and n on the same chair. Also the intersection $A_{2h-1} \cap A_{2h}$ is empty since any permutation in this intersection is such that the husband h sits in two different chairs. We can summarize these facts by saying that $A_J = \emptyset$ if J contains two consecutive integers in the sequence $(1, 2, \dots, 2n, 1)$. In particular, $A_J = \emptyset$ if $j = |J| > n$. If J does not contain two consecutive integers in the sequence $(1, 2, \dots, 2n, 1)$, then every permutation in A_J, $J = (k_1, \dots, k_j)$, fixes the chairs of the husbands $k_1, \dots, k_j$. Hence A_J contains as many permutations as the free permutations of the remaining $n - |J|$ husbands, i.e. $|A_J| = (n - |J|)!$ Summarizing, if $j := |J|$, then

$$|A_J| = \begin{cases} (n-j)! & \text{if } J \text{ does not contain two consecutive integers in the sequence } (1, \dots, 2n, 1), \\ 0 & \text{otherwise.} \end{cases}$$

Denote by $f_{2n,j}$ the number of subsets $J \subset \{1, \dots, 2n\}$ such that $|J| = j$ and which do not contain two consecutive integers in the sequence $(1, 2, \dots, 2n, 1)$. Then

$$U_n = \sum_{j=0}^{2n} (-1)^j \Big(\sum_{\substack{J \subset \{1,\dots,2n\} \\ |J|=j}} |A_J| \Big) = \sum_{j=0}^{n} (-1)^j \Big(\sum_{\substack{J \subset \{1,\dots,2n\} \\ |J|=j}} |A_J| \Big)$$

$$= \sum_{j=0}^{n} (-1)^j (n-j)!\, f_{2n,j}. \tag{2.26}$$

In the next lemma, see (2.27), we will compute $f_{2n,j}$:

$$f_{2n,j} = \frac{2n}{2n-j}\binom{2n-j}{j}.$$

Substituting in (2.26) we obtain the claim.

Lemma *The number $f_{n,k}$ of subsets of $\{1, \dots, n\}$ having cardinality k and that do not contain two consecutive integers in the sequence $(1, 2, \dots, n, 1)$ is*

$$f_{n,k} = \frac{n}{n-k}\binom{n-k}{k}. \tag{2.27}$$

Proof. It can be shown, see Exercise 1.30, that the number $g_{n,k}$ of subsets of cardinality k that do not contain two consecutive integers in $(1, \dots, n)$ is

$$g_{n,k} = \binom{n-k+1}{k}.$$

Let S be the family of subsets of $\{1, \dots, n\}$ having cardinality k and that do not contain two consecutive integers in the sequence $(1, 2, \dots, n, 1)$. Let S_1 be the family of the subsets in S that do not contain 1 and let S_2 be its complement in S, so that $|S| = |S_1| + |S_2|$. Each subset in S_1 has cardinality k and its elements are integers in $\{2, \dots, n\}$ which are not pairwise consecutive. Thus $|S_1| = g_{n-1,k}$. Each subset in S_2 has cardinality k and since it contains 1, it contains neither 2 nor n, and is detected by its remaining $k-1$ elements, Thus $|S_2| = g_{n-3,k-1}$. Consequently,

$$\begin{aligned} f_{n,k} &= |S| = |S_1| + |S_2| = g_{n-1,k} + g_{n-3,k-1} \\ &= \binom{n-1-k+1}{k} + \binom{n-3-k+1+1}{k-1} = \frac{n}{n-k}\binom{n-k}{k}. \end{aligned}$$

Exercise 2.74 (Euler function) *For any positive integer n compute the number of primes in $\{1, \dots, n-1\}$ which are relative prime to n, i.e.*

$$\phi(n) := \left|\left\{d \,\middle|\, 1 \le d \le n,\ \gcd(d, n) = 1\right\}\right|.$$

Solution. Write n as the product of prime numbers, $n = p_1^{\alpha_1} \dots p_k^{\alpha_k}$ with $p_1, p_2, \dots, p_k$ pairwise different prime numbers. If d and n are not coprime, then there exists $i \in \{1, \dots, k\}$ such that p_i divides d. For any $i = 1, \dots, k$ let

$$A_i := \left\{ d \,\middle|\, 1 \le d \le n, \ p_i \text{ divides } d \right\}$$
$$= \left\{ p_i, 2p_i, 3p_i, \dots \right\} \cap \{1, \dots, n\}.$$

We want to compute

$$\phi(n) = \left| (A_1 \cup A_2 \cup \dots A_k)^c \right|$$

by means of the inclusion-exclusion principle.

We first compute $|A_J|$ for every $J \subset \{1, \dots, n\}$. By (2.17) we have $A_\emptyset = \{1, \dots, n\}$, so that $|A_\emptyset| = n$. For any i, $A_i = \left\{ p_i, 2p_i, 3p_i, \dots \right\}$ so that $|A_i| = \frac{n}{p_i}$. Moreover,

$$A_{\{i,j\}} = A_i \cap A_j = \left\{ p_i p_j, 2p_i p_j, \dots, n \right\},$$

so that $|A_{\{i,j\}}| = \frac{n}{p_i p_j}$ and, in general,

$$|A_J| = \frac{n}{\prod_{i \in J} p_i} \qquad \forall J \subset \{1, \dots, k\}.$$

Applying the inclusion-exclusion principle we get

$$\begin{aligned}
\phi(n) &= \sum_{J \subset \{1,\dots,n\}} (-1)^{|J|} |A_J| \\
&= n \left(1 - \sum_i \frac{1}{p_i} + \sum_{i<j} \frac{1}{p_i p_j} - \sum_{i<j<k} \frac{1}{p_i p_j p_k} + \dots + (-1)^n \frac{1}{p_1 p_2 \cdots p_k} \right) \\
&= n \left(1 - \frac{1}{p_1} \right) \left(1 - \frac{1}{p_2} \right) \dots \left(1 - \frac{1}{p_k} \right).
\end{aligned} \tag{2.28}$$

3

Random variables

Events and their probabilities are important objects in probability theory, but not the only ones. Consider for instance the waiting time at a traffic light. It is impossible to evaluate the probability that the waiting time is, say, between 10 s and 20 s, simply because the answer depends on the arrival time probabilities. For instance, if everybody arrives during the green phase, then the waiting time is always zero! Waiting time probabilities depend in a definite manner on the arrival time. This is a typical example of *random variable*, i.e. of maps that link probability spaces. In Section 3.1 we first introduce real valued random variables and the basic related objects of value *distribution*, *law* and *generated events*, and the standard two descriptors of *expected value* and *variance*. We also state a few useful formulas to compute the distribution and law of the composition of random variables. A few classical discrete and absolutely continuous distributions on $\mathbb{R}$ are then presented in Section 3.2 and Section 3.3, respectively. Each section is complemented with examples and exercises. Other examples and exercises can be found e.g. in [3].

3.1 Random variables

Suppose $(\Omega, \mathcal{E}, \mathbb{P})$ is a probability space. Heuristically, a random variable is the result of an experiment (that is a well-defined deterministic experience) on the probability space $(\Omega, \mathcal{E}, \mathbb{P})$. The result of such an experiment is nevertheless random since the probability measure on Ω induces a probability measure on the set of the possible results of the experiment. As an example, consider a dice, for which the base space is the set $\Omega = \{1, \dots, 6\}$. If the experiment is 'What number is on the top face?', then the set of possible results is the set $\{1, 2, 3, 4, 5, 6\}$ and the result is random since the throwing of a dice is random. If the experiment is 'Is the number on the top face an even number?' then the set of possible results is {'True', 'False'}, and again the answer is random since the throwing of

A First Course in Probability and Markov Chains, First Edition. Giuseppe Modica and Laura Poggiolini.

a dice is random. Notice that in both cases the question is deterministic. Another example we have already presented is the waiting time at a traffic light: also in this case, the waiting time depends in a deterministic way on the time of arrival and randomness of the waiting time reflects the randomness of arrival times. As a further example, one may observe a certain feature of a physical system, say the temperature, a parameter that is known to vary with the state of the system in a well-defined way. If Ω is the base set of all possible states of the system, then the experiment of 'measuring the temperature' is a given map $T : \Omega \to \mathbb{R}$ from Ω to $\mathbb{R}$. What we observe is the value of such a mapping, not the state of the system. A certain degree of randomness in the system may induce randomness in the observed value of the temperature.

Summing up:

- A *random variable* is a map $X : \Omega \to \mathbb{R}$ defined on the base space Ω of a probability space $(\Omega, \mathcal{E}, \mathbb{P})$. We use the name 'variable' to stress the importance of the image of X.
- The probability measure $\mathbb{P}$ on Ω induces a probability measure on the image $X(\Omega)$, hence the name 'random variable'. The emphasis will be on the properties of the measure induced on $X(\Omega)$.
- In general, a random variable 'sees' only some events in Ω, as well as an observable on a physical system gives only partial information on the state of the system.

To give a formal presentation, we must furthermore keep in mind that, in general, not every subset of Ω is an event.

3.1.1 Definitions

3.1.1.1 Real valued random variables

Definition 3.1 *A real valued* random variable *on a probability space* $(\Omega, \mathcal{E}, \mathbb{P})$ *is an* $\mathcal{E}$*-measurable function* $X : \Omega \to \overline{\mathbb{R}}$*, i.e. a real-valued function defined on* Ω *such that for any* $t \in \mathbb{R}$ *the* t*-sublevel set of* X *belongs to* $\mathcal{E}$*,*

$$\left\{x \in \Omega \,\middle|\, X(x) \le t\right\} \in \mathcal{E}.$$

Here $\overline{\mathbb{R}} := \mathbb{R} \cup \{+\infty\} \cup \{-\infty\}$.

As previously noticed, we tacitly assume that $\mathcal{E} = \mathcal{P}(\Omega)$ if $X(\Omega)$ is finite or denumerable. In this case, any function $X : \Omega \to \overline{\mathbb{R}}$ is a random variable.

Example 3.2 Let $(\Omega, \mathcal{E}, \mathbb{P})$ be a probability space and let $E \in \mathcal{E}$. Then he characteristic function of E,

$$\mathbb{1}_E(x) = \begin{cases} 1 & \text{if } x \in E, \\ 0 & \text{otherwise} \end{cases}$$

is a random variable since for any $t \in \mathbb{R}$

$$\left\{x \in \Omega \,\middle|\, \mathbb{1}_E(x) \le t\right\} = \begin{cases} \Omega & \text{if } t \ge 1, \\ E^c & \text{if } 0 \le t < 1, \\ \emptyset & \text{if } t < 0 \end{cases}$$

and, of course, $\emptyset$, E^c and Ω are events.

3.1.1.2 Distribution of a random variable

Let $X : \Omega \to \overline{\mathbb{R}}$ be a random variable on the probability space $(\Omega, \mathcal{E}, \mathbb{P})$. Starting from

$$X^{-1}([-\infty, t]) = \left\{x \in \Omega \,\middle|\, X(x) \le t\right\} \in \mathcal{E}$$

and taking into account that $\mathcal{E}$ is a σ-algebra, one can show that

$$X^{-1}(A) = \left\{x \in \Omega \,\middle|\, X(x) \in A\right\}$$

is an event, $X^{-1}(A) \in \mathcal{E}$, for every Borel set $A \in \mathcal{B}(\mathbb{R})$, see also Appendix B. Consequently, we may define

$$\mathbb{P}_X(A) := \mathbb{P}(X \in A) := \mathbb{P}(X^{-1}(A)) \qquad \forall A \in \mathcal{B}(\mathbb{R}). \tag{3.1}$$

We will write $\mathbb{P}(X \in A)$ for $\mathbb{P}(\{x \in \Omega \mid X(x) \in A\})$.

Using De Morgan formulas, it is then easy to show that $(\mathcal{B}(\mathbb{R}), \mathbb{P}_X)$ is a measure on $\mathbb{R}$ which is called the *distribution* of X induced by $(\mathcal{E}, \mathbb{P})$. Since $\mathbb{P}_X(\mathbb{R}) = 1 - \mathbb{P}(X = \pm\infty)$, $\mathbb{P}_X$ is a probability measure on $\mathbb{R}$ if and only if $\mathbb{P}(X = \pm\infty) = 0$. If μ is a measure and $\mathbb{P}_X = \mu$ we also say that X *is distributed as* μ or that X follows μ.

We emphasize that the measure $\mathbb{P}_X$ depends on X as a function *and* on the measure $\mathbb{P}$, see Example 3.3. However, as usual, we still refer in the following to $\mathbb{P}_X$ as to the distribution of the random variable X, *thinking of the domain probability space as already understood in the definition of the random variable* X.

3.1.1.3 Law or probability distribution function

An efficient way to describe the distribution of a random variable $X : \Omega \to \mathbb{R}$ is through the so-called *probability distribution function* or *law* of X. It is the function $F_X : \mathbb{R} \to [0, 1]$ defined by

$$F_X(t) := \mathbb{P}(X \le t), \qquad t \in \mathbb{R}.$$

Example 3.3 In Figure 3.1 the laws of two random variables relative to the function $X(x) = x$ are drawn. Figure 3.1(a) shows the law of X driven by

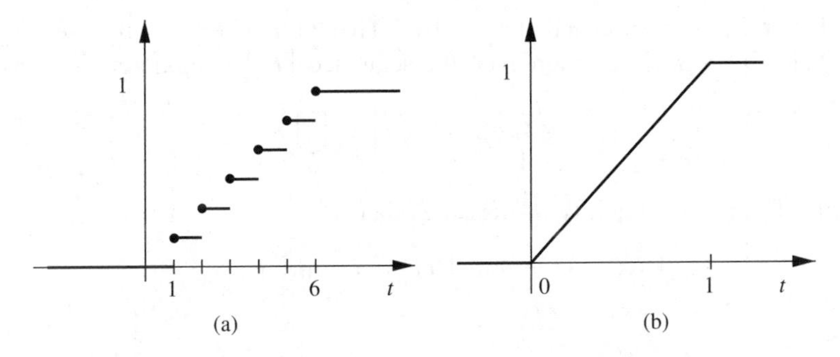

Figure 3.1 The laws in Example 3.3.

the probability measure associated with the throwing of a dice; Figure 3.1(b) shows the law of X driven by the uniform probability on the interval $[0, 1]$. Notice that the law, as well as the probability distribution, depends both on the function X *and* on the domain probability space.

The law F_X need not be continuous even if X is smooth, see e.g. Figure 3.1 where $X(x) = x$. However, we have the following.

Proposition 3.4 *Let $F_X : \mathbb{R} \to \mathbb{R}$ be the law of a random variable X. Then we have:*

(i) F_X is monotone increasing.

(ii) $F_X(t) \to \mathbb{P}(X = -\infty)$ as $t \to -\infty$.

(iii) $F_X(t) \to 1 - \mathbb{P}(X = +\infty)$ as $t \to +\infty$.

(iv) F_X is continuous from the right, i.e. $F_X(t) = \lim_{s \to t^+} F_X(s)$ $\forall t \in \mathbb{R}$.

(v) $F_X(t) = \mathbb{P}_X([-\infty, t])$ $\forall t \in \mathbb{R}$.

(vi) $\mathbb{P}(\{x \in \Omega \mid X(x) = t\}) = F_X(t) - \lim_{s \to t^-} F_X(s)$ $\forall t \in \mathbb{R}$.

Proof. Properties (i), (ii), (iii) and (v) are trivial. Let us prove (iv) and (vi).

(iv) Let $t \in [-\infty, \infty[$ and let $s_n \downarrow t$. For every n set $E_n := \{x \in \Omega \mid X(x) \le s_n\}$ $\in \mathcal{E}$ so that $\{E_n\}$ is a monotone decreasing sequence in $\mathcal{E}$. Since

$$\left\{x \in \Omega \,\middle|\, X(x) \le t\right\} = \bigcap_n E_n,$$

we infer from the continuity of measures that

$$F_X(t) = \mathbb{P}(X \le t) = \lim_{n \to \infty} \mathbb{P}(E_n) = \lim_{n \to +\infty} F_X(s_n).$$

Claim (iv) then follows, since $\{s_n\}$ is arbitrary.

(vi) Let $t \in]-\infty, \infty]$ and let $s_n \uparrow t$. Then for every n the set $F_n := \{x \in \Omega \,|\, X(x) \leq s_n\} \in \mathcal{E}$; moreover, the sequence $\{F_n\}$ is monotone increasing. Since

$$\Big\{x \in \Omega \,\Big|\, X(x) < t\Big\} = \bigcup_n F_n,$$

we infer from the continuity of measures that

$$\mathbb{P}(X < t) = \lim_{n\to\infty} \mathbb{P}(F_n) = \lim_{n\to+\infty} F_X(s_n),$$

hence

$$\mathbb{P}(X < t) = \lim_{s\to t^-} F_X(s) =: F_X(t^-).$$

Claim (vi) follows.

Although F_X need not be continuous, it is worth recalling that any monotone function f has at most a countable set of discontinuity points. In fact, for any integer k there are at most a countable number of points (a finite number if f is bounded) where the jump is larger than $1/k$.

The law F_X and the measure distribution $\mathbb{P}_X$ are two equivalent ways to describe the distribution of the random variable X. One can obtain F_X from $\mathbb{P}_X$ just by definition. On the other hand, from (v) of Proposition 3.4 and taking into account that $\mathbb{P}_X$ is a measure, one gets

$$\mathbb{P}_X(]a, b]) = F_X(b) - F_X(a) \qquad \forall a, b \in \mathbb{R},$$

i.e. given F_X, one recovers the value of $\mathbb{P}_X$ on intervals that are left-open and right-closed. Since any open set $A \in \mathbb{R}$ is an at most denumerable and disjoint union of such intervals, we can recover $\mathbb{P}_X(A)$ from F_X for any open set $A \subset \mathbb{R}$ and then for every Borel set because of the σ-additivity and continuity properties of measures. In this way, the law $t \to F_X(t)$ completely describes the measure $(\mathcal{B}(\mathbb{R}), \mathbb{P}_X)$, see also Appendix B.

3.1.1.4 Events generated by a random variable

Let $X : \Omega \to \mathbb{R}$ be a random variable on a probability space $(\Omega, \mathcal{E}, \mathbb{P})$. As already stated, for every $A \in \mathcal{B}(\mathbb{R})$ the preimage of A with respect to the random variable X,

$$X^{-1}(A) := \Big\{x \in \Omega \,\Big|\, X(x) \in A\Big\}, \tag{3.2}$$

belongs to $\mathcal{E}$. Using De Morgan formulas, it is easy to check that the family of such events

$$\mathcal{E}_X := \Big\{E \in \mathcal{E} \,\Big|\, E = X^{-1}(A) \text{ for some } A \in \mathcal{B}(\mathbb{R})\Big\}$$

is a σ-algebra, hence $\mathcal{E}_X$ is called the family of the events *detectable* or *generated* by X. Trivially, $\mathcal{E}_X$ is the smallest σ-algebra that contains all the events of the type $X^{-1}(A)$ for $A \in \mathcal{B}(\mathbb{R})$.

Example 3.5 Let $X : \Omega \to \mathbb{R}$ be a constant c. Then

$$X^{-1}(A) = \begin{cases} \Omega & \text{if } c \in A, \\ \emptyset & \text{if } c \notin A. \end{cases}$$

Therefore, the only events detectable by X are $\emptyset$ and Ω, i.e. $\mathcal{E}_X = \{\emptyset, \Omega\}$.

Example 3.6 Suppose that the image of $X : \Omega \to \mathbb{R}$ is finite or denumerable, $X(\Omega) = \{y_i\}_{i\in\mathbb{N}}$. Then $\mathcal{E}(X)$ is the σ-algebra generated by the denumerable family of events E_i, $i \in \mathbb{N}$, given for every $i \in \mathbb{N}$ by $E_i := \{x \in \Omega \mid X(x) = y_i\}$, $i \in \mathbb{N}$.

3.1.1.5 Special classes of distributions

The distributions of many random variables that appear in applications either fit in one of the two classes of measures described below or are linear combinations of them.

- *Discrete distribution*. We recall that a measure μ on $\mathbb{R}$ is *atomic* or *discrete* if it is concentrated on a finite or denumerable set of points $\{t_j\} \subset \overline{\mathbb{R}}$, that is, $\mu(A) = 0$ if and only if $A \cap \{t_j\} = \emptyset$. The vector (or sequence if the number of points is denumerable) $p = (p_j)$, $p_j := \mu(\{t_j\})$ is called the *density* (with respect to the counting measure), or *mass density function* of the measure μ. Moreover, the number p_j is called the *density* of μ at t_j.

  If X is a random variable with a discrete distribution $\mathbb{P}_X$ concentrated on $\{t_j\}$, we can express the law of X in terms of the density (p_j) of $\mathbb{P}_X$; in fact,

  $$F_X(t) = \mathbb{P}_X([-\infty, t]) = \sum_{t_j \le t} \mathbb{P}_X(\{t_j\}) = \sum_{t_j \le t} p_j.$$

  Notice that $F_X(t)$ is monotone increasing and piecewise constant with jumps at $\{t_j\}$, and the jump at t_j has size $\mathbb{P}_X(\{t_j\})$.

- *Absolutely continuous distribution*. Recall that a measure $(\mathcal{B}(\mathbb{R}), \mu)$ on $\mathbb{R}$ is said to be *absolutely continuous* if there exists a non-negative and Lebesgue summable function $f : \mathbb{R} \to \mathbb{R}$ such that

  $$\mu(A) = \int_A f(s)\, ds \qquad \forall A \in \mathcal{B}(\mathbb{R}). \tag{3.3}$$

  The function $f(s)$ is called the *density* of μ (with respect to the Lebesgue measure $\mathcal{L}^1$). In particular, for $F(t) := \mu(]-\infty, t])$ we have

  $$F(t) := \int_{-\infty}^{t} f(s)\, ds \qquad \forall t \in \mathbb{R}. \tag{3.4}$$

  Observe that $F(t) \to 0$ as $t \to -\infty$ and $F(t) \to \mu(\mathbb{R})$ as $t \to +\infty$.

The reader recognizes the relation (3.4) as a classical point in mathematical analysis. For instance, if f is piecewise continuous, then F is continuous and piecewise C^1 by the fundamental theorem of calculus. Moreover, $F'(t) = f(t)$ if f is continuous at t. If f is merely summable, it can be shown, see the absolute continuity theorem for the integral, Theorem B.58, that $F(t)$ is an absolutely continuous function, a condition strictly stronger than continuity, see Definition B.66. A celebrated theorem by Vitali, Theorem B.64, yields the converse: if F is an absolutely continuous function, then there exists a summable function f such that (3.4) holds. In conclusion, the following two claims agree:

— μ is an absolutely continuous measure, (3.3);

— $F(t) := \mu(]-\infty, t])$ is an absolutely continuous function.

If either of them is true, then the derivative of $F(t)$ exists for almost all t and

$$F'(t) = f(t) \qquad \text{for a.e. } t \in \mathbb{R}:$$

The density function agrees almost everywhere with the derivative of the law.

In particular, if X is a real valued random variable with an absolutely continuous distribution of density f_X, i.e.

$$\mathbb{P}_X(A) = \int_A f_X(t)\,dt \qquad \forall A \subset \mathcal{B}(\mathbb{R}),$$

then trivially,

$$F_X(t) = \mathbb{P}(X \le t) = \mathbb{P}(-\infty < X \le t) = \int_{-\infty}^{t} f_X(s)\,ds \qquad \forall t \in \mathbb{R}.$$

Observe that $\int_{\mathbb{R}} f_X(s)\,ds = 1$.

A typical example in this class is a random variable X *uniformly distributed* on an interval $]a, b[$, that is, such that $\mathbb{P}_X(A) := \int_A f_X(t)\,dt \ \forall A \in \mathcal{B}(\mathbb{R})$ where

$$f_X(t) = \begin{cases} \dfrac{1}{b-a} & \text{if } t \in]a, b[, \\ 0 & \text{otherwise.} \end{cases}$$

Distributions of random variables may also be a combination of the two types above, as in the following example.

Example 3.7 (Waiting time at a traffic light) Consider for simplicity a traffic light with two states, red and green. The red and green lights are on for r and g seconds, respectively. The cycle is iterated every $r + g$ seconds. Here the certain event is the time interval $[0, r + g]$. A car randomly reaches the traffic light according to the uniform distribution on $[0, r + g]$, i.e. the probability that a car driver arrives in a time interval $I = [a, b]$, $0 \le a \le b \le r + g$ is $\mathbb{P}(I) := (b - a)/(r + g)$. Thus the probability associated with arrival time is $(\mathcal{B}([0, r + g]), \mathbb{P})$ where $\mathbb{P}(A) := \frac{1}{r+g}\mathcal{L}^1(A)$.

For every arrival time $x \in [0, r + g]$ the waiting time at the traffic light is

$$T(x) := \begin{cases} r - x & \text{if } 0 \le x \le r, \\ 0 & \text{if } r \le x \le r + g. \end{cases}$$

Since T is continuous, T is a random variable on the probability space $([0, r + g], \mathcal{B}([0, r + g]), \mathbb{P})$ with law

$$F_T(t) := \mathbb{P}(T \le t) = \begin{cases} 0 & \text{if } t < 0, \\ \frac{t+g}{r+g} & \text{if } 0 < t < r, \\ 1 & \text{if } t \ge r, \end{cases}$$

see Figure 3.2. We point out that the law F_T is neither piecewise constant nor continuous. It has a jump at time $t = 0$ of amplitude $g/(r + g)$. Of course, F_T is the sum of a continuous piecewise linear function and of a piecewise constant function.

3.1.2 Expected value

Let $X : \Omega \to \overline{\mathbb{R}}$ be a random variable on $(\Omega, \mathcal{E}, \mathbb{P})$. The *integral*

$$\mathbb{E}[X] = \int_\Omega X(x)\,\mathbb{P}(dx), \tag{3.5}$$

if it exists, is also called the *expected value* or the *expectation* of the random variable X. Notice that $\mathbb{E}[X]$ is simply the *mean value* of X on Ω

$$\mathbb{E}[X] = \frac{1}{\mathbb{P}(\Omega)} \int_\Omega X(x)\,\mathbb{P}(dx)$$

since $\mathbb{P}(\Omega) = 1$.

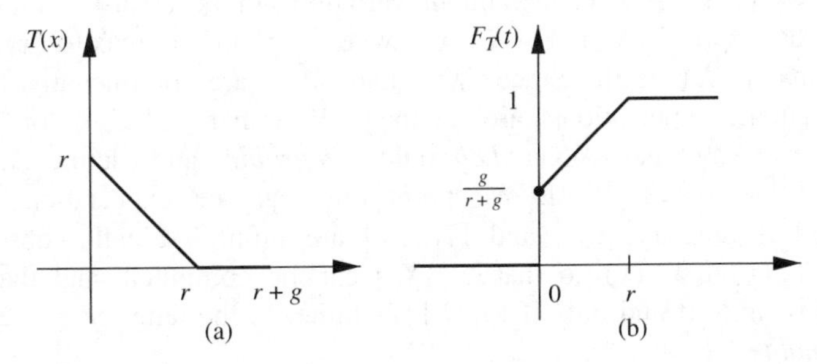

Figure 3.2 The traffic light. (a) The waiting time $T(x)$; (b) the law according to the uniform probability on $[0, r + g]$.

The integral in (3.5) is the *Lebesgue integral*, which we now briefly describe in this contest, even though the assumption $\mathbb{P}(\Omega) = 1$ is not necessary. The reader may refer to Appendix B for details.

We have already mentioned the definition of the Lebesgue integral of a simple function, i.e. of a random variable with a finite range, i.e. of a function

$$\varphi(x) = \sum_{i=1}^{n} c_i \mathbb{1}_{E_i}(x)$$

where $c_1, \dots, c_n$ are pairwise distinct and the sets $E_i := \{x \in \Omega \mid \varphi(x) = c_i\}$ are events of the σ-algebra $\mathcal{E}$. The integral of φ (with respect to the measure $\mathbb{P}$) has been defined as

$$\mathbb{E}[\varphi] = \int_\Omega \varphi(x)\, \mathbb{P}(dx) := \sum_{i=1}^{n} c_i \mathbb{P}(E_i), \tag{3.6}$$

as intuition suggests.

Let $X : \Omega \to \overline{\mathbb{R}}_+$ be a non-negative random variable (X may take the value $+\infty$). One can show that there exists a sequence $\{\varphi_n\}$ of simple functions such that $\varphi_1(x) \le \varphi_2(x) \le \dots$ and $\varphi_n(x) \uparrow X(x)$ $\forall x \in \Omega$. For each n the integral $I_n := \int_\Omega \varphi_n(x)\, \mathbb{P}(dx)$ is defined in (3.6) and $\{I_n\}$ is an increasing sequence of real numbers, so that $\lim_{n\to\infty} I_n$ exists and is in $[0, +\infty]$. It is an easy consequence of the Beppo Levi theorem that the limit does not depend on the choice of the approximating sequence $\{\varphi_n\}$. We then define the Lebesgue integral, equivalently the expectation of X, by

$$\mathbb{E}[X] = \int_\Omega X(x)\, \mathbb{P}(dx) := \lim_{n\to\infty} \int_\Omega \varphi_n(x)\, \mathbb{P}(dx).$$

Notice that $\mathbb{E}[X] = +\infty$ whenever $\mathbb{P}(X = +\infty) > 0$, and that $\mathbb{E}[X] = 0$ if and only if $\mathbb{P}(X > 0) = 0$; in the latter case, one says that $X = 0$ $\mathbb{P}$-*almost everywhere* or, when no confusion may arise we say that $X = 0$ *almost everywhere*, which is usually shortened to $X = 0$ a.e.

Let now $X : \Omega \to \overline{\mathbb{R}}$ be a random variable, not necessarily non-negative. Split X as $X(x) = X_+(x) - X_-(x)$ where $X_+(x) := \max\{X(x), 0\}$ and $X_-(x) = \max\{-X(x), 0\}$. Since X_+ and X_- are non-negative random variables, their expectations are defined. If either $\mathbb{E}[X_+]$ or $\mathbb{E}[X_-]$ is finite, we say that X is *Lebesgue integrable* and define $\mathbb{E}[X]$ as $\mathbb{E}[X] := \mathbb{E}[X_+] - \mathbb{E}[X_-]$. We point out that the expectation of X is not defined if both $\mathbb{E}[X_+]$ and $\mathbb{E}[X_-]$ are infinite. Finally, observe that $|X(x)| = X_+(x) + X_-(x)$ so that $\mathbb{E}[|X|]$ can be computed, and that $\mathbb{E}[X]$ exists and is finite if and only if $\mathbb{E}[|X|]$ is finite. In the latter case, we say that X is *summable*.

The usual properties of the integral or expectation are collected in Theorem B.21. There the reader may find the standard properties of the integral:

linearity, monotonicity and additivity, and the main result about limits, known as the Beppo Levi theorem.

Since the above holds whatever $\mathbb{P}(\Omega)$ is, this procedure defines the integral of random variables X on a probability space $(\Omega, \mathcal{E}, \mathbb{P})$ (with respect to $\mathbb{P}$) and the integral of $\mathcal{B}(\mathbb{R})$-measurable functions with respect to $\mathbb{P}_X$. Therefore, the notations

$$\int_\Omega X(x)\,\mathbb{P}(dx) \qquad \text{and} \qquad \int_\mathbb{R} \varphi(t)\mathbb{P}_X(dt)$$

make sense if X is a non-negative random variable and if $\varphi : \mathbb{R} \to \mathbb{R}$ is non-negative and $\mathcal{B}(\mathbb{R})$-measurable, respectively.

As already shown, the distribution function $F_X(t) : \mathbb{R} \to \mathbb{R}$ completely describes the distribution of the values of the random variable X. Thus, one also writes

$$\int_\Omega \varphi(t)\,dF_X(t) \qquad \text{instead of} \qquad \int_\mathbb{R} \varphi(t)\,\mathbb{P}_X(dt), \tag{3.7}$$

see also Appendix B and the exercises therein for explicit formulas of the integral of random variables with respect to several measures. Here we only quote the following two facts.

- Let X be a random variable with a discrete distribution $\mathbb{P}_X$ concentrated on $\{x_i\}$ and of density p_i at x_i. Then

$$\int_\mathbb{R} \varphi(t)\,\mathbb{P}_X(dt) = \sum_i \varphi(x_i)p_i.$$

- Let $X : \Omega \to \overline{\mathbb{R}}$ be a random variable with absolutely continuous distribution of density f_X. Then

$$\int_\mathbb{R} \varphi(t)\,dF_X(t) = \int_\mathbb{R} \varphi(t)\,\mathbb{P}_X(dt) = \int_\mathbb{R} \varphi(t)f_X(t)\,dt.$$

  Because of this last formula, the absolute continuity property (3.3) is usually written as

$$\mathbb{P}_X(dt) = dF_X(t) = f_X(t)\,dt.$$

3.1.3 Functions of random variables

Let X be a real valued random variable on the probability space $(\Omega, \mathcal{E}, \mathbb{P})$, let $(\mathcal{B}(\mathbb{R}), \mathbb{P}_X)$ be its distribution and let $\varphi : \mathbb{R} \to \mathbb{R}$ be a Borel-measurable function on $\mathbb{R}$. The composition $Y := \varphi \circ X : \Omega \to \mathbb{R}$ is again a random variable on $(\Omega, \mathcal{E}, \mathbb{P})$. In fact, for any open set $A \subset \mathbb{R}$ the set $\varphi^{-1}(A) \subset \mathbb{R}$ is a Borel set, hence $(\varphi \circ X)^{-1}(A) = X^{-1}(\varphi^{-1}(A))$ is an event. The following formulas for $Y = \varphi \circ X$ trivially hold:

$$\mathbb{P}_Y(A) = \mathbb{P}(Y \in A) = \mathbb{P}(X \in \varphi^{-1}(A)) = \mathbb{P}_X(\varphi^{-1}(A)), \tag{3.8}$$

$$F_Y(s) = \mathbb{P}_X(\{t \in \mathbb{R} \mid \varphi(t) \leq s\}). \tag{3.9}$$

From (3.9) we trivially infer that $F_Y(s) = \mathbb{P}(\varphi \circ X \leq s) = 0$ if s is below the image of φ and $F_Y(s) = 1$ if s is above the image of φ. However, there is no easy formula that follows from (3.9) as the following example shows.

Example 3.8 Let X be a real valued random variable and, for $0 < a \leq 1$, let

$$\varphi(t) := \begin{cases} 0 & \text{if } t < 0, \\ a & \text{if } t = 0, \\ 1 & \text{if } t > 0. \end{cases}$$

Then

$$\left\{x \in \Omega \,\middle|\, \varphi(X(x)) \leq s\right\} = \begin{cases} \emptyset & \text{if } s < 0, \\ \{x \mid X(x) < 0\} & \text{if } 0 \leq s < a, \\ \{x \mid X(x) \leq 0\} & \text{if } a \leq s < 1, \\ \Omega & \text{if } s \geq 1. \end{cases}$$

Hence $F_Y(0) = \mathbb{P}(\varphi(X(x)) \leq 0) = \mathbb{P}(X < 0) = F_X(0^-)$.

A few special cases of (3.9) are described below.

- If $\varphi : \mathbb{R} \to \mathbb{R}$ is continuous and strictly monotone increasing then for any $s \in \varphi(\mathbb{R})$ we have $\varphi(t) \leq s$ if and only if $t \leq \varphi^{-1}(s)$, hence

$$F_Y(s) = \mathbb{P}_X(y \leq \varphi^{-1}(s)) = F_X(\varphi^{-1}(s)).$$

- If φ is continuous and strictly monotone decreasing then for any $s \in \varphi(\mathbb{R})$ we have $\varphi(t) \leq s$ if and only if $t \geq \varphi^{-1}(s)$, so that

$$F_Y(s) = P_X(y \geq \varphi^{-1}(s)) = 1 - F_X(t^-), \qquad t := \varphi^{-1}(s).$$

- Assume $\varphi : \mathbb{R} \to \mathbb{R}$ is right-continuous and monotone increasing, possibly with jumps. For any $s \in \mathbb{R}$ the set

$$E_s := \left\{x \,\middle|\, s \leq \varphi(x)\right\}$$

is either empty or a *closed* half-line or $\mathbb{R}$. Accordingly, let $\psi(s) := \min E_s$. Trivially, $\psi(s) \leq t$ if and only if $t \in E_s$ i.e. if and only if $s \leq \varphi(t)$. Consequently,

$$\left\{t \,\middle|\, t \leq \psi(s)\right\} = \left\{t \,\middle|\, s \leq \varphi(t)\right\}$$

and we conclude that

$$F_Y(s) = F_{\varphi \circ X}(s) = F_X(\psi(s)). \tag{3.10}$$

The graph of $s \to \psi(s)$ is obtained from the graph of φ by adding vertical segments at the jump points and then looking at the resulting curve from the vertical axis, see Figure 3.3. Clearly, if φ is continuous and invertible, then $\psi(s)$ is finite on the range of φ and coincides with the inverse of φ.

The links between the distributions of X and $Y := \varphi \circ X$ given by (3.8) and (3.9) may be cumbersome, as we have seen, if either φ has jumps or is not strictly monotone. On the contrary, the relation between the relative integrals is much more convenient. Let us state the main property.

Theorem 3.9 *Let $X : \Omega \to \mathbb{R}$ be a random variable on a probability space $(\Omega, \mathcal{E}, \mathbb{P})$. Then for every non-negative Borel-measurable function $\varphi : \mathbb{R} \to \mathbb{R}$ we have*

$$\mathbb{E}[\varphi \circ X] = \int_\Omega \varphi(X(x))\,\mathbb{P}(dx) = \int_\mathbb{R} \varphi(t)\,\mathbb{P}_X(dt). \tag{3.11}$$

Proof. Assume first that φ is a simple function, $\varphi(t) = \sum_{i=1}^n c_i \mathbb{1}_{E_i}(t)$, where $c_1, \dots, c_n$ are pairwise distinct real numbers and the sets $E_i := \{t \in \mathbb{R} \mid \varphi(t) = c_i\}$ are events. Since $\{x \in \Omega \mid X(x) \in E_i\} = \{x \in \Omega \mid \varphi \circ X(x) = c_i\}$ and by the definition of integral, we have

$$\begin{aligned}\int_\mathbb{R} \varphi(t)\,\mathbb{P}_X(dt) &= \sum_{i=1}^n c_i \mathbb{P}_X(E_i) = \sum_{i=1}^n c_i \mathbb{P}(X \in E_i)\\ &= \sum_{i=1}^n c_i\,\mathbb{P}(\varphi \circ X = c_i) = \int_\Omega \varphi \circ X(x)\,\mathbb{P}(dx),\end{aligned}$$

hence the claim if φ is simple.

In the general case, let $\{\varphi_k\}$ be a monotone increasing sequence of simple functions such that $\varphi_k \uparrow \varphi$ pointwise. Then, from the above, for any integer k we have

$$\int_\mathbb{R} \varphi_k(t)\,\mathbb{P}_X(dt) = \int_\Omega \varphi_k(X(x))\,\mathbb{P}(dx).$$

Beppo Levi theorem allows us to pass to the limit in the previous equality as $k \to \infty$, thus proving (3.11) for φ.

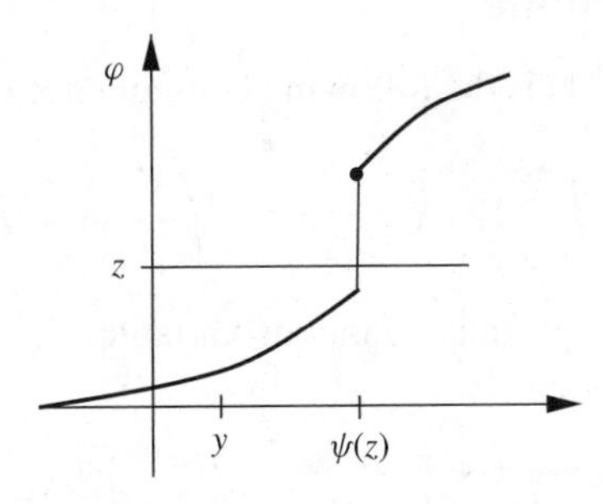

Figure 3.3 The generalized inverse of φ.

Corollary 3.10 *Let X be a random variable on $(\Omega, \mathcal{E}, \mathbb{P})$. Then $\mathbb{E}[X]$ is well-defined if and only if $\int_{\mathbb{R}} t\, \mathbb{P}_X(dt)$ exists and, in this case,*

$$\mathbb{E}[X] = \int X(x)\, \mathbb{P}(dx) = \int_{-\infty}^{+\infty} t\, \mathbb{P}_X(dt). \tag{3.12}$$

Proof. The claim trivially follows by applying (3.11) to X separately with $\varphi(t) = \max\{t, 0\}$ and $\varphi(t) = \max\{-t, 0\}$.

Corollary 3.10 shows in particular that the expectation of X depends only on the distribution of values of X. There is no need to go back to the probability space $(\Omega, \mathcal{E}, \mathbb{P})$ in order to compute the expectation of X. In particular, *if two random variables X and Y have the same distribution, $\mathbb{P}_X = \mathbb{P}_Y$, then $\mathbb{E}[X] = \mathbb{E}[Y]$.*

Theorem 3.9 yields a convenient formula for the distribution of a composition.

Corollary 3.11 (Composition) *Let X be a random variable on $(\Omega, \mathcal{E}, \mathbb{P})$ and let $\varphi : \mathbb{R} \to \mathbb{R}$ be a $\mathcal{B}(\mathbb{R})$-measurable function. Then for every non-negative $\mathcal{B}(\mathbb{R})$-measurable function $\psi : \mathbb{R} \to \mathbb{R}$ we have*

$$\int_{\mathbb{R}} \psi(s)\, \mathbb{P}_{\varphi \circ X}(ds) = \int_{\mathbb{R}} \psi(\varphi(t)) \mathbb{P}_X(dt). \tag{3.13}$$

Proof. By applying (3.11) to X with test function $\psi \circ \varphi$ and to $\varphi \circ X$ with test function ψ we get

$$\int_{\mathbb{R}} \psi(s)\, \mathbb{P}_{\varphi \circ X}(ds) = \int_{\Omega} \psi(\varphi(X(x))\, \mathbb{P}(dx)$$

and

$$\int_{\mathbb{R}} \psi(\varphi(t)) \mathbb{P}_X(dt) = \int_{\Omega} \psi(\varphi(X(x))\, \mathbb{P}(dx),$$

respectively. The claim follows at once.

3.1.4 Cavalieri formula

Another interesting formula is the following consequence of the *Cavalieri formula* (B.22):

$$\mathbb{E}[X] = \int_0^{+\infty} \mathbb{P}(X > t)\, dt = \int_0^{+\infty} (1 - F_X(t))\, dt \tag{3.14}$$

which holds for any non-negative random variable X. Applying (3.14) to X_+ and X_- and summing, one gets

$$\mathbb{E}[X] = \int_0^{+\infty} (1 - F_X(t))\, dt - \int_{-\infty}^{0} F_X(t)\, dt \tag{3.15}$$

provided X is summable, see Corollary B.24. Formula (3.15) shows again that the expected value of a random variable X depends only on the distribution $\mathbb{P}_X$; actually it shows how the expected value depends on the law of X. In particular (3.15) yields $\mathbb{E}[X] = \mathbb{E}[Y]$ if $F_X = F_Y$.

Theorem 3.12 (Composition) *Let $X : \Omega \to \mathbb{R}$ be a random variable on $(\Omega, \mathcal{E}, \mathbb{P})$, let I be an interval such that $I \supset X(\Omega)$ and let $\varphi : I \to \mathbb{R}$ be a non-negative and monotone increasing function of class C^1. Then*

$$\mathbb{E}[\varphi \circ X] = m\mathbb{P}(\Omega) + \int_I \varphi'(t)(1 - F_X(t))\,dt \tag{3.16}$$

where $m := \inf_{t\in I} \varphi(t)$.

Proof. (i) Assume φ is strictly increasing. Let $M := \sup_{t\in I} \varphi(t)$ so that $]m, M[\subset \varphi(I) \subset [m, M]$. Notice that

$$\Big\{x \in \Omega \,\Big|\, \varphi \circ X(x) > s\Big\} = \begin{cases} \Omega & \text{if } s \le m,\ s \notin \varphi(I), \\ \Big\{x \in \Omega \,\Big|\, X(x) > \varphi^{-1}(s)\Big\} & \text{if } s \in \varphi(I), \\ \emptyset & \text{if } s \ge M,\ s \notin \varphi(I). \end{cases}$$

Therefore, applying the Cavalieri formula (3.15) to the random variable $\varphi \circ X$ and then, changing variable with $s = \varphi(t)$, $t \in I$, we get

$$\begin{aligned}\mathbb{E}[\varphi \circ X] &= \int_0^\infty \mathbb{P}(\varphi \circ X > s)\,ds = m\mathbb{P}(\Omega) + \int_m^M \mathbb{P}(X > \varphi^{-1}(s))\,ds \\ &= m\mathbb{P}(\Omega) + \int_I \varphi'(t)\mathbb{P}(X > t)\,dt \\ &= m\,\mathbb{P}(\Omega) + \int_I \varphi'(t)(1 - F_X(t))\,dt.\end{aligned}$$

(ii) When φ is not strictly monotone, consider a non-negative bounded, strictly monotone function $\psi : I \to \mathbb{R}$ of class $C^1(I)$ with $\inf_{t\in I} \psi(t) = 0$. Then $\varphi + \psi$ is strictly monotone and (i) yields

$$\begin{aligned}\mathbb{E}[\varphi \circ X] + \mathbb{E}[\psi \circ X] &= \mathbb{E}[(\varphi + \psi) \circ X] \\ &= m\mathbb{P}(\Omega) + \int_I (\varphi'(t) + \psi'(t))(1 - F_X(t))\,dt,\end{aligned}$$

$$\mathbb{E}[\psi \circ X] = \int_I \psi'(t)(1 - F_X(t))\,dt.$$

The claim then follows by subtracting the second equality from the first one since $\mathbb{E}[\psi \circ X]$ is finite.

3.1.5 Variance

The expectation of a random variable X is a 'central' value around which the values of the range of X are distributed. Several indicators can be defined to roughly describe the distribution of X around the expected value. For instance, to evaluate how much of the range of X is dispersed (or concentrated) around the expected value, a relevant indicator is the so-called *variance* of X defined as

$$\mathrm{Var}\,[\,X\,] := \mathbb{E}\big[\,(X - \mathbb{E}\,[X])^2\,\big].$$

Of course, the variance is defined if and only if $\mathbb{E}\,[\,|X|\,] < +\infty$; moreover, in this case, it may happen that $\mathrm{Var}\,[\,X\,] = +\infty$. The following formulas are easily proven by using the linearity property of the expected value:

$$\mathrm{Var}\,[\,X\,] = 0 \text{ if and only if } X(x) = \mathbb{E}\,[\,X\,] \text{ for } \mathbb{P}\text{-a.e. } x, \tag{3.17}$$

$$\mathrm{Var}\,[\,X\,] = \mathbb{E}\big[\,X^2 - 2\mathbb{E}\,[X]\,X + \mathbb{E}\big[X^2\big]\big] = \mathbb{E}\big[\,X^2\,\big] - (\mathbb{E}\,[\,X\,])^2, \tag{3.18}$$

$$\mathrm{Var}\,[\alpha X + \beta] = \alpha^2 \mathrm{Var}\,[\,X\,] \qquad \forall \alpha, \beta \in \mathbb{R}. \tag{3.19}$$

Finally, the number

$$\sigma(X) := \sqrt{\mathrm{Var}\,[\,X\,]}. \tag{3.20}$$

is called the *standard deviation* of the random variable X. Notice that $\sigma(X)$ is positively homogeneous of degree 1, $\sigma(\lambda X) = |\lambda|\,\sigma(X)$ $\forall \lambda \in \mathbb{R}$.

3.1.6 Markov and Chebyshev inequalities

Proposition 3.13 *Let $X : \Omega \to \mathbb{R}$ be a random variable on the probability space $(\Omega, \mathcal{E}, \mathbb{P})$. Let I be an interval such that $X(\Omega) \subset I$ and let $\varphi : I \to \mathbb{R}$ be a non-negative monotone increasing function of class C^1. Then $\forall t \in I$*

$$\varphi(t)\,\mathbb{P}(X \geq t) \leq \mathbb{E}\,[\,\varphi \circ X\,]. \tag{3.21}$$

The inequality (3.21) is called Markov inequality. *Moreover, if $\mathbb{E}[\,|X|\,]$ is finite, then $\forall t > 0$ the* Chebyshev inequality

$$\mathbb{P}\Big(\Big|X - \mathbb{E}\,[\,X\,]\Big| \geq t\Big) \leq \frac{\mathrm{Var}\,[\,X\,]}{t^2} \tag{3.22}$$

holds.

Proof. Let $t \in \mathbb{R}$. Since

$$\{x \in \Omega \,|\, X(x) \geq t\} \subset \{x \in \Omega \,|\, \varphi(X(x)) \geq \varphi(t)\}$$

we have

$$\begin{aligned}\varphi(t)\mathbb{P}(X \geq t) &\leq \varphi(t)\mathbb{P}\Big(\varphi \circ X \geq \varphi(t)\Big) \\ &\leq \int_{\{\varphi\circ X \geq \varphi(t)\}} \varphi(X(x))\,\mathbb{P}(dx) \leq \int \varphi(X(x))\,\mathbb{P}(dx),\end{aligned}$$

i.e. Markov inequality.

Assume $\mathbb{E}[\,|X|\,]$ is finite. By applying Markov inequality to the random variable $x \mapsto |X(x) - \mathbb{E}[\,X\,]|$ and the map $\varphi(t) := t^2$ we get

$$\mathbb{P}\Big(|X - \mathbb{E}[\,X\,]| \geq t\Big) \leq \frac{1}{t^2}\mathbb{E}\big[\,(X - \mathbb{E}[X])^2\,\big] = \frac{\mathrm{Var}[\,X\,]}{t^2},$$

that is, Chebyshev inequality.

3.1.7 Variational characterization of the median and of the expected value

Definition 3.14 *Let $X : \Omega \to \mathbb{R}$ be a random variable on a probability space $(\Omega, \mathcal{E}, \mathbb{P})$. The real number*

$$t_M := \inf\Big\{t \,\Big|\, F_X(t) \geq 1/2\Big\}.$$

is called the median *of the distribution* $\mathbb{P}_X$.

From the definition and the right-continuity of F_X, we get

$$F_X(t_M) \geq 1/2 \qquad \text{and} \qquad \mathbb{P}(X < t_M) = \lim_{t \to t_M^-} F_X(t) \leq 1/2.$$

Trivially, $F_X(t_M) = 1/2$ if F_X is continuous at t_M.

Proposition 3.15 *Let $X : \Omega \to \mathbb{R}$ be a random variable on a probability space $(\Omega, \mathcal{E}, \mathbb{P})$ with $\mathbb{E}[\,|X|\,] < +\infty$. Then t_M is a minimum point of the function*

$$\phi(s) := \int_\Omega |X(x) - s|\,\mathbb{P}(dx), \qquad s \in \mathbb{R}.$$

Proof. Let $s < t_M$. Let us prove that $\phi(s) \geq \phi(t_M)$.

$$\begin{aligned}
\phi(s) &= \int_{\{X(x) \leq s\}} (s - X(x))\,\mathbb{P}(dx) + \int_{\{X(x) > s\}} (X(x) - s)\,\mathbb{P}(dx) \\
&= \int_{\{X(x) \leq s\}} (t_M - X(x))\,\mathbb{P}(dx) + \int_{\{X(x) \leq s\}} (s - t_M)\,\mathbb{P}(dx) \\
&\quad + \int_{\{s < X(x) \leq t_M\}} (X(x) - t_M)\,\mathbb{P}(dx) + \int_{\{s < X(x) \leq t_M\}} (t_M - s)\,\mathbb{P}(dx) \\
&\quad + \int_{\{X(x) > t_M\}} (X(x) - t_M)\,\mathbb{P}(dx) + \int_{\{X(x) > t_M\}} (t_M - s)\,\mathbb{P}(dx)
\end{aligned}$$

$$
\begin{aligned}
&= \int |X(x) - t_M|\,\mathbb{P}(dx) - 2\int_{\{s<X(x)\le t_M\}} (t_M - X(x))\,\mathbb{P}(dx) \\
&\qquad + (t_M - s)\Big(- \mathbb{P}(X \le s) + \mathbb{P}(s < X \le t_M) + \mathbb{P}(X > s)\Big) \\
&\ge \phi(t_M) - 2(t_M - s)\mathbb{P}(s < X \le t_M) \\
&\qquad + (t_M - s)\Big(- \mathbb{P}(X \le s) + \mathbb{P}(s < X \le t_M) + \mathbb{P}(X > t_M)\Big) \\
&= \phi(t_M) + (t_M - s)\Big(- \mathbb{P}(X \le s) - \mathbb{P}(s < X \le t_M) + 1 - \mathbb{P}(X \le t_M)\Big) \\
&= \phi(t_M) + (t_M - s)\Big(1 - 2\mathbb{P}(X \le t_M)\Big) \\
&\ge \phi(t_M).
\end{aligned}
$$

Similarly, one shows that $\phi(s) \ge \phi(t_M)$ if $s > t_M$.

Proposition 3.16 *Let $X : \Omega \to \mathbb{R}$ be a random variable on the probability space $(\Omega, \mathcal{E}, \mathbb{P})$ with $\mathbb{E}[\,|X|\,]) < +\infty$ and* Var $[\,X\,] < +\infty$. *Then $\mathbb{E}[\,X\,]$ is the unique minimizer of the function*

$$\phi(s) := \int_\Omega |X(x) - s|^2\,\mathbb{P}(dx), \qquad s \in \mathbb{R}.$$

Proof. The function $s \mapsto \phi(s) = \mathbb{E}\big[\,X^2\,\big] - 2s\,\mathbb{E}[\,X\,] + s^2$ is a polynomial of degree two. Therefore, $\phi(s)$ has a unique minimizer at $\mathbb{E}[\,X\,]$.

3.1.8 Exercises

For the reader's convenience the main definitions and formulas given in this chapter are summarized in Figures 3.4–3.6.

Exercise 3.17 *The life of a bulb (measured in days) is a random variable X with an absolute continuous distribution of density*

$$f(x) = \begin{cases} 0 & \text{if } x < 100, \\ \dfrac{3 \cdot 10^6}{x^4} & \text{if } x \ge 100. \end{cases}$$

Compute the expected value of X. [150]

Exercise 3.18 *Let X be a random variable. Compute the law F_{X^2} of the random variable X^2. Moreover, assuming that $\mathbb{P}_X$ is absolutely continuous with density $\rho_X(t)$, show that $\mathbb{P}_{X^2}$ is absolutely continuous by computing its density.*

Random variables

A random variable X on the probability space $(\Omega, \mathcal{E}, \mathbb{P})$ is an $\mathcal{E}$-measurable function $X : \Omega \to \overline{\mathbb{R}}$. Then its distribution, law, expected value and variance are defined by

$$\begin{aligned}
&\mathbb{P}_X(A) := \mathbb{P}(X \in A) \qquad \forall A \in \mathcal{B}(\mathbb{R}),\\
&F_X(t) := \mathbb{P}(X \le t) = \mathbb{P}_X([-\infty, t]),\\
&\mathbb{E}[\,X\,] := \int_\Omega X(x)\,\mathbb{P}(dx) = \int_{-\infty}^{+\infty} t\,\mathbb{P}_X(dt),\\
&\qquad = \int_0^{+\infty} (1 - F_X(t))\,dt - \int_{-\infty}^0 F_X(t)\,dt,\\
&\mathrm{Var}[X] := \mathbb{E}\big[\,(X - \mathbb{E}[X])^2\,\big] = \mathbb{E}\big[\,X^2\,\big] - \mathbb{E}[\,X\,]^2,
\end{aligned}$$

respectively.

Composition formulas

Let X be a random variable on the probability space $(\Omega, \mathcal{E}, \mathbb{P})$, let $\varphi : \mathbb{R} \to \mathbb{R}$ be a continuous function and let $Y = \varphi \circ X$. Then

- $\mathbb{P}_Y(A) = \mathbb{P}_X(\varphi^{-1}(A)) \quad \forall A \in \mathcal{B}(\mathbb{R})$,
- $F_Y(s) = \begin{cases} F_X(t) & \text{if } \varphi \text{ is strictly increasing,} \\ 1 - F_X(t^-) & \text{if } \varphi \text{ is strictly decreasing,} \end{cases} \quad t = \varphi^{-1}(s)$
- $\mathbb{E}[\,Y\,] = \int_{-\infty}^{+\infty} s\,\mathbb{P}_Y(ds) = \int_{-\infty}^{+\infty} \varphi(t)\,\mathbb{P}_X(dt)$,
- $\int \psi(s)\,\mathbb{P}_Y(ds) = \int \psi(\varphi(t))\,\mathbb{P}_X(dt)$.

If $\varphi \in C^1(\mathbb{R})$ is strictly monotone and $\mathbb{P}_X$ is absolutely continuous, $\mathbb{P}_X(ds) = \rho_X(s)\,ds$, then also $\mathbb{P}_Y$ is absolutely continuous and

$$\mathbb{P}_Y(ds) = \rho_Y(s)\,dx, \qquad \text{where} \qquad \rho_Y(s) = \frac{\rho_X(t)}{|\varphi'(t)|}, \quad t = \varphi^{-1}(s).$$

Figure 3.4 Random variables.

Solution. Since

$$F_{X^2}(t) = \mathbb{P}(X^2 \le t) = \begin{cases} 0 & \text{if } t < 0, \\ \mathbb{P}(-\sqrt{t} \le X \le \sqrt{t}) & \text{if } t \ge 0. \end{cases}$$

Random variables with discrete distribution

A random variable X is in this class if $\mathbb{P}_X$ is a discrete measure concentrated at $\{t_j\}$. The sequence $p = (p_j)$ where

$$p_j := \mathbb{P}_X(\{t_j\}) = \mathbb{P}(X = t_j), \qquad j = 1, \dots,$$

is the *density*, or *mass density function*, of $\mathbb{P}_X$, from which the basic quantities of distribution $\mathbb{P}_X$, law $F_X(t)$, expected value $\mathbb{E}[\,X\,]$ and variance $\mathrm{Var}[X]$ can be computed as follows:

$$\mathbb{P}_X(A) = \sum_{t_j \in A} \mathbb{P}(X = t_j),$$

$$F_X(t) = \sum_{t_j \le t} \mathbb{P}(X = t_j),$$

$$\int_{\mathbb{R}} \varphi(t)\, \mathbb{P}_X(dt) = \sum_j \varphi(t_j)\mathbb{P}(X = t_j),$$

$$\mathbb{E}[\,X\,] = \sum_j t_j \mathbb{P}(X = t_j),$$

$$\mathrm{Var}[X] = \sum_j (t_j - \mathbb{E}[\,X\,])^2 \mathbb{P}(X = t_j).$$

Moreover, if the range of X is integer valued, then

$$\mathbb{E}[\,X\,] = \sum_{k=0}^{\infty} \mathbb{P}(X > k) - \sum_{k-\infty}^{0} \mathbb{P}(X < k).$$

Figure 3.5 *A few formulas relative to discrete random variables.*

For every $t > 0$ we have

$$F_{X^2}(t) = \mathbb{P}(-\sqrt{t} \le X \le \sqrt{t}) = F_X(\sqrt{t}) - F_X(-\sqrt{t}) + \mathbb{P}(X = \sqrt{t}).$$

Assume now that $\mathbb{P}_X(dt) = \rho_X(t)\, dt$. Then for every $t \ge 0$ we have

$$\begin{aligned} F_{X^2}(t) &= \int_{-\infty}^{\sqrt{t}} \rho(s)\, ds - \int_{-\infty}^{-\sqrt{t}} \rho(s)\, ds = \int_{-\sqrt{t}}^{0} \rho(s)\, ds + \int_0^{\sqrt{t}} \rho(s)\, ds \\ &= \int_0^t \frac{\rho(\sqrt{\sigma})}{2\sqrt{\sigma}} d\sigma + \int_0^t \frac{\rho(-\sqrt{\sigma})}{2\sqrt{\sigma}} d\sigma, \end{aligned}$$

Random variables with absolutely continuous distribution
The distribution $\mathbb{P}_X$ of a random variable $X : \Omega \to \mathbb{R}$ defined on a probability space $(\Omega, \mathcal{E}, \mathbb{P})$ is said to be *absolutely continuous* if there exists a non-negative $\mathcal{L}^1$-summable function $f_X : \mathbb{R} \to \mathbb{R}$ such that

$$\mathbb{P}_X(A) = \int_A f_X(t)\,dt \qquad \forall A \in \mathcal{B}(\mathbb{R})$$

or, equivalently, if the law F_X is such that

$$F_X(t) = \int_{-\infty}^{t} f_X(s)\,ds \qquad \forall t \in \mathbb{R}.$$

The function f_X is called the *density* of $\mathbb{P}_X$ (with respect to the Lebesgue measure). If $\mathbb{P}_X(ds) = f_X(s)\,ds$, then

- F_X is an absolutely continuous function,
- $F_X(t)$ is differentiable at a.e. $t \in \mathbb{R}$ and $F'_X(t) = f_X(t)$ at a.e. t,
- if $f_X \in C^0(\mathbb{R})$ then $F_X \in C^1(\mathbb{R})$ and $F'_X(t) = f_X(t)\ \forall t$,
- $\mathbb{E}[X] = \int_{\mathbb{R}} t\, f_X(t)\,dt$,
- $\mathrm{Var}[X] = \int_{\mathbb{R}} (t - \mathbb{E}[X])^2 f_X(t)\,dt = \int_{\mathbb{R}} t^2 f_X(t)\,dt - (\mathbb{E}[X])^2$,
- for every Borel-measurable function $\varphi : \mathbb{R} \to \mathbb{R}$

$$\int_{\mathbb{R}} \varphi(t)\,\mathbb{P}_X(dt) = \int_{\mathbb{R}} \varphi(t) f_X(t)\,dt.$$

Figure 3.6 A few formulas relative to absolutely continuous random variables.

hence $\mathbb{P}_{X^2}$ is absolutely continuous, $\mathbb{P}_{X^2}(dt) = \tau(t)\,dt$, where

$$\tau(t) := \begin{cases} 0 & \text{if } t < 0, \\ \frac{\rho(\sqrt{t}) + \rho(-\sqrt{t})}{2\sqrt{t}} & \text{if } t > 0. \end{cases}$$

Exercise 3.19 *Let X be a random variable. Compute the law of the random variable $Y := \alpha X + \beta$ where $\alpha, \beta \in \mathbb{R}$, $\alpha \neq 0$. Assuming $\mathbb{P}_X$ to be absolutely continuous of density ρ, $\mathbb{P}_X(dt) = \rho(t)\,dt$, show that $\mathbb{P}_Y$ is also absolutely continuous by computing its density.*

Solution. We proceed by computing the law of Y. We have

$$F_Y(t) = \mathbb{P}(\alpha X + \beta \le t) = \begin{cases} \mathbb{P}\Big(X \le \frac{t-\beta}{\alpha}\Big) & \text{if } \alpha > 0, \\ \mathbb{P}\Big(X \ge \frac{t-\beta}{\alpha}\Big) & \text{if } \alpha < 0, \end{cases}$$

hence

$$F_Y(t) = \begin{cases} F_X\left(\frac{t-\beta}{\alpha}\right) & \text{if } \alpha > 0, \\ 1 - F_X\left(\frac{t-\beta}{\alpha}\right) + \mathbb{P}\left(X = \frac{t-\beta}{\alpha}\right) & \text{if } \alpha < 0. \end{cases}$$

Assume now $\mathbb{P}_X(dt) = \rho(t)\,dt$. If $\alpha > 0$, then changing variable with $u = \alpha s + \beta$, we get

$$F_Y(t) = F_X\Big(\frac{t-\beta}{\alpha}\Big) = \int_{-\infty}^{\frac{t-\beta}{\alpha}} \rho(s)\,ds = \int_{-\infty}^{t} \rho\Big(\frac{u-\beta}{\alpha}\Big)\frac{1}{\alpha}\,du$$

i.e. $\mathbb{P}_Y(dt) = \frac{1}{\alpha}\rho\left(\frac{t-\beta}{\alpha}\right)dt$. If $\alpha < 0$, then $\alpha X + \beta \le t$ if and only if $X \ge \frac{t-\beta}{\alpha}$, consequently, changing variable with $u = \alpha s + \beta$

$$\begin{aligned} F_Y(t) = 1 - F_X\Big(\frac{t-\beta}{\alpha}\Big) &= \int_{\frac{t-\beta}{\alpha}}^{+\infty} \rho(s)\,ds \\ &= \int_t^{-\infty} \Big(\frac{u-\beta}{\alpha}\Big)\frac{1}{\alpha}\,du = -\int_{-\infty}^{t} \Big(\frac{u-\beta}{\alpha}\Big)\frac{1}{\alpha}\,du \end{aligned}$$

i.e. $\mathbb{P}_Y(dt) = -\frac{1}{\alpha}\,\rho\left(\frac{t-\beta}{\alpha}\right)dt$. The two cases above can be summarized to

$$\mathbb{P}_Y(dt) = \frac{1}{|\alpha|}\rho\Big(\frac{t-\beta}{\alpha}\Big)\,dt \qquad \text{if} \qquad \mathbb{P}_X(dt) = \rho(t)\,dt.$$

Exercise 3.20 *Let X be a random variable with $\mathbb{P}_X(dt) = r(t)\,dt$ and let $f : \mathbb{R} \to \mathbb{R}$ be a strictly monotone function of class $C^1(\mathbb{R})$. Show that $f \circ X$ is also absolutely continuous, $\mathbb{P}_{f\circ X}(dt) = s(t)\,dt$, with density $s(t) = r(w)/|f'(w)|$, $w = f^{-1}(t)$.*

Solution. For any non-negative Borel-measurable function $\varphi : \mathbb{R} \to \mathbb{R}$ we have

$$\begin{aligned} \int_{\mathbb{R}} \varphi(s)\,\mathbb{P}_{f\circ X}(ds) &= \int_{\mathbb{R}} \varphi(f(t))\,\mathbb{P}_X(dt) \\ &= \int_{\mathbb{R}} \varphi(f(t))\,r(t)\,dt = \int_{\mathbb{R}} \varphi(s) r(f^{-1}(s))\frac{1}{|f'(f^{-1}(s))|}\,ds. \end{aligned}$$

In order to get the last equality, we substituted $s = f(t)$, so that $t = f^{-1}(s)$ and $dt = (f^{-1})'(s)\,ds = 1/|f'(f^{-1}(s))|\,ds$. Therefore,

$$\mathbb{P}_{f\circ X}(ds) = \frac{r(w)}{|f'(w)|}\,ds, \qquad w = f^{-1}(s).$$

since φ is arbitrary.

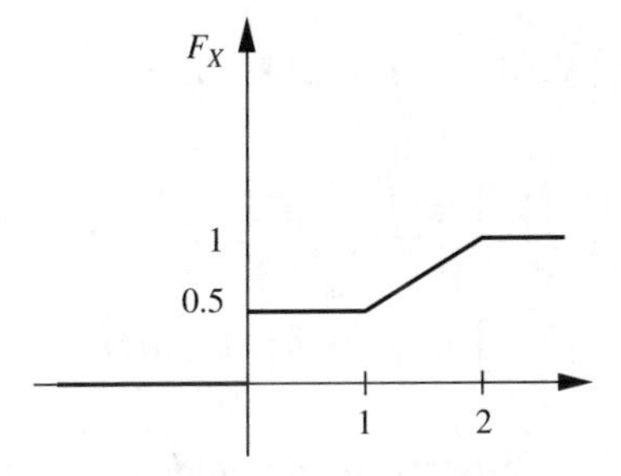

Figure 3.7 The law in Exercise 3.22.

Exercise 3.21 *Let $X(x)$ be the life (measured in years) of a battery, rounded to an integer. Assume that the density of $\mathbb{P}_X$ is given by*

$$p_X(k) = \mathbb{P}_X(\{k\}) = \begin{cases} 0.2 & \text{if } 3 \le k \le 7, \\ 0 & \text{otherwise.} \end{cases}$$

Compute:

(i) The probability that a 3-years-old battery is still working.

(ii) The probability that a battery which is at least 3 years old is still working.

(iii) The probability that the battery is working a further 3 years even if it has already been working for 5 years.

Exercise 3.22 *Assume Figure 3.7 represents the graph of the law of a random variable X. Compute $\mathbb{E}[X]$, $\mathrm{Var}[X]$ and $\mathbb{P}(X \le 0.8)$.*

Exercise 3.23 *Let the real-valued random variable X be uniformly distributed in $[0, 1]$. Compute the law of $Y = \max\{X - 0.5, 0\}$.*

Exercise 3.24 *Show that for any real valued random variable X*

$$\mathbb{E}[X_+] = \int_0^{+\infty} (1 - F_X(s))\, ds,$$

$$\mathbb{E}[|X|^p] = p \int_0^{+\infty} t^{p-1}(1 - F_X(s))\, ds.$$

As usual, $X_+ := \max\{X, 0\}$.

Exercise 3.25 *Show that for any random variable $X : \Omega \to \mathbb{R}$ the following equalities hold:*

$$\mathbb{E}[\max\{X, 0\}] = \int_0^{+\infty} |t|\, \mathbb{P}_X(dt),$$

$$\mathbb{E}[\min\{X,0\}] = \int_{-\infty}^{0} |t|\, \mathbb{P}_X(dt),$$
$$\mathbb{E}[\,|X|\,] = \int_{-\infty}^{+\infty} |t|\, \mathbb{P}_X(dt), \tag{3.23}$$
$$\mathbb{E}\big[\,|X|^p\,\big] = \int_{-\infty}^{+\infty} |t|^p\, \mathbb{P}_X(dt) \qquad \forall p \geq 0.$$

Exercise 3.26 *Let $X : \Omega \to \mathbb{R}$ be a discrete random variable. Prove the following inequalities.*

$$\mathbb{E}[\max\{X,0\}] = \sum_{y \in X(\Omega),\, y \geq 0} y\mathbb{P}(X = y),$$
$$\mathbb{E}[\min\{X,0\}] = \sum_{y \in X(\Omega),\, y \leq 0} y\mathbb{P}(X = y),$$
$$\mathbb{E}[\,|X|\,] = \sum_{y \in X(\Omega)} |y|\mathbb{P}(X = y), \tag{3.24}$$
$$\mathbb{E}\big[\,|X|^p\,\big] = \sum_{y \in X(\Omega)} |y|^p\mathbb{P}(X = y).$$

Exercise 3.27 *Let $X : \Omega \to \mathbb{R}$ be a random variable and let $\varphi : [a,b] \to \mathbb{R}$ be a strictly monotone increasing function of class $C^1([a,b])$. Prove the integration by parts formula:*

$$\int_a^b \varphi'(t) F_X(t)\, dt + \int_{]a,b]} \varphi(t)\, \mathbb{P}_X(dt) = \varphi(b)F_X(b) - \varphi(a)F_X(a). \tag{3.25}$$

Solution. Let $E_\tau := \Big\{t \in \mathbb{R} \,\Big|\, \varphi(t)\mathbb{1}_{]a,b]}(t) > \tau\Big\}$. Cavalieri formula (B.22) yields

$$\int_{]a,b]} \varphi(t)\, \mathbb{P}_X(dt) = \int_{\mathbb{R}} \varphi(t)\mathbb{1}_{]a,b]}(t)\, \mathbb{P}_X(dt) = \int_0^{+\infty} \mathbb{P}_X(E_\tau)\, d\tau. \tag{3.26}$$

Since

$$E_\tau = \Big\{t \in]a,b] \,\Big|\, \varphi(t) > \tau\Big\} = \begin{cases}]a,b] & \text{if } \tau \leq \varphi(a), \\ ]\varphi^{-1}(\tau), b] & \text{if } \varphi(a) < \tau \leq b, \\ \emptyset & \text{if } \tau > \varphi(b), \end{cases}$$

we get

$$\mathbb{P}_X(E_\tau) = \begin{cases} F_X(b) - F_X(a) & \text{if } 0 < \tau \leq \varphi(a), \\ F_X(b) - F_X(\varphi^{-1}(\tau)) & \text{if } \varphi(a) < \tau \leq \varphi(b), \\ 0 & \text{if } \tau > \varphi(b). \end{cases}$$

Thus

$$\begin{aligned}\int_0^\infty \mathbb{P}_X(E_\tau)\,d\tau &= \int_0^{\varphi(a)} \Big(F_X(b) - F_X(a)\Big)\,d\tau \\ &\quad + \int_{\varphi(a)}^{\varphi(b)} \Big(F_X(b) - F_X(\varphi^{-1}(\tau))\Big)\,d\tau \\ &= \varphi(a)(F_X(b) - F_X(a)) \\ &\quad + F_X(b)(\varphi(b) - \varphi(a)) - \int_a^b \varphi'(s)F_X(s)\,ds \\ &= F_X(b)\varphi(b) - \varphi(a)F_X(a) - \int_a^b \varphi'(s)F_X(s)\,ds.\end{aligned}$$

The last equality and (3.26) yield the claim.

3.2 A few discrete distributions

3.2.1 Bernoulli distribution

The *Bernoulli distribution* of parameter p, $0 \le p \le 1$, is the probability measure $B(p)$ supported on the two points set $\{0, 1\}$ whose density is

$$B(p)(\{0\}) = 1 - p, \qquad B(p)(\{1\}) = p.$$

The expectation and the variance of random variables X such that $\mathbb{P}_X = B(p)$ are

$$\mathbb{E}[\,X\,] = p \qquad \text{and} \qquad \mathrm{Var}[\,X\,] = p(1-p), \tag{3.27}$$

respectively.

Any random variable X whose range is the two points set $\{0, 1\}$ has a Bernoulli type distribution $B(p)$ with $p := \mathbb{P}(X = 1)$. These random variables are called *Bernoulli trials* of parameter p. For instance let $(\Omega, \mathcal{E}, \mathbb{P})$ be a probability space and $E \in \mathcal{E}$. The characteristic function $\mathbb{1}_E(x)$ is a Bernoulli trial of parameter $\mathbb{P}(E)$ since $\big\{x \in \Omega \,|\, \mathbb{1}_E(x) = 1\big\} = E$.

3.2.2 Binomial distribution

The *binomial distribution* $B(n, p)$ of parameters n and p, $n \ge 1$, $0 \le p \le 1$, is the measure supported on the $n + 1$ points set $\{0, \dots, n\}$ of density

$$B(n, p)(\{k\}) = \binom{n}{k} p^k (1-p)^{n-k}, \qquad 0 \le k \le n.$$

$B(n, p)$ is a probability measure since the Newton binomial formula yields

$$\sum_k B(n, p)(\{k\}) = \sum_{k=0}^{n} \binom{n}{k} p^k (1-p)^{n-k} = (p + 1 - p)^n = 1.$$

The expected value and the variance of a random variable X such that $\mathbb{P}_X = B(n, p)$ are

$$\mathbb{E}[X] = np \qquad \text{and} \qquad \mathrm{Var}[X] = np(1-p), \tag{3.28}$$

respectively.

Let us prove (3.28). For $p = 0$ or 1, (3.28) is trivial. Assume $0 < p < 1$ and let $z := p/(1-p)$. From (iv) of Figure 1.4, we infer

$$\begin{aligned}\mathbb{E}[X] &= \sum_{k=0}^{n} k\binom{n}{k} p^k (1-p)^{n-k} \\ &= (1-p)^n \sum_{k=0}^{n} k\binom{n}{k} z^k = (1-p)^n nz(1+z)^{n-1} = np;\end{aligned}$$

moreover, (v) of Figure 1.4 yields

$$\begin{aligned}\mathbb{E}\left[X^2\right] &= \sum_{k=0}^{n} k^2\binom{n}{k} p^k (1-p)^{n-k} = (1-p)^n \sum_{k=0}^{n} k^2\binom{n}{k} z^k \\ &= (1-p)^n nz(1+z)^{n-2}(1+nz) = np(1-p+np),\end{aligned}$$

hence

$$\mathrm{Var}[X] = \mathbb{E}\left[X^2\right] - \mathbb{E}[X]^2 = np(1-p+np) - n^2p^2 = np(1-p).$$

Example 3.28 Consider n 'independent' random trials of parameter p, $0 \le p \le 1$. Each sequence x of outcomes can be arranged as n-tuple of zeros and ones, a n-byte, $x = (x_1, \dots, x_n)$. We already know that any such sequence has probability $p^k(1-p)^{n-k}$, see Section 2.2.5, therefore we model n 'independent' trials as the probability space $(\Omega, \mathcal{E}, \mathbb{P})$ where the sample space is $\Omega = \{0, 1\}^n$, $\mathcal{E} = \mathcal{P}(\Omega)$ and $\mathbb{P}$ is defined by its density, $\mathbb{P}(\{x\}) := p^k(1-p)^{n-k}$ if x has k ones and $n-k$ zeros.

Let $X : \{0, 1\}^n \to \mathbb{N}$ be the random variable counting the number of successes, i.e. the function that for each n-byte x returns the number of ones in it. Then for every $k = 0, 1, \dots,$

$$\mathbb{P}(X = k) = \binom{n}{k} p^k (1-p)^{n-k} \tag{3.29}$$

since there are $\binom{n}{k}$ n-bytes with exactly k ones and all of them have probability $p^k(1-p)^{n-k}$. Therefore, the distribution of X is the binomial one of parameters n and p, $\mathbb{P}_X = B(n, p)$ for short.

3.2.2.1 Graph of the binomial distribution

As k increases from 0 to $n-1$, the quotient

$$\frac{B(n, p)(\{k+1\})}{B(n, p)(\{k\})} = \frac{p}{1-p}\frac{n-k}{k+1} \tag{3.30}$$

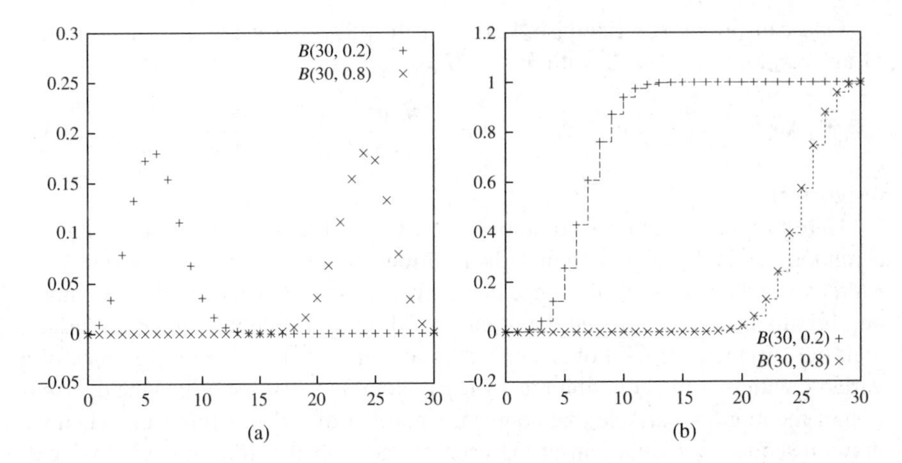

Figure 3.8 (a) The densities and (b) the laws of two binomial distributions.

decreases from $\frac{p}{1-p}n$ to $\frac{p}{1-p}\frac{1}{n}$. Thus,

(i) if $n\frac{p}{1-p} < 1$, then $k \to B(n, p)(\{k\})$ is monotone decreasing,

(ii) if $\frac{1}{n}\frac{p}{1-p} > 1$, then $k \to B(n, p)(\{k\})$ is monotone increasing,

(iii) if $\frac{1}{n}\frac{p}{1-p} < 1 < \frac{p}{1-p}n$, then there exists $\overline{k}$ such that $k \to B(n, p)(\{k\})$ is strictly increasing on $0, \dots, \overline{k} - 1$ and strictly decreasing on $\overline{k}, \dots, n$. By (3.30), $\overline{k}$ is the smallest integer greater than the solution of $\frac{p}{1-p}\frac{n-k}{k+1} = 1$, that is the smallest integer greater than $np - 1 + p$.

The densities and the laws of two binomial distributions are plotted in Figure 3.8.

3.2.3 Hypergeometric distribution

Let w, r be non-negative and let $0 \leq n \leq w + r$. The *hypergeometric distribution* $H(w, r, n)$ of parameters w, r and n is the discrete measure whose density is

$$p(k) = H(w, r, n)(\{k\}) := \frac{\binom{w}{k}\binom{r}{n-k}}{\binom{w+r}{n}}, \qquad k = 0, 1, \dots, n.$$

$H(w, r, n)$ is a probability measure since

$$\sum_k H(w, r, n)(\{k\}) = \sum_{k=0}^{n} \frac{\binom{w}{k}\binom{r}{n-k}}{\binom{w+r}{n}} = 1$$

by the *Vandermonde formula* (A.5).

Of course, $p(k) \neq 0$ only if $0 \leq k \leq w$ and $n - r \leq k \leq n$, so that $H(w, r, n)$ is supported on the integers between $\max\{0, n - r\}$ and $\min\{w, n\}$.

One can prove, see Exercise 4.10, that the expected value and the variance of any random variable X with $\mathbb{P}_X = H(w, r, n)$ are

$$\mathbb{E}[X] = \frac{nw}{w+r} \quad \text{and} \quad \mathrm{Var}[X] = \frac{nwr}{(w+r)^2}\left(1 - \frac{n-1}{w+r-1}\right), \tag{3.31}$$

respectively.

Here we only want to point out that X has the same expected value as a random variable with binomial distribution $B(n, p)$ where $p := w/(w+r)$. Moreover, when $(w+r)/n \to \infty$, that is, when the ratio between the total number of balls inside the urn and the number of drawings becomes larger and larger, then the variance $\mathrm{Var}[X]$ of X converges to $np(1-p)$, i.e. to the variance of a variable with binomial distribution $B(n, p)$. Note that $B(n, p)$ is also the distribution of the random variable that counts the number of drawn white balls when we draw in sequence n balls reinserting each extracted ball before drawing the next.

Example 3.29 Assume we have an urn containing w distinguishable white balls and r distinguishable red balls. We extract (without replacement) n balls. Consider the probability space $(\Omega, \mathcal{E}, \mathbb{P})$ where the sample space Ω is the set of all admissible drawings, i.e. of the n-words made of symbols chosen among w white balls and r red balls, $\mathcal{E} = \mathcal{P}(\Omega)$ and $\mathbb{P}$ is the uniform probability on Ω of density $1/\binom{w+r}{n}$.

Trivially, the random variable $X(x)$ from Ω into $\mathbb{N}$ that counts the number of white balls in the drawn sequence x of n balls follows the distribution $\mathbb{P}_X = H(w, r, n)$. In fact, for any k, the number of drawings with k white balls and $n-k$ red balls is $\binom{w}{k}\binom{r}{n-k}$.

3.2.4 Negative binomial distribution

Let $n \geq 1$ and $p \in [0, 1]$. The *negative binomial* or *Pascal distribution* $B(-n, p)$ is the measure supported on non-negative integers of density

$$B(-n, p)(\{k\}) := \binom{n+k-1}{k} p^n (1-p)^k, \qquad \forall k \geq 0.$$

Since

$$\begin{aligned}\sum_{k=0}^{\infty} B(-n, p)(\{k\}) &= \sum_{k=0}^{\infty} \binom{n+k-1}{k} p^n (1-p)^k = p^n \sum_{k=0}^{\infty} \binom{-n}{k} (p-1)^k \\ &= \frac{p^n}{(1+p-1)^n} = 1,\end{aligned}$$

see Proposition 1.2, $B(-n, p)$ is a probability measure.

The expected value and the variance of a random variable X with $\mathbb{P}_X = B(-n, p)$ are given by

$$\mathbb{E}[X] = n\frac{1-p}{p} \quad \text{and} \quad \mathrm{Var}[X] = n\frac{1-p}{p^2}, \tag{3.32}$$

respectively. We have

$$\mathbb{E}[X] = p^n \sum_{k=0}^{\infty} k \binom{-n}{k} (p-1)^k$$
$$= -np^n z(1+z)^{n-1} = -np^n(p-1)p^{-n-1} = n\frac{1-p}{p}$$

from (iv) of Figure 1.4 with $z := p-1$; moreover, from (v) of Figure 1.4 again with $z := p-1$, we get

$$\mathbb{E}[X^2] = p^n \sum_{k=0}^{\infty} k^2 \binom{-n}{k} (p-1)^k$$
$$= -np^n z(1-nz)(1+z)^{-n-2} = \frac{1-p}{p^2}(1+n-np),$$

hence

$$\mathrm{Var}[X] = \mathbb{E}[X^2] - \mathbb{E}[X]^2 = \frac{1-p}{p^2}(1+n-np) - n^2\frac{(1-p)^2}{p^2} = n\frac{1-p}{p^2}.$$

Example 3.30 Consider a denumerable sequence of Bernoulli trials. Each outcome is a binary sequence. Consider the function T that counts the number of failures before the nth success. T is a discrete random variable with values in the set of non-negative integers. We claim that $\mathbb{P}_T = B(-n, p)$.

Let $E_{n,k}$ be the set of sequences of $n+k$ trials with k zeroes and n ones, and with a one at the $(n+k)$th trial. Then, trivially, $\mathbb{P}(T = k) = \mathbb{P}(E_{n,k})$. The cardinality of $E_{n,k}$ is $\binom{n+k-1}{n-1} = \binom{n+k-1}{k}$, i.e. the number of $n+k-1$ drawings with $n-1$ ones and k zeroes, and each sequence in $E_{n,k}$ has probability $(1-p)^k p^n$, see Section 2.2.5. Therefore,

$$\mathbb{P}_T(\{k\}) := \binom{n+k-1}{k}(1-p)^k p^n = B(-n, p)(\{k\}),$$

thus concluding that $\mathbb{P}_T = B(-n, p)$.

The densities and the laws of two negative binomial distributions are plotted in Figure 3.9.

3.2.5 Poisson distribution

The *Poisson distribution* $P(\lambda)$ of parameter $\lambda > 0$ is the measure supported on the non-negative integers of mass density

$$P(\lambda)(\{k\}) := \frac{\lambda^k}{k!}e^{-\lambda}. \tag{3.33}$$

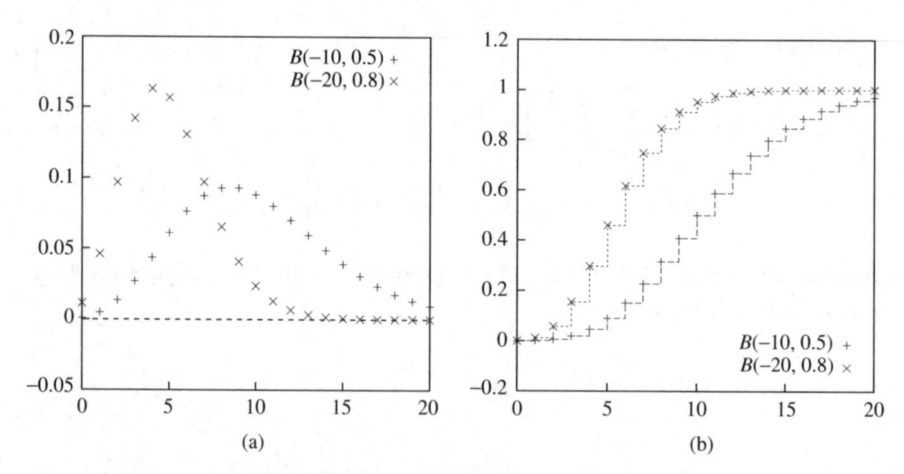

Figure 3.9 *(a) The densities and (b) the laws of two negative binomial distributions.*

$P(\lambda)$ is a probability measure since

$$\sum_{k=0}^{\infty} P(\lambda)(\{k\}) = e^{-\lambda} \sum_{k=0}^{\infty} \frac{\lambda^k}{k!} = e^{-\lambda} e^{\lambda} = 1.$$

The mean and the variance of a random variable X such that $\mathbb{P}_X = P(\lambda)$ are given by

$$\mathbb{E}[\,X\,] = \lambda \qquad \text{and} \qquad \mathrm{Var}[\,X\,] = \lambda, \tag{3.34}$$

respectively. The proof is elementary.

Poisson distribution approximates $B(n, \lambda/n)$ for large n. More precisely, the following holds.

Proposition 3.31 (Rare events property) *Let $\lambda \in \overline{\mathbb{R}}_+$ and let $\{p_n\}$ be a non-negative sequence such that $np_n \to \lambda$. Then, for any $k \in \mathbb{N}$*

$$B(n, p_n)(\{k\}) \longmapsto \begin{cases} P(\lambda)(\{k\}) & \text{if } 0 < \lambda < +\infty, \\ 0 & \text{if } \lambda = 0, +\infty \end{cases} \qquad \text{as } n \to \infty. \tag{3.35}$$

Proof. Since

$$\begin{aligned} B(n, p_n)(\{k\}) &= \binom{n}{k} p_n^k (1-p_n)^{n-k} \\ &= \frac{n(n-1)\cdots(n-k+1)}{k!n^k}(np_n)^k(1-p_n)^{n-k} \\ &=: \frac{n(n-1)\cdots(n-k+1)}{k!n^k} h_n \end{aligned} \tag{3.36}$$

and $\frac{n(n-1)\cdots(n-k+1)}{n^k} \to 1$ as $n \to \infty$, it suffices to compute the limit of h_n, or, equivalently, of

$$\log h_n = k\log(np_n) + (n-k)\log(1-p_n)$$
$$= np_n\Big(k\frac{\log(np_n)}{np_n} + \Big(1-\frac{k}{n}\Big)\frac{\log(1-p_n)}{p_n}\Big).$$

If $\lambda < +\infty$, then $p_n \to 0$. Since $\frac{\log(1-x)}{x} \to -1$ as $x \to 0$, we get

$$\log h_n \to k\log\lambda - \lambda, \qquad \text{i.e.} \qquad h^n \to \lambda^k e^{-\lambda}$$

as $n \to \infty$. This proves (3.36) for finite λ. If $\lambda = +\infty$, recalling that $\log x/x \to 0$ as $x \to +\infty$ and $\log(1-x)/x \le -1$ for any $0 < x < 1$, we get that $\log h_n \to -\infty$, so that $h_n \to 0$.

Remark 3.32 Actually, we have just proved that if $n_j p_{n_j} \to \lambda$ for some subsequence $\{p_{n_j}\}$ of $\{p_n\}$, then

$$B(n_j, p_{n_j})(\{k\}) \longmapsto \begin{cases} P(\lambda)(k) & \text{if } 0 < \lambda < +\infty, \\ 0 & \text{if } \lambda = 0, +\infty \end{cases} \qquad \text{as } j \to \infty.$$

Remark 3.33 Using McLaurin expansions

$$\frac{n(n-1)\cdots(n-k+1)}{n^k} = 1 - \frac{k(k-1)}{2n} + O\Big(\frac{1}{n^2}\Big),$$
$$\Big(1-\frac{\lambda}{n}\Big)^n = e^{-\lambda} + e^{-\lambda}\frac{\lambda^2}{n} + O\Big(\frac{1}{n^2}\Big)$$

one easily estimates the relative error: we have

$$\Big|B(n,\lambda/n)(\{k\}) - P(\lambda)(\{k\})\Big| \le P(\lambda)(\{k\})\frac{(k+\lambda)^2}{n}.$$

$P(\lambda)$ is known as the distribution of rare events. In fact, many counting problems can be seen as the counting of the number of successes in n 'independent' trials with n large and the probability of success in each trial, p, small.

Example 3.34 Assume a public service (such as the emergency services) receives – on average– λ calls per day. Assume there are n possible users and that they call the service independently of the others. The number of daily calls thus follows a binomial distribution $B(n, p)$, where p is unknown. Since the expected value of $B(n, p)$ is np, we can assume $p = \lambda/n$. Thus, for large n, the probability that the service gets k daily calls is approximately $P(\lambda)(\{k\}) = \frac{\lambda^k}{k!}e^{-\lambda}$.

Events randomly distributed in time or space that can be modelled by means of a Poisson distribution with good approximation appear in many different contexts.

For instance, the number of calls per hour a callcentre receives, the number of cars passing through a toll gate at a certain time of the day, the number of car accidents that occur on a certain stretch of road on a public holiday, the number of flaws in a certain length of yarn produced by a textile machine, and the number of wrong notes played per hour by a musician during a concert are possible examples.

Example 3.35 Assume a machine is subject to an average number λ of breakdowns per unit time; assume also the average number of breakdowns is proportional to the observation time. Thus, in an arbitrary time interval $[s, s+t]$, the average number of breakdowns is λt.

Let us divide the interval $[0, t]$ in to N intervals $I_1, \dots, I_N$ of equal length. Assume:

- the probability that a breakdown occurs in the interval I_i is $p = \lambda t/N$;
- the probability that two or more breakdowns occur in any of the intervals I_i is negligible with respect to p;
- the events X_i, 'a breakdown occurs during the interval I_i' are independent, so that the distribution of the random variable $\sum_{i=1}^{N} X_i$ that counts the number of breakdowns in the interval $[0, t]$ is the binomial one, $B(N, \lambda t/N)$.

If the assumptions stated above hold for any large enough N, then the probability that k breakdowns occur in the time interval $[0, t]$ is approximately

$$P(\lambda t)(\{k\}) = \frac{\lambda^k t^k}{k!} e^{-\lambda t}.$$

For a precise statement and further details see Section 6.1.

The densities and the laws of a binomial distribution and of a Poisson one are plotted in Figure 3.10.

3.2.6 Geometric distribution

The geometric distribution of parameter p, $0 < p < 1$, is the probability measure $G(p)$ supported on positive integers of density

$$G(p)(\{k\}) := (1-p)^{k-1} p, \qquad \forall k \geq 1.$$

Thus

$$F_{G(p)}(t) = \sum_{1 \leq k \leq t} (1-p)^{k-1} p = 1 - (1-p)^{[t]}.$$

The expected value and the variance of a random variable X with geometric distribution, $\mathbb{P}_X = G(p)$ are

$$\mathbb{E}[X] = \frac{1}{p} \qquad \text{and} \qquad \mathrm{Var}[X] = \frac{(1-p)}{p^2}, \tag{3.37}$$

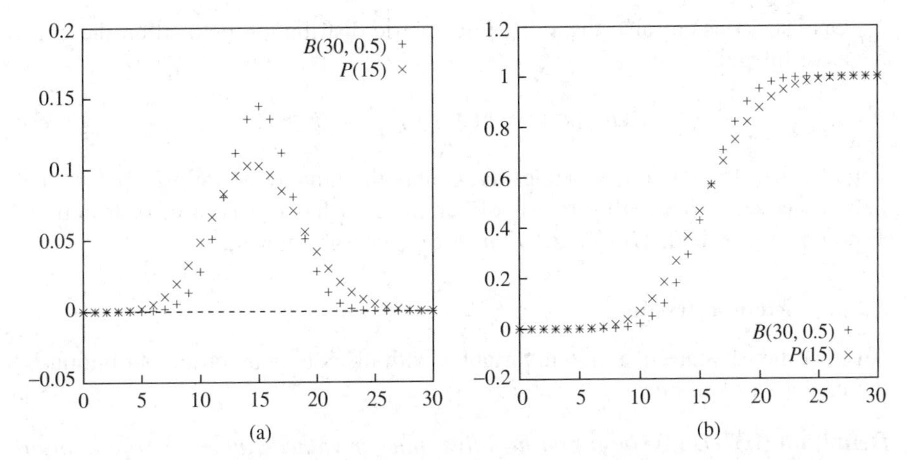

Figure 3.10 (a) The densities and (b) the laws of the binomial distribution $B(30, 0.5)$ and of the Poisson distribution $P(15)$, respectively.

respectively. To show the former, apply formula (i) of Figure 1.4 with $z = 1 - p$; we get

$$\mathbb{E}[X] = \sum_{k=1}^{\infty} kG(p)(\{k\}) = \frac{p}{1-p}\sum_{k=1}^{\infty} k(1-p)^k = \frac{p}{1-p}\frac{1-p}{p^2} = \frac{1}{p}.$$

For the latter, apply formula (ii) of Figure 1.4 with $z = 1 - p$ to get

$$\mathbb{E}[X^2] = \sum_{k=0}^{\infty} k^2 p(1-p)^{k-1} = p\frac{2-p}{p^3},$$

hence

$$\mathrm{Var}[X] = \mathbb{E}[X^2] - \mathbb{E}[X]^2 = \frac{2-p}{p^2} - \frac{1}{p^2} = \frac{1-p}{p^2}.$$

Example 3.36 Let the random variable T describe the waiting time for the first success in a Bernoulli process of parameter p, $0 < p < 1$,

$$T(x) := \min\left\{n \,\middle|\, x = (x_k),\ x_1 = \cdots = x_{n-1} = 0,\ x_n = 1\right\},$$

where $T(x) = +\infty$ if success never occurs, i.e. if $x = (x_n)$ is such that $x_n = 0$ $\forall n$. The distribution of the random variable X is the geometric one $G(p)$. In fact, in Section 2.2.5 we observed that

$$\mathbb{P}(T = +\infty) = \mathbb{P}(\{0, 0, \dots\}) = 0$$

and

$$\mathbb{P}(T = k) = \mathbb{P}(\{(\underbrace{0, 0, \dots, 0}_{k-1}, 1)\}) = (1-p)^{k-1}p.$$

One can consider also the shifted geometric distribution defined on the non-negative integers

$$G'(p)(\{k\}) := p(1-p)^k \qquad \forall k \geq 0. \tag{3.38}$$

For instance, the random variable that counts the number of failures before the first success in a Bernoulli process of parameter p has this type of distribution. The densities of both $G(p)$ and $G'(p)$ are geometric sequences.

3.2.6.1 Memorylessness

An important feature of random variables with discrete geometric distribution is the *memoryless property*.

Definition 3.37 *Let X be a non-negative, integer valued random variable on a probability space $(\Omega, \mathcal{E}, \mathbb{P})$. We say that X has the* memoryless property *if*

$$\mathbb{P}\Big(X \leq i+j \,\Big|\, X \geq j\Big) = \mathbb{P}(X \leq i) \qquad \forall i, j \in \mathbb{N}. \tag{3.39}$$

We have the following.

Theorem 3.38 *Let X be a non-negative random variable with discrete distribution supported on $\mathbb{N}$ and assume that $p := \mathbb{P}_X(X=0)$ is such that $0 < p < 1$. Then X is a discrete memoryless random variable if and only if $\mathbb{P}_X = G'(p)$.*

Proof. Suppose that $\mathbb{P}_X = G'(p)$ with $0 < p < 1$. Then for every $k \geq 0$, $\mathbb{P}(X = k) = p(1-p)^k$, hence we have $\mathbb{P}(X = 0) = p$ and, moreover, for every $i, j \geq 0$

$$\mathbb{P}(X \leq i) = p\sum_{k=0}^{i}(1-p)^k,$$

$$\mathbb{P}(j \leq X \leq i+j) = \sum_{k=j}^{j+i} p(1-p)^k$$

$$= p(1-p)^j \sum_{k=0}^{i}(1-p)^k = (1-p)^j \mathbb{P}(X \leq i)$$

$$\mathbb{P}(X \geq j) = p\sum_{k=j}^{\infty}(1-p)^k = p(1-p)^j \frac{1}{1-(1-p)} = (1-q)^j,$$

from which (3.39) follows.

Let us prove the converse. For $k = 0, 1, \ldots$, let $x_k := \mathbb{P}(X = k)$ be the density of $\mathbb{P}_X$ at k. We first infer from (3.39) that $\mathbb{P}(X = 0) < 1$ since $P(X \geq j) > 0$

for all non-negative integers j, and that $\mathbb{P}(X = 0) > 0$ since otherwise $\mathbb{P}(X = j) = 0\ \forall j$. Memorylessness property (3.39) with $i = 0$ yields

$$\frac{\mathbb{P}(X = j)}{\mathbb{P}(X \geq j)} = \mathbb{P}(X = 0) \qquad \forall j$$

i.e.

$$x_j = x_0 \sum_{k=j}^{\infty} x_k \qquad \forall j \geq 0.$$

Subtracting from the previous formula a similar formula with j replaced by $j + 1$, we get $x_j - x_{j+1} = x_0 x_j$, i.e.

$$x_{j+1} = x_j(1 - x_0) \qquad \forall j \geq 0,$$

hence by induction $x_j = x_0(1 - x_0)^j\ \forall j \geq 0$, as required.

The density and the law of a geometric distribution are plotted in Figure 3.11.

3.2.7 Exercises

Exercise 3.39 *Let X be the number of defective parts produced each day in a production chain. Assume X is a random variable that follows the binomial distribution, with mean 6 and variance 5.88:*

- *Compute the density of X.*
- *Compute the probability that no defective part is produced today. How can this value be approximated?* $\left[50^{-300} \simeq e^{-6}\right]$

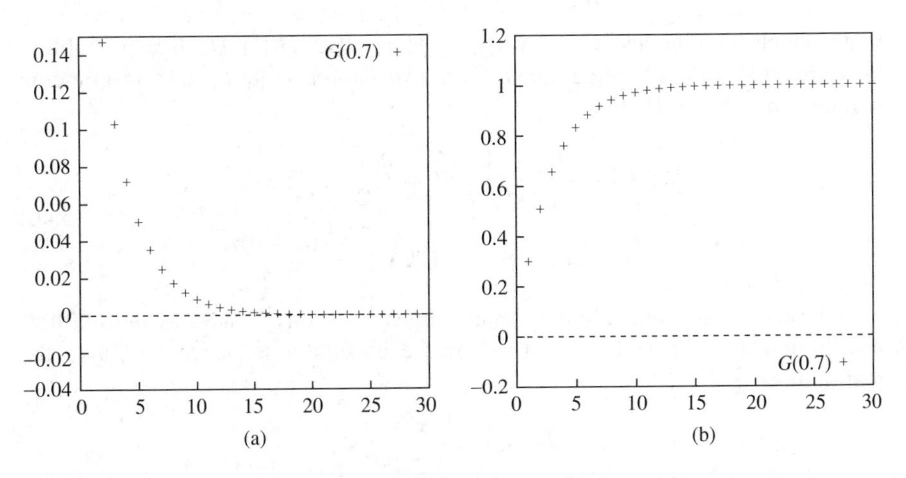

Figure 3.11 (a) The density and (b) the law of a geometric distribution.

Exercise 3.40 *A random variable X follows the binomial distribution. The mean and the variance are* 6 *and* 2.4, *respectively. Compute* $\mathbb{P}(X = 5)$, $\mathbb{P}(X = 6)$ *and* $\mathbb{P}(X = 7)$. $[\mathbb{P}(X = 5) \simeq 0.2, \mathbb{P}(X = 6) \simeq 0.25, \mathbb{P}(X = 7) \simeq 0.215]$

Exercise 3.41 *Show that the memoryless property (3.39) is equivalent to*

$$\mathbb{P}(X = 0) = \frac{\mathbb{P}(X = j)}{\mathbb{P}(X \geq j)} \qquad \forall j \geq 0.$$

Exercise 3.42 *Let X be a geometrically distributed random variable, that is* $\mathbb{P}(X = k) = (1 - p)^{k-1} p \ \forall k \geq 1$. *Show that*

$$\mathbb{P}\Big(X < i + j \,\Big|\, X \geq j\Big) = \mathbb{P}(X \leq i) \qquad \forall i, j \geq 1.$$

Infer that the above equality is equivalent to

$$\mathbb{P}(X = 1) = \frac{\mathbb{P}(X = j)}{\mathbb{P}(X \geq j)} \qquad \forall j \geq 1.$$

3.3 Some absolutely continuous distributions

3.3.1 Uniform distribution

Let $]a, b[\subset \mathbb{R}$, $a < b$, be an interval of $\mathbb{R}$ and let

$$\rho(x) = \begin{cases} 0 & \text{if } x \notin]a, b[, \\ \frac{1}{b-a} & \text{if } x \in]a, b[. \end{cases}$$

The measure $(\mathcal{B}(\mathbb{R}), U)$ defined as

$$U(A) := \frac{\mathcal{L}^1(A \cap]a, b[)}{b - a} = \int_A \rho(x)\, dx, \qquad A \in \mathcal{B}(\mathbb{R}),$$

is absolutely continuous with density ρ. Moreover, $(\mathcal{B}(\mathbb{R}), U)$ is a probability measure on $\Omega :=]a, b[$ called the *uniform distribution* on $]a, b[$. If X is a random variable with $\mathbb{P}_X = U$, then

$$\begin{aligned} \mathbb{E}[X] &= \int_a^b t \frac{1}{b - a} dt = \frac{a + b}{2}, \\ \text{Var}[X] &= \mathbb{E}\big[X^2\big] - \mathbb{E}[X]^2 = \frac{(a - b)^2}{12}. \end{aligned} \tag{3.40}$$

Flipping a fair coin infinitely many times is strictly related to the uniform distribution. Let $T\colon \{0, 1\}^\infty \to [0, 1]$ map each binary sequence $\{\alpha_k\}$ into the real number $\sum_{k=1}^\infty \frac{\alpha_k}{2^k}$:

$$T(\{\alpha_k\}) := \sum_{k=1}^\infty \frac{\alpha_k}{2^k}.$$

Theorem 3.43 *Let* $(\{0, 1\}^\infty, \mathcal{E}, Ber(\infty, \frac{1}{2}))$ *denote the probability space generated by an infinite Bernoulli process of parameter* $p = 1/2$. *Then* T *is a random variable on* $(\{0, 1\}^\infty, \mathcal{E}, Ber(\infty, \frac{1}{2}))$ *which is uniformly distributed on the interval* $[0, 1]$. *In other words,* $\mathbb{P}_T$ *is the Lebesgue measure on the interval* $[0, 1]$.

Proof. Let $\mathcal{E}$ be the family of events associated with the infinite Bernoulli process. We have to prove the following:

- T is a random variable, i.e. $\Big\{\{\beta_k\} \,|\, T(\{\beta_k\}) \le x\Big\} \in \mathcal{E}\ \forall x \in [0, 1]$.
- $Ber\Big(\infty, \frac{1}{2}\Big)(T \le x) = \mathcal{L}^1([0, x]) = x\ \forall x \in [0, 1]$.

As previously stated, see the proof of Proposition 2.32, the map T is surjective but it is not injective. In fact if a sequence $\{\alpha_k\}$ is definitely 0 but not identically 0 (definitely 1 but not identically 1), then there exists a sequence $\{\beta_k\}$ which is definitely 1 (0, respectively) such that $T(\{\alpha_k\}) = T(\{\beta_k\})$. Thus, if N is the subset of the sequences which are definitely 1, T is a one-to-one map between $\{0, 1\}^\infty \setminus N$ and $[0, 1[$. Notice that N is denumerable, hence a zero-probability event in $\mathcal{E}$, see Example 2.33.

Let $x, y \in [0, 1[$ and $\alpha = \{\alpha_k\}, \beta = \{\beta_k\} \in \{0, 1\}^\infty \setminus N$ such that $T(\alpha) = x$, $T(\beta) = y$. One can easily check that $y \le x$ if and only if $\beta \le \alpha$ with respect to the *lexicographic order*, i.e. $\beta_1 < \alpha_1$, or $\beta_1 = \alpha_1$ and $\beta_2 < \alpha_2$, or $\beta_1 = \alpha_1$, $\beta_2 = \alpha_2$ and $\beta_3 < \alpha_3$ and so on.

For any $\alpha = \{\alpha_k\} \in \{0, 1\}^\infty \setminus N$ and for any non-negative integer $k \ge 1$ define

$$A_k := \Big\{\beta \in \{0, 1\}^\infty \,\Big|\, \beta_i = \alpha_i\ \forall i = 1, \dots, k-1,\ \beta_k < \alpha_k\Big\}$$

so that $T(\beta) = y \le x = T(\alpha)$ if and only if $\beta \in A_k$ for some k, that is

$$\cup_k A_k \subset \Big\{\beta \,\Big|\, T(\beta) \le x\Big\} \subset \cup_k A_k \cup N.$$

Thus, there exists a zero-probability event $M \subset N$ such that

$$\Big\{\beta \,\Big|\, T(\beta) \le x\Big\} = \cup_k A_k \cup M.$$

Notice that A_k is empty whenever $\alpha_k = 0$. If $\alpha_k = 1$, then the elements of A_k are the sequences $\beta = \{\beta_j\}$ such that $\beta_j = \alpha_j$ for every $j = 1, \dots, k-1$ and $\beta_k = 0$. In particular each set A_k is a cylinder, hence $A_k \in \mathcal{E}$ and therefore, $\{\beta \,|\, T(\beta) \le x\} \in \mathcal{E}$ being a denumerable union of $\mathcal{E}$-measurable sets. Since $x \in [0, 1[$ is arbitrary, we have proven that T is a random variable.

Moreover, the previous description proves that for every $k \ge 1$

$$Ber\Big(\infty, \frac{1}{2}\Big)(A_k) = \begin{cases} 0 & \text{if } \alpha_k = 0, \\ Ber\left(k, \frac{1}{2}\right)(\{\alpha_1, \dots, \alpha_{k-1}, 0\}) = \dfrac{1}{2^k} & \text{if } \alpha_k = 1, \end{cases}$$

i.e. $Ber(\infty, \frac{1}{2})(A_k) = \frac{\alpha_k}{2^k}$. Since $Ber(\infty, \frac{1}{2})(M) = 0$ and the sets A_k are pairwise disjoint, we get

$$Ber\Big(\infty, \frac{1}{2}\Big)(T \leq x) = \sum_{k=1}^{\infty} Ber\Big(\infty, \frac{1}{2}\Big)(A_k) = \sum_{k=1}^{\infty} \frac{\alpha_k}{2^k} = x.$$

Both claims are proven.

3.3.2 Normal distribution

The *normal* or *Gaussian distribution* of parameters $m \in \mathbb{R}$ and $\sigma > 0$ is the measure $(\mathcal{B}(\mathbb{R}), N(m, \sigma^2))$ with

$$N(m, \sigma^2)(A) := \int_A f(t)\, dt, \qquad A \in \mathcal{B}(\mathbb{R})$$

where

$$f(t) := \frac{1}{\sqrt{2\pi\sigma^2}} \exp\Big(\frac{-(t-m)^2}{2\sigma^2}\Big).$$

Since

$$\int_{-\infty}^{+\infty} e^{-s^2}\, ds = \sqrt{\pi},$$

$N(m, \sigma^2)$ is a probability measure on $\mathbb{R}$. Any random variable X such that $\mathbb{P}_X = N(m, \sigma^2)$ has mean m and variance σ^2, that is

$$\mathbb{E}[X] = m \qquad \text{and} \qquad \mathrm{Var}[X] = \sigma^2,$$

as an easy computation shows. The law of the normal distribution $N(0, 1)$ of parameters $m = 0$ and $\sigma^2 = 1$ is usually denoted by $\Phi(t)$ instead of $F_{N(0,1)}(t)$,

$$\Phi(t) := \frac{1}{\sqrt{2\pi}} \int_{-\infty}^{t} e^{-\frac{x^2}{2}}\, dx.$$

Clearly $\Phi(-\infty) = 0$, $\Phi(0) = 0.5$ and $\Phi(+\infty) = 1$, see Figures 3.12 and 3.13.

In practice, 'independent' measurements are often 'distributed' according to a normal distribution. As we shall see, the Gaussian distribution plays a central role among distributions associated with 'independent' random variables.

Exercise 3.44 *The densities of two normal distributions are rescaled versions of each other since*

$$f(t) = \frac{1}{\sqrt{2\pi\sigma^2}} \exp\Big(\frac{-(t-m)^2}{2\sigma^2}\Big) = \frac{1}{\sigma}\Phi'\Big(\frac{t-m}{\sigma}\Big).$$

Infer that if X is a random variable such that $\mathbb{P}_X = N(m, \sigma^2)$, then the random variable $Y = (X - m)/\sigma$ follows the Gaussian distribution $N(0, 1)$.

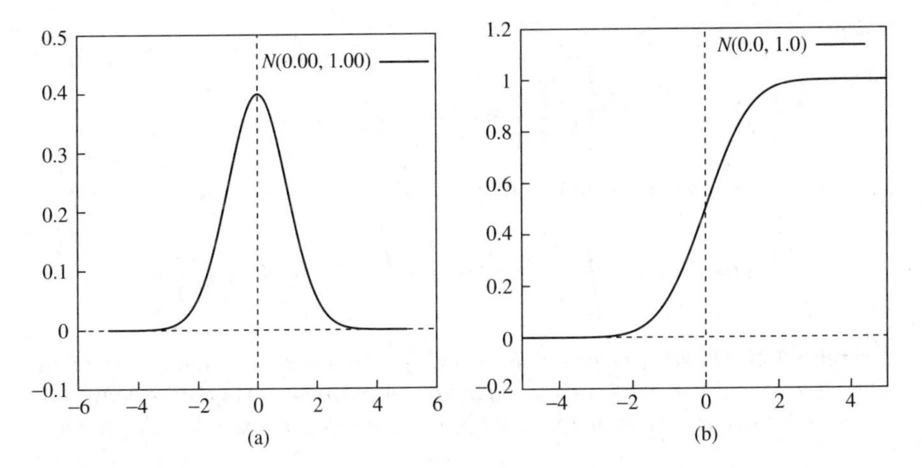

Figure 3.12 (a) The density and (b) the law of the normal distribution $N(0, 1)$.

t	$F_{N(0,1)}(t)$	t	$F_{N(0,1)}(t)$	$F_{N(0,1)}(t)$	t	$F_{N(0,1)}(t)$	t
0.0	0.5	2.0	0.9772	0.5	0.0	0.8	0.8416
0.25	0.5987	2.25	0.9878	0.53	0.0753	0.85	1.0364
0.50	0.6915	2.5	0.9938	0.57	0.1764	0.9	1.2816
0.75	0.7734	2.75	0.9970	0.6	0.2533	0.95	1.6449
1.0	0.8413	3.0	0.9987	0.62	0.3055	0.99	2.3263
1.25	0.8944	3.25	0.9994	0.67	0.4399	0.999	3.0902
1.5	0.9332	3.5	0.9998	0.7	0.5244	0.9999	3.7190
1.75	0.9599	3.75	0.9999	0.75	0.6745	0.99999	4.2649

Figure 3.13 A few values of the normal law and of its inverse.

Exercise 3.45 *Assume X is a Gaussian random variable whose mean and variance are m and σ^2, respectively. Compute the distribution and the expected value of the random variable $|X - m|$.*

Solution. We first compute the law of $|X - m|$. Clearly, $\mathbb{P}(|X - m| \leq t) = 0$ if $t \leq 0$. Since the random variable $(X - m)/\sigma$ follows the distribution $N(0, 1)$, see Exercise 3.44, for $t > 0$ we have

$$\mathbb{P}\Big(|X - m| \leq t\Big) = \mathbb{P}\left(-\frac{t}{\sigma} \leq \frac{X - m}{\sigma} \leq \frac{t}{\sigma}\right) = \frac{2}{\sqrt{2\pi}} \int_0^{\frac{t}{\sigma}} \exp\left(-\frac{x^2}{2}\right) dx,$$

from which we infer that $|X - m|$ is absolutely continuous and

$$\mathbb{P}_{|X-m|}(A) = \int_A g(t)\, dt$$

where

$$g(t) = \begin{cases} 0 & \text{if } t \le 0, \\ \sqrt{\dfrac{2}{\pi}} \dfrac{1}{\sigma} \exp\left(-\dfrac{t^2}{2\sigma^2}\right) & \text{if } t > 0. \end{cases}$$

We now compute the expected value:

$$\mathbb{E}[\,|X - m|\,] = \int_{-\infty}^{+\infty} |t|\, g(t)\, dt = \int_0^{+\infty} t\, g(t)\, dt = \sqrt{\frac{2}{\pi}}\, \sigma.$$

Exercise 3.46 *The life (measured in hours) of a bulb is a Gaussian random variable of expected value* 200 *and variance* σ^2. *A customer asks that at least 90% of the bulbs live longer than 150 h. What is the maximum possible value for* σ^2*?*

[$\simeq 39$]

3.3.3 Exponential distribution

For $\lambda > 0$, set

$$\rho_\lambda(t) = \begin{cases} 0 & \text{if } t < 0, \\ \lambda e^{-\lambda t} & \text{if } t \ge 0, \end{cases} \tag{3.41}$$

and for any Borel set $A \subset \mathbb{R}$ let

$$\exp(\lambda)(A) := \int_A \rho_\lambda(t)\, dt.$$

Since $\exp(\lambda)(\mathbb{R}) = \int_0^{+\infty} \lambda e^{-\lambda t}\, dt = 1$, $\exp(\lambda)$ is a probability measure on $\mathbb{R}$ supported on $[0, \infty[$, called the *exponential distribution* of parameter λ. The expected value and the variance of an exponentially distributed random variable X are

$$\mathbb{E}[\,X\,] = \frac{1}{\lambda} \quad \text{and} \quad \operatorname{Var}[\,X\,] = \frac{1}{\lambda^2}, \tag{3.42}$$

respectively.

Definition 3.47 *A non-negative random variable* X *is* memoryless *if*

$$\mathbb{P}\Big(X \le t + s \,\Big|\, X \ge t\Big) = \mathbb{P}(X \le s) \qquad \forall t, s > 0. \tag{3.43}$$

The meaning of the above property is clear: X is memoryless if the distribution of X does not change when we shift the observation interval from $[0, s]$ to $[t, t + s]$, whatever t. The (conditional) distribution remains the same independently of the beginning of observation. Observe that $\mathbb{P}(X \le t + s \mid X \ge t)$ and $\mathbb{P}(X \le t + s \mid X > t)$ differ if and only if $\mathbb{P}(X = t) \ne 0$, i.e. if and only if F_X has a jump at t.

Example 3.48 Random variables that describe 'time' often follow a memoryless distribution. Assume, for instance, that a random variable X describes the time needed for completing a certain task. Then $\mathbb{P}(X \leq t+s \mid X \geq s)$ is the probability that the task is completed by time $t+s$ knowing that it was still going on immediately before time s.

Exponentially distributed random variables are the only random variables where the distribution takes values in $[0, +\infty]$ and which are 'memoryless'.

Theorem 3.49 *Assume that the range of a random variable X is in $[0, +\infty]$ and $F_X(0) < 1$. Thus X is memoryless if and only if X follows the exponential law $\exp(\mu)$ where $\mu = -\log(1 - F_X(1)) = F'_X(0^+)$.*

In order to prove the theorem, we need the following.

Lemma 3.50 *Let $\alpha : \mathbb{R}_+ \to \mathbb{R}$ be either continuous or monotone and such that $\alpha(t+s) = \alpha(t) + \alpha(s)$ for any $t, s \geq 0$. Then $\alpha(t) = t\,\alpha(1)$ $\forall t \geq 0$.*

Proof. Choosing $t = s = 0$ we immediately get $\alpha(0) = 0$. By induction, we get $\alpha(k) = \alpha(1 + \cdots + 1) = k\,\alpha(1)$ and $n\alpha(k/n) = \alpha(k/n) + \cdots + \alpha(k/n) = \alpha(k) = k\,\alpha(1)$ for any two positive integers k and n. Thus $\alpha(q) = q\,\alpha(1)$ for any positive rational number q. If α is continuous the claim follows. If α is monotone, then for any positive t let $\{p_n\}$ and $\{q_n\}$ be two sequences of rational numbers approximating t from below and from the above, respectively. Assuming, for instance, α monotone increasing, we have

$$p_n\alpha(1) = \alpha(p_n) \leq \alpha(t) \leq \alpha(q_n) = q_n\alpha(1)$$

for any n, hence the claim, letting $n \to \infty$.

Proof of Theorem 3.49. If $\mathbb{P}_X = \exp(\mu)$, then $F_X(t) = 1 - e^{-\mu t}$ for any $t \geq 0$, so that, for any $t, s > 0$ we get

$$\begin{aligned}\mathbb{P}(X \leq t+s,\ X \geq t) &= F_X(t+s) - \lim_{s \to t^-} F_X(s) = e^{-\mu t} - e^{-\mu(t+s)} \\ &= e^{-\mu t}(1 - e^{-\mu s}) = (1 - F_X(t))F_X(s) \\ &= \mathbb{P}(X \geq t)\mathbb{P}(X \leq s),\end{aligned}$$

i.e. (3.43) is proven.

Conversely, define $F_X(t^-) := \lim_{s \to t^-} F_X(s)$ and $\delta_X(t) := F_X(t) - F_X(t^-)$. Property (3.43) amounts to the equality

$$(1 - F_X(t^-))F_X(s) = F_X(t+s) - F_X(t^-) \qquad \forall t, s > 0.$$

i.e.

$$(1 - F_X(t) + \delta_X(t))F_X(s) = F_X(t+s) - F_X(t) + \delta_X(t) \qquad \forall t, s > 0. \quad (3.44)$$

From the right continuity of F_X and letting $s \to 0^+$ in (3.44) we get

$$(1 - F_X(t) + \delta_X(t))F_X(0) = \delta_X(t) \qquad \forall t > 0. \tag{3.45}$$

We now prove that $F_X(0) = 0$. Assume, by contradiction, that $F_X(0) > 0$. Since F_X is right-continuous and, by assumption $F_X(0) < 1$, then $F_X(t) \in]0, 1[$ for any t in a right-hand neighbourhood of 0, so that by (3.45) the jump $\delta_X(t)$ would be positive for any t in such a neighbourhood. This is a contradiction, since F_X can be discontinuous in at most a denumerable set of points.

Since $F_X(0) = 0$, again from (3.45) we infer $\delta_X(t) = 0$ for any $t \geq 0$, i.e. F_X is continuous on $\mathbb{R}$, hence (3.44) reduces to

$$(1 - F_X(t))F_X(s) = F_X(t + s) - F_X(t) \qquad \forall t, s > 0,$$

or, equivalently, to

$$(1 - F_X(t))(1 - F_X(s)) = (1 - F_X(t + s)) \qquad \forall t, s > 0. \tag{3.46}$$

Set $\alpha(t) := \log(1 - F_X(t))$, $t > 0$. Then $\alpha(t + s) = \alpha(t) + \alpha(s)$ $\forall t, s > 0$. Since $\alpha(t)$ is monotone decreasing, Lemma 3.50 yields $\alpha(t) = t\alpha(1)$ $\forall t > 0$, i.e.

$$1 - F_X(t) = e^{-\mu t} \qquad \forall t > 0$$

where $\mu := -\alpha(1) > 0$. Finally, by continuity we infer $1 - F_X(t) = e^{-\mu t}$ $\forall t \geq 0$ and, consequently, the right-hand side derivative of F_X at 0 exists and is equal to μ

$$F'_{X+}(0) = \lim_{t \to 0+} \frac{F_X(t)}{t} = \mu.$$

Remark 3.51 Looking at the proof of Theorem 3.49, one sees that the exponentially distributed random variables are the only random variables X with range of the distribution in $[0, +\infty[$ such that $F_X(0) < 1$ and

$$\mathbb{P}\Big(X \leq t + s \,\Big|\, X > t\Big) = \mathbb{P}(X \leq s) \qquad \forall t, s > 0. \tag{3.47}$$

Therefore (3.43) and (3.47) are equivalent formulations of the memoryless property. We need a proof, since we have not assumed that the law of X is continuous.

The densities and the laws of three exponential distributions are plotted in Figure 3.14.

3.3.4 Gamma distributions

Let α and λ be positive real parameters, and for $t \in \mathbb{R}$, let

$$f(t) = f_{\alpha,\lambda}(t) := \begin{cases} 0 & \text{if } t \leq 0, \\ \frac{\lambda^\alpha t^{\alpha-1} e^{-\lambda x}}{\Gamma(\alpha)} & \text{if } t > 0, \end{cases} \tag{3.48}$$

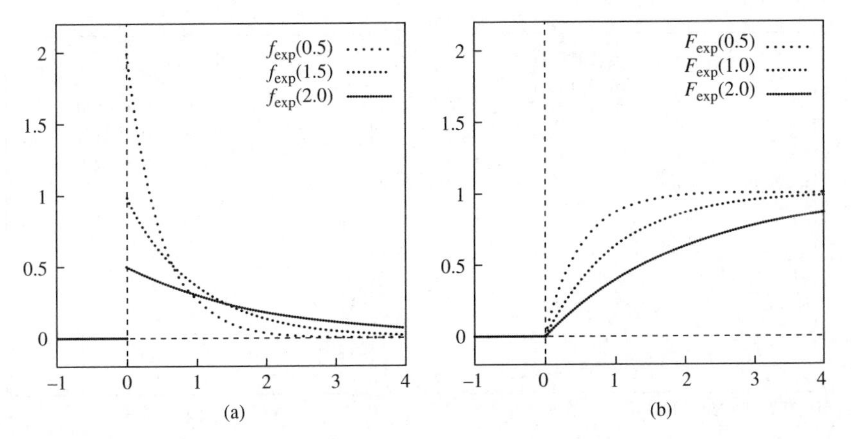

Figure 3.14 Three exponential distributions. On the left, their densities; on the right, their laws.

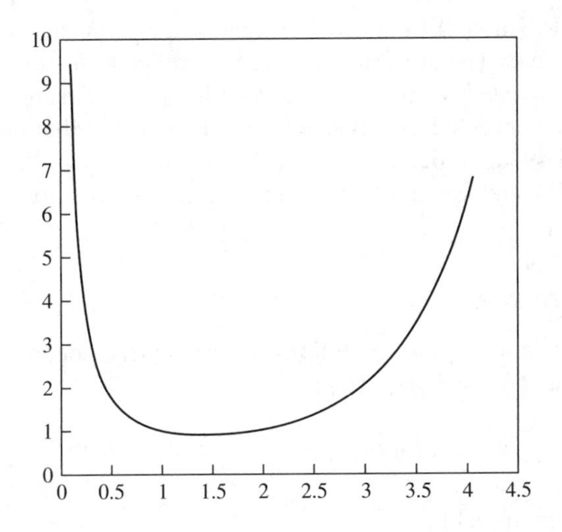

Figure 3.15 The $\Gamma(\alpha)$ function.

where $\Gamma(\alpha) := \int_0^{+\infty} s^{\alpha-1} e^{-s}\, ds$, see Figure 3.15. For any Borel set $A \subset \mathbb{R}$ let

$$\Gamma(\alpha, \lambda)(A) := \int_A f(t)\, dt.$$

It can be easily shown that $\Gamma(\alpha, \lambda)$ is a probability measure on $\mathbb{R}$ supported on $\mathbb{R}_+$, called the *Gamma distribution* of *shape* α and *scale* $1/\lambda$. Observe that $\Gamma(1, \lambda)$ is the exponential distribution $\exp(\lambda)$.

If X is a Gamma distributed random variable with shape α and scale $1/\lambda$, then the expected value and the variance of X are

$$\mathbb{E}[X] = \frac{\alpha}{\lambda} \qquad \text{and} \qquad \mathrm{Var}[X] = \frac{\alpha}{\lambda^2}, \tag{3.49}$$

respectively.

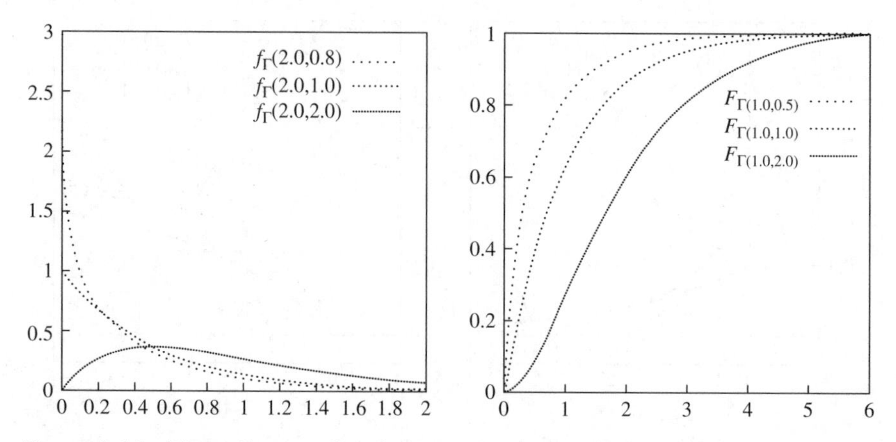

Figure 3.16 Three Gamma distributions. On the left, their densities; on the right, their laws.

Gamma distributions $\Gamma(\alpha, \lambda)$ with integer shape parameter α are also called *Erlang distributions*. The distribution $\Gamma(\frac{n}{2}, \frac{1}{2})$ is called a *chi-squared distribution with n degrees of freedom*; it is also denoted by $\chi^2(n)$. Erlang and chi-squared distributions are connected to exponential and normal distributions, respectively, see Exercises 4.49 and 4.62.

The densities and the laws of three Gamma distributions are plotted in Figure 3.16.

3.3.5 Failure rate

Assume the random variable X follows an absolutely continuous distribution whose density is $f(t) = F'_X(t)$. Define

$$w(\tau; t) := \frac{f(t+\tau)}{1 - F_X(t)}.$$

Then Vitali theorem yields

$$\mathbb{P}\Big(X \le t + s \,\Big|\, X \ge t\Big) = \frac{F_X(t+s) - F_X(t)}{1 - F_X(t)} = \int_0^s w(\tau; t)\, d\tau.$$

The function $w(0; t)$,

$$w(0; t) = \frac{f(t)}{1 - F_X(t)} = \frac{F'_X(t)}{1 - F_X(t)} = -\frac{d}{dt}\log(1 - F_X(t)),$$

is called the *hazard function* or *failure rate* of the distribution of X. In fact, assuming that X represents a failure time, then

$$W(A) := \mathbb{P}\Big(X \in A \,\Big|\, X \ge t\Big) = \frac{1}{1 - F_X(t)} \int_{A \cap [t,\infty[} f(s)\, ds$$

is the distribution probability of the failure time, assuming that no failure has occurred before time t. Notice that W is supported on $[t, +\infty]$. In particular, for the probability of a failure during a small time interval $[t, t + \delta t]$, we have

$$\mathbb{P}(X \leq t + \delta t \mid X \geq t) = \frac{1}{1 - F_X(t)} \int_t^{t+\delta t} f(s)\, ds \simeq w(0; t)\, \delta t$$

if f is continuous at t. In this sense $w(0; t)$ models the *urgency* of completing a task at time t before the equipment devoted to it breaks down.

It can be easily checked that the failure rate function of a random variable that follows the exponential distribution $\exp(\lambda)$ is constant and equal to λ.

3.3.6 Exercises

Exercise 3.52 *The temperature T (measured in degrees Celsius) of a certain volume of gas follows the Gaussian distribution of mean $\mu = 40$ and variance $\sigma^2 = 100$. Compute:*

- $\mathbb{P}(T \leq 50)$.
- $\mathbb{P}(|T - \mu| \leq 5)$.
- $\mathbb{P}(|T - \mu| \leq 10 | T \geq 35)$.

$$\left[\Phi(1),\, \Phi(0.5),\, \frac{\Phi(1)+\Phi(0.5)-1}{\Phi(0.5)}\right]$$

Exercise 3.53 *Assume X is a Gaussian random variable with mean and variance equal to* 1 *and* 4*, respectively. Compute $\mathbb{P}(|X - 1| \leq 2)$ in terms of the law Φ of the standard Gaussian distribution.*

$$\left[\Phi\left(\frac{1}{2}\right) - \Phi\left(\frac{-3}{2}\right) = \Phi\left(\frac{1}{2}\right) + \Phi\left(\frac{3}{2}\right) - 1\right]$$

Exercise 3.54 *Assume X is a random variable with absolutely continuous distribution of density*

$$f(x) = \begin{cases} |1 - x| & \text{if } 0 \leq x \leq 2, \\ 0 & \text{otherwise.} \end{cases}$$

Compute $\mathrm{Var}[\, X\,]$.

$$\left[\frac{1}{2}\right]$$

Exercise 3.55 *Let*

$$f(x) = \begin{cases} c & \text{if } -2 \leq x \leq -1, \\ 2c & \text{if } 1 \leq x \leq 2, \\ 0 & \text{otherwise,} \end{cases}$$

be the density of an absolutely continuously distributed random variable X. Compute the density of $Y := X^2$.

Exercise 3.56 *Let $\theta > 0$ and assume that*

$$f(x,\theta) = \begin{cases} \theta x^{-\theta-1} & \text{if } x > 1, \\ 0 & \text{otherwise,} \end{cases}$$

is the density of an absolutely continuously distributed random variable X. Compute $\mathbb{E}[X]$.

Exercise 3.57 *John thinks that the life (measured in kilometers) of a car is an exponentially distributed random variable with mean $\mu = 300\,000$. He buys a used car that has already done $100\,000$ km. What is the probability that the car does another $25\,000$ km before it breaks down?* [$\simeq 0.435$]

Exercise 3.58 *The daily production (measured in kilograms) of cheese from a certain dairy farm is a Gaussian random variable of mean $\mu = 1000$ and standard deviation $\sigma = 50$. Compute the probability that the daily production is at least 975 kg.* [$\Phi(0.50) \simeq 0.69$]

Exercise 3.59 *The fuel consumption (measured in litres per/100 km) of an AAA car is a Gaussian random variable with mean $\mu_1 = 6.5$ and standard deviation $\sigma_1 = 0.8$. The fuel consumption of a BBB car is another Gaussian random variable with mean $\mu_2 = 7$ and standard deviation $\sigma_2 = 0.5$. Compute the probability that the AAA car consumes more fuel than the BBB car.*

[$\simeq 1 - \Phi(0.53) \simeq 0.30$]

Exercise 3.60 *The weight of a box of spaghetti (measured in grams) is a Gaussian random variable whose expected value and standard deviation are 510 and 12, respectively. Compute the probability that a box weighs more than 500 g.*

[$\simeq 0.80$]

Exercise 3.61 *Assume X is an exponentially distributed random variable whose expected value is $1/\lambda$. Compute the laws of the random variables e^X and $\min\{X, 3\}$.*

Exercise 3.62 *Let X be a normally distributed random variable, $\mathbb{P}_X = N(10, 9)$. Compute for every $t > 0$ $\mathbb{P}(X \geq t)$ and $\mathbb{P}(X^2 > t)$ in terms of the density of the normalized Gaussian distribution $N(0, 1)$.*

4

Vector valued random variables

4.1 Joint distribution

Let X and Y be two random variables. The distribution vector $(\mathbb{P}_X, \mathbb{P}_Y)$ does not bear enough information about the relations occurring between X and Y, for instance to compute $\mathbb{P}(A \cap B)$ where A and B are generated by X and Y, respectively; one needs to introduce a new object called the *joint (probability) distribution* of X and Y.

Example 4.1 (Discrete random variables) Let X and Y be two discrete random variables on the same probability space $(\Omega, \mathcal{E}, \mathbb{P})$. Let $X(\Omega) = \{x_1, \dots, x_n\}$, $Y(\Omega) = \{y_1, \dots, y_m\}$ and let $(p_1, \dots, p_n)$ and $(q_1, \dots, q_m)$ be the mass density vectors of $\mathbb{P}_X$ and $\mathbb{P}_Y$, respectively.

The image of the map $(X, Y) : \Omega \to \mathbb{R}^2$ is a subset of $\{(x_i, y_j), i = 1, \dots, n, j = 1, \dots, m\}$ (whose cardinality is nm) and its distribution is concentrated at this set. Let $\{p_{ij}\}$ be the nm-tuple defined by

$$p_{ij} := \mathbb{P}(X = x_i, Y = y_j).$$

Clearly, the mass densities (p_i) and (q_j) can be computed in terms of the p_{ij}'s: in fact,

$$p_i = \mathbb{P}(X = x_i) = \sum_{j=1}^{m} \mathbb{P}(X = x_i, Y = y_j) = \sum_j p_{ij}$$

A First Course in Probability and Markov Chains, First Edition. Giuseppe Modica and Laura Poggiolini.

$$q_j = \mathbb{P}(Y = y_j) = \sum_{i=1}^{n} \mathbb{P}(X = x_i, Y = y_j) = \sum_i p_{ij}.$$

But, since (p_{ij}) has nm components while the components of (p_i) and (q_j) are only $n + m$, one cannot expect in general to recover (p_{ij}) from (p_i) and (q_j): if $n \wedge m > 2$, the vector (p_{ij}) is likely to bear more information than the (p_i)'s and the (q_j)'s. Example 4.5 shows that, in general, even in the simplest case $n = m = 2$, the knowledge of (p_1, p_2) and (q_1, q_2) does not yield the possibility to compute (p_{ij}).

4.1.1 Joint and marginal distributions

4.1.1.1 Vector valued random variables

Definition 4.2 *Let $X_1, \dots, X_N : \Omega \to \mathbb{R}$ be real-valued maps defined on a probability space $(\Omega, \mathcal{E}, \mathbb{P})$ and let $X : \Omega \to \mathbb{R}^N$, $X = (X_1, \dots, X_N)$. If*

$$\Big\{x \in \Omega \,\Big|\, X(x) \in A\Big\} \in \mathcal{E}$$

for any Borel set $A \subset \mathbb{R}^N$, then X is said to be an $\mathcal{E}$-measurable map, *or a* $\mathbb{R}^N$-valued random variable *or a* multivariate random variable.

The following proposition can be easily proved.

Proposition 4.3 *$X = (X_1, \dots, X_N)$ is a $\mathbb{R}^N$-valued random variable if and only if all the X_i's are real-valued random variables.*

4.1.1.2 Joint distribution

Let $X = (X_1, \dots, X_N) : \Omega \to \mathbb{R}^N$ be a random variable on the probability space $(\Omega, \mathcal{E}, \mathbb{P})$. We can define a new function on the Borel sets of $\mathbb{R}^N$, $\mathbb{P}_X : \mathcal{B}(\mathbb{R}^N) \to [0, 1]$, as

$$\mathbb{P}_X(A) := \mathbb{P}(X \in A) = \mathbb{P}(X^{-1}(A)) \qquad \forall A \subset \mathcal{B}(\mathbb{R}^N).$$

Following the same line as in the scalar case, see Section 3.1.1, one can use De Morgan formulas to prove that $(\mathcal{B}(\mathbb{R}^N), \mathbb{P}_X)$ is a probability measure on $\mathbb{R}^N$. $(\mathcal{B}(\mathbb{R}^N), \mathbb{P}_X)$ is called the *joint distribution* of $X = (X_1, \dots, X_N)$.

Also, we define the *joint distribution law* $F_X : \mathbb{R}^N \to \mathbb{R}$ of $X = (X_1, \dots, X_N)$ as

$$F_X(t_1, \dots, t_N) = \mathbb{P}(X_1 \le t_1,\ X_2 \le t_2, \dots, X_N \le t_N).$$

Namely, if for each $t = (t_1, \dots, t_N) \in \mathbb{R}^N$, we introduce the set

$$E_t :=]-\infty, t_1] \times]-\infty, t_2] \times \cdots \times]-\infty, t_N] \subset \mathbb{R}^N; \tag{4.1}$$

then

$$F_X(t) := \mathbb{P}(X_1 \le t_1,\ X_2 \le t_2, \dots, X_N \le t_N) = \mathbb{P}_X(E_t).$$

Thus, knowing $\mathbb{P}_X$ one can compute F_X and, as in the scalar case, $\mathbb{P}_X$ can be recovered from F_X, see Section 3.1.1 and Appendix B.

4.1.1.3 Marginal random variables and marginal distributions

Let $X = (X_1, X_2) : \Omega \to \mathbb{R}^2$ be a random variable on a probability space $(\Omega, \mathcal{E}, \mathbb{P})$, Both X_1 and X_2 are real valued random variables and their distributions $\mathbb{P}_{X_1}$ and $\mathbb{P}_{X_2}$ can be computed from the joint distribution $\mathbb{P}_X$ of X_1 and X_2. In fact, for any Borel set $A \subset \mathcal{B}(\mathbb{R})$, the sets $A \times \mathbb{R}$ and $\mathbb{R} \times A$ are Borel sets in $\mathbb{R}^2$,

$$\begin{aligned}\mathbb{P}_{X_1}(A) &= \mathbb{P}(X_1 \in A) = \mathbb{P}(X_1 \in A,\ X_2 \in \mathbb{R})\\ &= \mathbb{P}((X_1, X_2) \in A \times \mathbb{R}) = \mathbb{P}_{(X_1,X_2)}(A \times \mathbb{R})\end{aligned}$$

and, analogously,

$$\mathbb{P}_{X_2}(A) = \mathbb{P}_{(X_1,X_2)}(\mathbb{R} \times A).$$

When the stress is on the distribution $\mathbb{P}_X$ of $X := (X_1, X_2)$, $\mathbb{P}_{X_1}$ and $\mathbb{P}_{X_2}$ are called the *marginal distributions of* $\mathbb{P}_X$.

Clearly, the above can be extended to random variables of arbitrary dimension. Assume, for example, that $X : \Omega \to \mathbb{R}^h$, $Y : \Omega \to \mathbb{R}^k$ and $Z := (X, Y) : \Omega \to \mathbb{R}^{h+k}$. Then X and Y are random variables if and only if Z is a random variable. Moreover, for any $A \in \mathcal{B}(\mathbb{R}^h)$ and any $B \in \mathcal{B}(\mathbb{R}^k)$ we have

$$\mathbb{P}_X(A) = \mathbb{P}_Z(A \times \mathbb{R}^k) \qquad \text{and} \qquad \mathbb{P}_Y(B) = \mathbb{P}_Z(\mathbb{R}^h \times B).$$

Remark 4.4 Random variables are defined only by means of the set theoretic notion of preimage. This fact suggests a possible generalization: let $(\Omega, \mathcal{E}, \mathbb{P})$ and $(\Delta, \mathcal{F}, \mathbb{Q})$ be two probability spaces. A map $X : \Omega \to \Delta$ is said to be a random variable between such spaces if $X^{-1}(F) \in \mathcal{E}$ for every $F \in \mathcal{F}$. If X is a random variable (according to this definition), then $\mathbb{P}_X(F) := \mathbb{P}(X^{-1}(F))$, $F \in \mathcal{F}$, is a set function defined on $\mathcal{F}$ and the couple $(\mathcal{F}, \mathbb{P}_X)$ is a probability measure on Δ, called the *distribution* of X. We are not going to treat this point any further.

Example 4.1 suggests that the joint distribution conveys more information than the marginal ones. Here we provide one explicit example.

Example 4.5 Let $E, F \subset [0, 1]$ be two intervals. Let $X = \mathbb{1}_E$, $Y = \mathbb{1}_F$ and let $\mathbb{P}$ be the uniform distribution on $[0, 1]$. Thus

$$\mathbb{P}(X = 0) = \mathbb{P}(E^c), \qquad \mathbb{P}(X = 1) = \mathbb{P}(E),$$

$$\mathbb{P}(Y = 0) = \mathbb{P}(F^c), \qquad \mathbb{P}(Y = 1) = \mathbb{P}(F),$$

and the matrix of joint densities is

$$\mathbb{P}_{\mathbb{1}_E,\mathbb{1}_F}(\{i,j\}) = \begin{pmatrix} \mathbb{P}(E^c \cap F^c) & \mathbb{P}(E \cap F^c) \\ \mathbb{P}(E^c \cap F) & \mathbb{P}(E \cap F) \end{pmatrix}$$

This clearly shows that moving E with respect to F changes $\mathbb{P}_{X,Y}$ but does not change $\mathbb{P}_X$ and $\mathbb{P}_Y$.

4.1.1.4 Composition

Let $X : \Omega \to \mathbb{R}^N$ be a random variable on the probability space $(\Omega, \mathcal{E}, \mathbb{P})$. The definition of joint distribution in terms of the probability $\mathbb{P}$,

$$\mathbb{P}_X(A) := \mathbb{P}(X \in A) \qquad \forall A \subset \mathbb{R}^N$$

can be conveniently written by means of integrals on Ω and $\mathbb{R}^N$.

Theorem 4.6 *Let $X_1, \ldots, X_N : \Omega \to \mathbb{R}$ be random variables on $(\Omega, \mathcal{E}, \mathbb{P})$. Then, for any $\mathcal{B}(\mathbb{R}^N)$-measurable non-negative function $\psi : \mathbb{R}^N \to \mathbb{R}$, we have*

$$\int_\Omega \psi(X_1(x), X_2(x), \ldots, X_N(x))\, \mathbb{P}(dx) \\ = \int_{\mathbb{R}^N} \psi(t_1, \ldots, t_N)\, \mathbb{P}_{(X_1,\ldots,X_N)}(dt_1 dt_2 \cdots dt_N). \tag{4.2}$$

Proof. The proof follows the same line of the proof of Theorem 3.9.

The *composition formula* easily follows, see Corollary 3.11.

Corollary 4.7 *Let $X : \Omega \to \mathbb{R}^N$ be a random variable on $(\Omega, \mathcal{E}, \mathbb{P})$ and let $\varphi : \mathbb{R}^N \to \mathbb{R}^k$ be a $\mathcal{B}(\mathbb{R}^N)$-measurable function. Then $\varphi \circ X : \Omega \to \mathbb{R}^k$ is a random variable on $(\Omega, \mathcal{E}, \mathbb{P})$ and for every non-negative $\mathcal{B}(\mathbb{R}^k)$-measurable function $\psi : \mathbb{R}^k \to \mathbb{R}$ we have*

$$\int_{\mathbb{R}^k} \psi(s_1, \ldots, s_k)\, \mathbb{P}_{\varphi \circ X}(ds_1 \cdots ds_k) \\ = \int_{\mathbb{R}^N} \psi(\varphi(t_1, \ldots, t_N))\, \mathbb{P}_X(dt_1 \cdots dt_N). \tag{4.3}$$

4.1.1.5 Sum of random variables

Formula (4.3) allows to compute the distribution of the sum of two random variables in terms of their joint distribution.

Proposition 4.8 *Let X and $Y : \Omega \to \mathbb{R}^N$ be two random variables on $(\Omega, \mathcal{E}, \mathbb{P})$. Then*

$$\int_{\mathbb{R}^N} \psi(s)\, \mathbb{P}_{X+Y}(ds) = \iint_{\mathbb{R}^{2N}} \psi(t+s)\, \mathbb{P}_{(X,Y)}(dtds) \tag{4.4}$$

for any non-negative $\mathcal{B}(\mathbb{R}^N)$-measurable function $\varphi : \mathbb{R}^N \to \mathbb{R}$. In particular,

$$\mathbb{P}_{X+Y}(A) = \mathbb{P}_{X,Y}\Big(\Big\{(x, y) \in \mathbb{R}^{2N} \,\Big|\, x + y \in A\Big\}\Big) \qquad \forall A \in \mathcal{B}(\mathbb{R})$$

and

$$F_{X+Y}(t) = \mathbb{P}_{X,Y}\Big(\Big\{(x, y) \in \mathbb{R}^{2N} \,\Big|\, x_1 + y_1 \le t_1, \; \ldots, \; x_N + y_N \le t_N\Big\}\Big)$$

for every $t = (t_1, \ldots, t_N) \in \mathbb{R}^N$.

Proof. Equality (4.4) is the claim of Corollary 4.7 for $\varphi : \mathbb{R}^N \times \mathbb{R}^N \to \mathbb{R}^N$, $\varphi(x, y) := x + y$. The other claims follow choosing $\psi = \mathbb{1}_A$ in (4.4) and then choosing $A = \prod_{i=1}^N]-\infty, t_i[$.

4.1.2 Exercises

Exercise 4.9 *Let X, Y be two random variables on $(\Omega, \mathcal{E}, \mathbb{P})$. Prove the following identities:*

$$\begin{aligned}
\mathbb{E}[X] &= \iint_{\mathbb{R}^2} x \, \mathbb{P}_{X,Y}(dxdy), \\
\mathbb{E}[Y] &= \iint_{\mathbb{R}^2} y \, \mathbb{P}_{X,Y}(dxdy), \\
\mathbb{E}[X+Y] &= \iint_{\mathbb{R}^2} (x+y) \, \mathbb{P}_{X,Y}(dxdy), \\
\mathbb{E}[XY] &= \iint_{\mathbb{R}^2} xy \, \mathbb{P}_{X,Y}(dxdy), \\
\mathbb{E}[\varphi \circ X] &= \iint_{\mathbb{R}^2} \varphi(x) \, \mathbb{P}_{X,Y}(dxdy), \\
\mathbb{E}[\varphi \circ Y] &= \iint_{\mathbb{R}^2} \varphi(y) \, \mathbb{P}_{X,Y}(dxdy), \\
\mathbb{E}[\max(X, Y)] &= \iint_{\mathbb{R}^2} \max(x, y) \, \mathbb{P}_{X,Y}(dxdy), \\
\mathbb{E}[\min(X, Y)] &= \iint_{\mathbb{R}^2} \min(x, y) \, \mathbb{P}_{X,Y}(dxdy).
\end{aligned} \tag{4.5}$$

Exercise 4.10 (Hypergeometric distribution) *Assume we are given an urn containing w white balls and r red balls. We draw n balls. Let X be the random variable that counts the number of drawn white balls. Compute $\mathbb{E}[X]$ and* $\mathrm{Var}[X]$.

Solution. X follows the hypergeometric distribution. Let X_j, $j = 1, \ldots, n$, be the random variables such that $X_j = 1$ if the jth drawn ball is white and $X_j = 0$

otherwise. Then $X = \sum_{j=1}^{n} X_j$ and each random variable X_j is a Bernoulli trial with distribution $B(1, p)$ where $p = w/(w+r)$ (see Section 3.2.3). Thus $\mathbb{E}[X_j] = w/(w+r)$ for every $j = 1, \dots, n$ so that

$$\mathbb{E}[X] = \sum_{j=1}^{n} \mathbb{E}[X_j] = \frac{wr}{w+r}.$$

Iterating formula (4.15) one gets

$$\mathrm{Var}[X] = \sum_{j=1}^{n} \mathrm{Var}[X_j] + 2\sum_{j=2}^{n}\sum_{i=1}^{j-1} \mathrm{Cov}(X_i, X_j).$$

Since $\mathrm{Cov}(X_i, X_j) = \mathbb{E}[X_i X_j] - \mathbb{E}[X_i]\mathbb{E}[X_j]$ and

$$\mathbb{E}[X_i X_j] = \mathbb{P}(X_i = 1, X_j = 1) = \mathbb{P}(X_j = 1 | X_i = 1)\mathbb{P}(X_i = 1)$$
$$= \frac{w-1}{w+r-1}\frac{w}{w+r},$$

we get

$$\mathrm{Cov}(X_i, X_j) = \frac{(w-1)w}{(w+r-1)(w+r)} - \frac{w^2}{(w+r)^2}$$
$$= \frac{-wr}{(w+r)^2(w+r-1)}.$$

Finally,

$$\mathrm{Var}[X] = \frac{nwr}{(w+r)^2} - \frac{n(n-1)}{2}\frac{wr}{(w+r)^2(w+r-1)} = \frac{wr}{(w+r)^2}\frac{w+r-n}{w+r-1}.$$

Exercise 4.11 *Assume the joint distribution $\mathbb{P}_{X,Y}$ of X and Y is uniformly distributed on the square $[0, 1]^2$. Compute the distribution of the random variable XY.*

Exercise 4.12 *Assume the joint distribution $\mathbb{P}_{X,Y}$ of X and Y is uniformly distributed on the parallelogram of vertices $(0, 0)$, $(1, 0)$, $(2, 1)$ and $(3, 1)$. Compute:*

(i) The distributions $\mathbb{P}_X$ and $\mathbb{P}_Y$.

(ii) Mean and variance of X and Y.

Exercise 4.13 *Assume that the joint distribution $\mathbb{P}_{X,Y}$ of two random variables X and Y is absolutely continuous with respect to $\mathcal{L}^2$ with density*

$$f(x, y) = \begin{cases} cx^2y^2 & \text{if } 0 \le x \le 1 \text{ and } 0 \le y \le 1, \\ 0 & \text{otherwise.} \end{cases}$$

Find c and compute the joint distribution of $3X$ and XY.

Exercise 4.14 *Assume that the joint distribution $\mathbb{P}_{X,Y}$ of two random variables X and Y is absolutely continuous with respect to $\mathcal{L}^2$ with density*

$$f(x,y) = \begin{cases} c(x-y)^2 & \text{if } 0 \le x, y \le 1, \\ 0 & \text{otherwise.} \end{cases}$$

Find c and compute the joint distribution of X^2 and X^2Y^2.

Exercise 4.15 *Assume that the joint distribution $\mathbb{P}_{X,Y}$ of two random variables X and Y is absolutely continuous with respect to $\mathcal{L}^2$ with density*

$$f(x,y) = \begin{cases} \dfrac{3\sqrt{x^2+y^2}}{14\pi} & \text{if } 1 \le x^2+y^2 \le 4, \\ 0 & \text{otherwise.} \end{cases}$$

Find $r > 0$ such that $\mathbb{P}(X^2 + Y^2 \ge r^2) = \dfrac{1}{2}$.

$$\left[\frac{\sqrt[3]{36}}{2}\right]$$

Exercise 4.16 *Assume that the joint distribution $\mathbb{P}_{X,Y}$ of two random variables X and Y is absolutely continuous with respect to $\mathcal{L}^2$ with density*

$$f(x,y) = \begin{cases} \dfrac{\sqrt{2}(x+y)}{36} & \text{if } x^2+y^2 \le 9,\ x+y > 0, \\ 0 & \text{otherwise.} \end{cases}$$

Compute $\mathbb{P}(XY \ge 0)$.

$$\left[\frac{\sqrt{2}}{2}\right]$$

Exercise 4.17 *Let a and $b > 0$ and let X and Y be two random variables with an absolutely continuous joint distribution $\mathbb{P}_{X,Y}$ of density*

$$f(x,y) = \begin{cases} Ce^{-ax}e^{-by} & \text{if } x > 0,\ y > 0, \\ 0 & \text{otherwise.} \end{cases}$$

Compute:

(i) The constant C.

(ii) The probability $\mathbb{P}(Y \le X)$.

(iii) The joint law of X and Y.

$$\left[(i)\ ab,\ (ii)\ \frac{a}{a+b},\ (iii)\ (1-e^{-as})(1-e^{-bt})\right]$$

4.2 Covariance

4.2.1 Random variables with finite expected value and variance

Proposition 4.18 *Let $X, Y : \Omega \to \mathbb{R}$ be two random variables on the probability space $(\Omega, \mathcal{E}, \mathbb{P})$. Then*

$$\mathbb{E}\left[\,|XY|\,\right] \leq \mathbb{E}\left[\,X^2\,\right]^{1/2} \mathbb{E}\left[\,Y^2\,\right]^{1/2}. \tag{4.6}$$

Proof. If either $\mathbb{E}\left[\,X^2\,\right] = +\infty$ or $\mathbb{E}\left[\,Y^2\,\right] = +\infty$, the claim is trivial. If $\mathbb{E}\left[\,X^2\,\right] = 0$, then $X = 0$ $\mathbb{P}$-a.e., so also $\mathbb{E}\left[\,|XY|\,\right] = 0$, hence (4.6) holds true. The same is true if $\mathbb{E}\left[\,Y^2\,\right] = 0$.

Let us assume that $\mathbb{E}\left[\,X^2\,\right]$ and $\mathbb{E}\left[\,Y^2\,\right]$ are both finite and positive. Consider the random variables

$$a(x) = \frac{1}{\mathbb{E}\left[\,X^2\,\right]^{1/2}} |X(x)|, \qquad b(x) = \frac{1}{(\mathbb{E}\left[\,Y^2\,\right])^{1/2}} |Y(x)|.$$

Since $2a(x)b(x) \leq a^2(x) + b^2(x)$ $\forall x \in \Omega$, integrating with respect to $\mathbb{P}$ we get

$$\begin{aligned} 2\,\mathbb{E}\left[\,ab\,\right] &= 2\int_\Omega a(x)b(x)\,\mathbb{P}(dx) \\ &\leq \mathbb{E}\left[\,a^2\,\right] + \mathbb{E}\left[\,b^2\,\right] = \frac{1}{\mathbb{E}\left[\,X^2\,\right]}\mathbb{E}\left[\,X^2\,\right] + \frac{1}{\mathbb{E}\left[\,Y^2\,\right]}\mathbb{E}\left[\,Y^2\,\right] = 2, \end{aligned}$$

i.e. $\mathbb{E}\left[\,ab\,\right] \leq 1$, which proves (4.6).

Let

$$\mathcal{L}^2(\Omega; \mathbb{P}) := \left\{ X : \Omega \to \mathbb{R} \text{ is a random variable} \,\middle|\, \mathbb{E}\left[\,X^2\,\right] < \infty \right\}. \tag{4.7}$$

When the context is clear, we shall write $\mathcal{L}^2(\Omega)$ or $\mathcal{L}^2$ in place of $\mathcal{L}^2(\Omega; \mathbb{P})$. If $X \in \mathcal{L}^2$ we say that X is a *square integrable* random variable.

Proposition 4.19 *If $X, Y \in \mathcal{L}^2$, then $\mathbb{E}\left[\,XY\,\right]$ is finite. Moreover, $\mathcal{L}^2$ is a linear space, the map $(X, Y) \mapsto \mathbb{E}\left[\,XY\,\right]$ is a positive semidefinite symmetric bilinear form on $\mathcal{L}^2$, and $X \mapsto \mathbb{E}\left[\,X^2\,\right]^{1/2}$ is a seminorm on $\mathcal{L}^2$.*

Proof. By inequality (4.6) $\left|\mathbb{E}\left[\,XY\,\right]\right|$ is finite whenever X and $Y \in \mathcal{L}^2$. Hence, also $\mathbb{E}\left[\,XY\,\right] \in \mathbb{R}$.

Let $X, Y \in \mathcal{L}^2$. Since $\mathbb{E}\left[\,XY\,\right]$ is bilinear and again by (4.6) we get

$$\begin{aligned} \mathbb{E}\left[\,(X+Y)^2\,\right] &= \mathbb{E}\left[\,X^2\,\right] + 2\mathbb{E}\left[\,XY\,\right] + \mathbb{E}\left[\,Y^2\,\right] \\ &\leq \left(\mathbb{E}\left[\,X^2\,\right]^{1/2} + \mathbb{E}\left[\,Y^2\,\right]^{1/2}\right)^2 < \infty; \end{aligned}$$

thus $X+Y \in \mathcal{L}^2$. Moreover, for any $\alpha \in \mathbb{R}$ and any $X \in \mathcal{L}^2$, trivially $\mathbb{E}\left[\,(\alpha X)^2\,\right] = \alpha^2 \mathbb{E}\left[\,X^2\,\right] < \infty$, i.e. $\alpha X \in \mathcal{L}^2$. This proves that $\mathcal{L}^2$ is a linear space. It is now obvious that $(X, Y) \to \mathbb{E}[\,XY\,]$ is real valued, and therefore a positive semidefinite symmetric bilinear form.

Remark 4.20 Notice the following:

- In general, the form $(X|Y) := \mathbb{E}[\,XY\,]$ is not positive definite and, consequently, $X \mapsto \mathbb{E}\left[\,X^2\,\right]^{1/2}$ is not a norm. In fact, $\mathbb{E}\left[\,X^2\,\right] = 0$ does not imply that $X(x) = 0\ \forall x$ but only that $X = 0$ $\mathbb{P}$-a.e.
- Inequality (4.6) is the Cauchy–Schwarz inequality for the positive semidefinite symmetric bilinear form $(X, Y) \mapsto \mathbb{E}[\,XY\,]$ on $\mathcal{L}^2$. In particular, for $X, Y \in \mathcal{L}^2$ we have

$$\mathbb{E}[\,XY\,] \le \mathbb{E}[\,|XY|\,] \le \mathbb{E}\left[\,X^2\,\right]^{1/2} \mathbb{E}\left[\,Y^2\,\right]^{1/2} \tag{4.8}$$

  and the equality holds if and only if either $Y = 0$ $\mathbb{P}$-a.e. or $X(x) = \lambda Y(x)$ $\mathbb{P}$-a.e. for some $\lambda \in \mathbb{R}$, $\lambda \ge 0$.

Applying inequality (4.6) with $Y = 1$ we get the classical inequality between mean and mean square of X

$$\mathbb{E}[\,|X|\,] \le \sqrt{\mathbb{E}\left[\,X^2\,\right]}. \tag{4.9}$$

Finally, (4.8) and (4.9) yield the following.

Proposition 4.21 *X is square integrable, i.e.* $\mathbb{E}\left[\,X^2\,\right] < \infty$ *if and only if both* $\mathbb{E}[\,X\,]$ *and* $\mathrm{Var}[\,X\,]$ *are finite. In particular*

$$\mathcal{L}^2 \subset \mathcal{L}^1. \tag{4.10}$$

Example 4.22 In general, $\mathcal{L}^2$ and $\mathcal{L}^1$ do not coincide. For example, let $(\mathcal{B}(\mathbb{R}), \mathbb{P})$ the Lebesgue measure on the interval $[0, 1]$ and let $X(t) = 1/\sqrt{t}$, $0 < t < 1$. Then

$$\mathbb{E}[\,X\,] = \int_0^1 \frac{1}{\sqrt{t}}\,dt = 2 \qquad \text{while} \qquad \mathbb{E}\left[\,X^2\,\right] = \int_0^1 \frac{1}{t}\,dt = +\infty.$$

Definition 4.23 *Let X, Y be two random variables in $\mathcal{L}^2$, equivalently two random variables whose expected value and variance are both finite. The number*

$$\mathrm{Cov}(X, Y) := \mathbb{E}\Big[\Big(X - \mathbb{E}[X]\Big)\Big(Y - \mathbb{E}[Y]\Big)\Big] \tag{4.11}$$

is called the covariance *of X and Y.*

Trivially:

- $(X, Y) \mapsto \mathrm{Cov}(X, Y)$ is a positive semidefinite symmetric bilinear form on the linear space $\mathcal{L}^2$.

- $\mathrm{Cov}(X, X) = 0$ if and only if X is $\mathbb{P}$-a.e. constant. In this case $X(x) = \mathbb{E}[X]$ for $\mathbb{P}$-a.e. x.
- The following formulas hold:

$$\begin{aligned}&\mathrm{Cov}(X, Y) = \mathbb{E}[XY] - \mathbb{E}[X]\mathbb{E}[Y],\\&\mathrm{Cov}(\alpha X + \beta, \gamma Y + \delta) = \alpha\gamma\,\mathrm{Cov}(X, Y),\\&\mathrm{Cov}(X, X) = \mathrm{Var}[X].\end{aligned} \tag{4.12}$$

- Theorem 4.6 yields

$$\begin{aligned}\mathrm{Cov}(X, Y) &= \iint_{\mathbb{R}^2} xy\,\mathbb{P}_{X,Y}(dxdy) - \mathbb{E}[X]\mathbb{E}[Y]\\&= \iint_{\mathbb{R}^2} (x - \mathbb{E}[X])\,(y - \mathbb{E}[Y])\,\mathbb{P}_{X,Y}(dxdy).\end{aligned} \tag{4.13}$$

- The Cauchy–Schwarz inequality for the form $(X, Y) \mapsto \mathrm{Cov}(X, Y)$ on $\mathcal{L}^2$ reads as follows:

$$\mathrm{Cov}(X, Y) \le \sqrt{\mathrm{Var}[X]}\sqrt{\mathrm{Var}[Y]}, \tag{4.14}$$

where the equality holds if and only if either Y is constant $\mathbb{P}$-a.e. or there exist $\alpha \ge 0$ and $\beta \in \mathbb{R}$ such that $X(x) = \alpha Y(x) + \beta$ $\mathbb{P}$-a.e.

- The *Carnot theorem* reads

$$\mathrm{Var}[X + Y] = \mathrm{Var}[X] + \mathrm{Var}[Y] + 2\,\mathrm{Cov}(X, Y) \tag{4.15}$$

from which the *Pythagoras formula* follows:

$$\mathrm{Var}[X + Y] = \mathrm{Var}[X] + \mathrm{Var}[Y] \tag{4.16}$$

if and only if $\mathrm{Cov}(X, Y) = 0$.

Definition 4.24 *If $X, Y \in \mathcal{L}^2$ and $\mathrm{Cov}(X, Y) = 0$, i.e.*

$$\mathbb{E}[XY] = \mathbb{E}[X]\mathbb{E}[Y],$$

we say that X and Y are uncorrelated.

Finally, iterating (4.16) we get that

$$\mathrm{Var}[X_1 + \cdots + X_n] = \mathrm{Var}[X_1] + \cdots + \mathrm{Var}[X_n] \tag{4.17}$$

if and only if the random variables $X_1, \ldots, X_n$ are pairwise uncorrelated.

4.2.2 Correlation coefficient

For $X, Y \in \mathcal{L}^2$, let $\sigma(X) := \mathrm{Var}[X]^{1/2}$ and $\sigma(Y) := \mathrm{Var}[Y]^{1/2}$. If X and Y are not constant $\mathbb{P}$-a.e., one defines the *correlation coefficient* of X and Y as

$$\mathrm{Corr}(X, Y) := \frac{\mathrm{Cov}(X, Y)}{\sigma(X)\sigma(Y)}. \tag{4.18}$$

$\mathrm{Corr}(X, Y)$ gives some measure of the 'similarities between the shapes' of the distributions of X and Y, regardless of their expected value and variance. In fact, $\mathrm{Corr}(X, Y)$ is invariant by change of scale either in X or in Y. Moreover, the Cauchy–Schwarz inequality reads

$$-1 \le \mathrm{Corr}(X, Y) \le 1, \tag{4.19}$$

and $\mathrm{Corr}(X, Y) = 1$ ($= -1$) if and only if there exists $\alpha > 0$ ($\alpha < 0$, respectively) and $\beta \in \mathbb{R}$ such that $X = \alpha Y + \beta$ $\mathbb{P}$-a.e.

4.2.3 Exercises

Exercise 4.25 *Let X be a random variable with finite expected value and variance. Let $Y = aX + b$, $a \neq 0$, $b \in \mathbb{R}$. Show that* $\mathrm{Corr}(X, Y) = \mathrm{sgn}(a)$.

Exercise 4.26 *Let X be a random variable uniformly distributed in $[0, b]$. Compute* $\mathrm{Cov}(X, X^2)$ *and the correlation coefficient* $\mathrm{Corr}(X, X^2)$.

Solution. We have

$$\mathbb{E}[X] = \int_0^b x\frac{1}{b}\,dx = \frac{b}{2}, \qquad \mathbb{E}[X^2] = \frac{1}{b}\int_0^b x^2\,dx = \frac{b^2}{3},$$

$$\mathbb{E}[X^3] = \frac{1}{b}\int_0^b x^3\,dx = \frac{b^3}{4}, \qquad \mathbb{E}[X^4] = \frac{1}{b}\int_0^b x^4\,dx = \frac{b^4}{4},$$

$$\mathrm{Var}[X] = \mathbb{E}[X^2] - \mathbb{E}[X]^2 = \frac{1}{12}b^2,$$

$$\mathrm{Var}[X^2] = \mathbb{E}[X^4] - \mathbb{E}[X^2]^2 = \frac{4}{45}b^4.$$

Hence

$$\mathrm{Cov}(X, X^2) = \mathbb{E}[X^3] - \mathbb{E}[X]\mathbb{E}[X^2] = \Big(\frac{1}{4} - \frac{1}{2}\frac{1}{3}\Big)b^3 = \frac{b^3}{12}.$$

and

$$\mathrm{Corr}(X, X^2) = \frac{\mathrm{Cov}(X, X^2)}{\sqrt{\mathrm{Var}[X]}\sqrt{\mathrm{Var}[X^2]}} = \frac{\frac{1}{12}}{\frac{1}{\sqrt{12}}\frac{2}{\sqrt{45}}} = \frac{\sqrt{15}}{4}.$$

Exercise 4.27 *Throw a fair dice. If the result is less than or equal to 3, flip a fair coin. Otherwise flip two fair coins. Let Y be the result of the dice and let X be the number of heads obtained in the flipping of the coins. Compute:*

1. *The distribution of X.*
2. *The expected value and the variance of X.*
3. *The correlation coefficient* $\mathrm{Corr}(X, Y)$.

Exercise 4.28 *An experiment has three possible results, 1, 2 and 3, whose probabilities are p_1, p_2 and p_3, respectively. Repeat the experiment n times and assume that the result of each trial is independent of the other trials.*

Let N_1 and N_2 be the number of times that you obtain the results 1 *and* 2, *respectively. Compute* $\mathrm{Corr}(N_1, N_2)$.

Taking into account the constraints $p_1 \geq 0$, $p_2 \geq 0$, $p_3 \geq 0$ and $p_1 + p_2 + p_3 = 1$ compute the maximum and the minimum possible values for ρ. Can $\mathrm{Corr}(N_1, N_2)$ *be zero?*

4.3 Independent random variables

4.3.1 Independent events

Definition 4.29 *Let $(\mathcal{E}, \mathbb{P})$ be a probability space and let $A, B \in \mathcal{E}$. If $\mathbb{P}(A \cap B) = \mathbb{P}(A)\mathbb{P}(B)$, then we say that A and B are* independent *events.*

It can easily checked that, if $\mathbb{P}(B) > 0$, then A and B are independent if and only if $\mathbb{P}(A) = \mathbb{P}(A \mid B)$: no knowledge on A is gained from the knowledge of B.

Example 4.30 Throw two dice. Here $\Omega = \{1, 2, \ldots, 6\}^2$ and, if the dice are fair, then each pair (i, j) in Ω has the same probability, i.e. $\mathbb{P}(\{(i, j)\}) = \frac{1}{36}$ $\forall (i, j) \in \Omega$. Let A be the event 'The number on the top face of the first dice is 6', i.e.

$$A = \Big\{(6, j) \Big| j = 1, \ldots, 6\Big\}$$

and let B be the event 'The number on the second dice is 4', i.e.

$$B = \Big\{(j, 4) \Big| j = 1, \ldots, 6\Big\}.$$

Then $\mathbb{P}(A) = \mathbb{P}(B) = \frac{1}{6}$. Since $A \cap B = \{(6, 4)\}$, then $\mathbb{P}(A \cap B) = \frac{1}{36} = \mathbb{P}(A)\mathbb{P}(B)$, thus A and B are independent events.

Definition 4.31 *Let $(\Omega, \mathcal{E}, \mathbb{P})$ be a probability space and let $A_1, \ldots, A_n \in \mathcal{E}$. $A_1, \ldots, A_n$ are said to be independent if for any $k = 2, \ldots, n$ and for any choice of k events $A_{j_1}, \ldots, A_{j_k}$ among them, then*

$$\mathbb{P}(A_{j_1} \cap A_{j_2} \cap \cdots \cap A_{j_k}) = \mathbb{P}(A_{j_1})\mathbb{P}(A_{j_2}) \cdots \mathbb{P}(A_{j_k}). \tag{4.20}$$

We say that a denumerable family of events $\{A_n\}$ is independent if for every n the events $A_1, \dots, A_n$ are independent.

Notice that the pairwise independence of the events $A_1, \dots, A_n$, $n \geq 3$, does not guarantee the independence of $A_1, \dots, A_n$. Here we provide an example.

Example 4.32 Consider the uniform probability on $\Omega = \{0, 1\}^2$. Each singleton has probability 1/4. Let $A_1 = \{(1, 0), (1, 1)\}$, $A_2 = \{(0, 1), (1, 1)\}$ and $A_3 = \{(i, j) \mid i + j = 1\} = \{(1, 0), (0, 1)\}$. Then

$$\mathbb{P}(A_1) = 1/2, \qquad \mathbb{P}(A_2) = 1/2, \qquad \mathbb{P}(A_3) = 1/2,$$
$$\mathbb{P}(A_1 \cap A_2) = \mathbb{P}(\{1, 1\}) = 1/4,$$
$$\mathbb{P}(A_2 \cap A_3) = \mathbb{P}(\{0, 1\}) = 1/4, \qquad \mathbb{P}(A_1 \cap A_3) = \mathbb{P}(\{(1, 0)\}) = 1/4.$$

Hence A_1, A_2 and A_3 are pairwise independent but they are not independent since

$$\mathbb{P}(A_1 \cap A_2 \cap A_3) = \mathbb{P}(\emptyset) = 0 \neq \frac{1}{8}.$$

If A and B are independent events, then also A^c and B are independent. In fact,

$$\mathbb{P}(A^c \cap B) = \mathbb{P}(B) - \mathbb{P}(A \cap B) = \mathbb{P}(B) - \mathbb{P}(A)\mathbb{P}(B) = \mathbb{P}(A^c)\mathbb{P}(B).$$

Similarly, one proves the independence of the couples (A, B^c) and (A^c, B^c). In general, if $A_1, \dots, A_n$ are independent, then all the n-tuples of events whose first component is either A_1 or A_1^c, the second component is either A_2 or A_2^c and so on, are independent.

The above shows that the notion of independence is actually the independence of the corresponding generated σ-algebras. This suggests the following extension: let $(\Omega, \mathcal{E}, \mathcal{P})$ be a probability space. Two σ-algebras $\mathcal{A}, \mathcal{B} \subset \mathcal{E}$ are *independent* if $\mathbb{P}(A \cap B) = \mathbb{P}(A)\mathbb{P}(B)$ for every $A \in \mathcal{A}$ and $B \in \mathcal{B}$.

Example 4.33 (Reliability: series and parallel systems) Assume you are given a system made of n subsystems working in *parallel*, see Figure 4.1. For each $i = 1, \dots, n$ let E_i be the event 'The ith subsystem is not working' and let E be the event 'The whole system does not work'. Then $E = E_1 \cap E_2 \cap \dots \cap E_n$, hence

$$\mathbb{P}(E) \leq \min(\mathbb{P}(E_1), \mathbb{P}(E_2), \dots, \mathbb{P}(E_n)).$$

The probability of a breakdown of the system is less than or equal to the probability of a breakdown of the strongest component. If we assume that the failures of the subsystems are independent, then we actually have

$$\mathbb{P}(E) = \mathbb{P}(E_1)\mathbb{P}(E_2) \cdots \mathbb{P}(E_n).$$

For instance, if the probability that each subsystem fails is 0.5, then the probability that the whole system is not working is $(0.5)^n$, a huge increase in

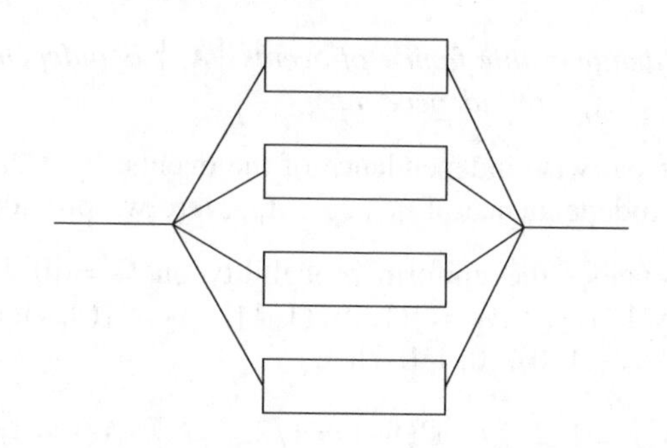

Figure 4.1 A parallel system.

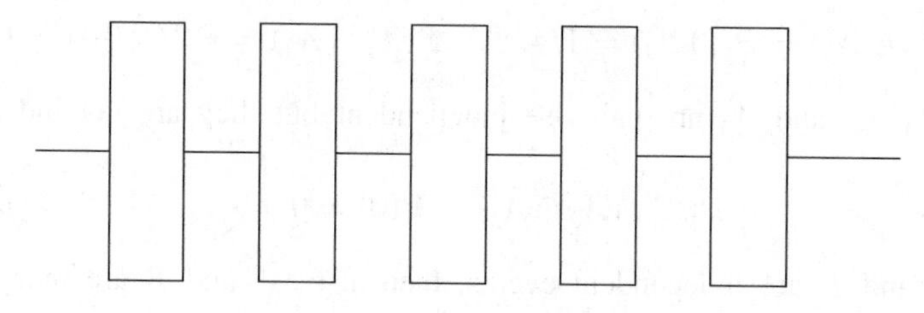

Figure 4.2 A series system.

the reliability of the whole system. However, it is worth recalling that this gain in the reliability of the system relies on the independence of the failures of the subsystems, which is not so easily granted!

Assume now that the subsystems are working in series, see Figure 4.2. The event E can now be written as:

$$E = E_1 \cup E_2 \cup \cdots \cup E_n;$$

therefore,

$$\mathbb{P}(E) \geq \max(\mathbb{P}(E_1), \ldots, \mathbb{P}(E_n)),$$

that is, the occurrence of the breakdown of the whole system is at least as probable as the occurrence of a breakdown of the weakest component of the system. If we assume that the failures of the subsystems $\{E_i\}$ are *independent*, since

$$E^c = E_1^c \cap E_2^c \cap \cdots \cap E_n^c.$$

then

$$\mathbb{P}(E^c) = \mathbb{P}(E_1^c)\mathbb{P}(E_2^c) \ldots \mathbb{P}(E_n^c)$$

or, equivalently,

$$\mathbb{P}(E) = 1 - (1 - \mathbb{P}(E_1))(1 - \mathbb{P}(E_2)) \ldots (1 - \mathbb{P}(E_n)).$$

For instance, if $\mathbb{P}(E_i) = 0.5$ then $\mathbb{P}(E) = 1 - 1/2^n$: the reliability is much worse in a series system if the failures on the subsystems are independent.

4.3.2 Independent random variables

Definition 4.34 *Let X and Y be two random variables on the probability space $(\Omega, \mathcal{E}, \mathbb{P})$. Assume X and Y take values on $\mathbb{R}^n$ and $\mathbb{R}^m$, respectively. X and Y are said to be* independent random variables *if*

$$\mathbb{P}(X \in A, Y \in B) = \mathbb{P}(X \in A)\,\mathbb{P}(Y \in B), \tag{4.21}$$

for any $A \in \mathcal{B}(\mathbb{R}^n)$ and $B \in \mathcal{B}(\mathbb{R}^m)$.

The independence of two random variables X and Y can be restated in several equivalent ways:

- The σ-algebras $\mathcal{E}_X$ and $\mathcal{E}_Y$ of the events detected by X and Y are independent.
- The distribution of X does not provide any piece of information on the distribution of Y and conversely: if A, B are such that $\mathbb{P}(X \in A)\mathbb{P}(Y \in B) > 0$, then

  $$\mathbb{P}\Big(X \in A \,\Big|\, Y \in B\Big) = \mathbb{P}(X \in A), \qquad \mathbb{P}\Big(Y \in B \,\Big|\, X \in A\Big) = \mathbb{P}(Y \in B).$$

- Equality (4.21) can be rewritten as

  $$\mathbb{P}_{X,Y}(A \times B) = \mathbb{P}_X(A)\mathbb{P}_Y(B) \qquad \forall A \in \mathcal{B}(\mathbb{R}^n),\ B \in \mathcal{B}(\mathbb{R}^m). \tag{4.22}$$

  Since the value of the joint distribution on 'rectangles'. $A \times B$ determines the joint distribution on all of $\mathcal{B}(\mathbb{R}^{n+m})$, we conclude that the joint distribution of X and Y, hence the probabilistic interaction between X and Y, is completely described by $\mathbb{P}_X$ and $\mathbb{P}_Y$.

- Since a law completely determines a measure, we also have that X and Y are independent if and only if $F_{(X,Y)}(t, s) = F_X(t)F_Y(s)$ $\forall (t, s) \in \mathbb{R}^n \times \mathbb{R}^m$.

In the jargon of measure theory, (4.22) says that $\mathbb{P}_{X,Y}$ is the *product measure* of $\mathbb{P}_X$ and $\mathbb{P}_Y$, $\mathbb{P}_{X,Y} = \mathbb{P}_X \times \mathbb{P}_Y$. From Fubini theorem, see Appendix B, we get the following.

Proposition 4.35 *Let X and Y be two random variables on the probability space $(\Omega, \mathcal{E}, \mathbb{P})$ with values in $\mathbb{R}^n$ and $\mathbb{R}^m$, respectively. Then X and Y are independent if and only if*

$$\begin{aligned}\iint_{\mathbb{R}^{n+m}} \varphi(x, y)\,\mathbb{P}_{X,Y}(dxdy) &= \int_{\mathbb{R}^m}\left(\int_{\mathbb{R}^n} \varphi(x, y)\,\mathbb{P}_X(dx)\right)\mathbb{P}_Y(dy)\\ &= \int_{\mathbb{R}^n}\left(\int_{\mathbb{R}^m} \varphi(x, y)\,\mathbb{P}_Y(dy)\right)\mathbb{P}_X(dx)\end{aligned} \tag{4.23}$$

for any $\mathcal{B}(\mathbb{R}^{n+m})$-measurable and non-negative function $\varphi : \mathbb{R}^{n+m} \to \mathbb{R}$.

Finally, from Definition 4.34 we get the following.

Proposition 4.36 *Let X and Y be two independent variables on $(\Omega, \mathcal{E}, \mathbb{P})$ with values in $\mathbb{R}^n$ and $\mathbb{R}^m$, respectively, and let $\alpha : \mathbb{R}^n \to \mathbb{R}^h$ and $\beta : \mathbb{R}^m \to \mathbb{R}^k$ be Borel-measurable. Then the random variables $\alpha \circ X$ and $\beta \circ Y$ are also independent. In particular, equality (4.23) holds true also for the random variables $\alpha \circ X$ and $\beta \circ Y$ (replacing n and m with h and k, of course).*

Theorem 4.37 (Expected value of a product) *Let X and Y be two independent random variables with finite expected value. Then X and Y are uncorrelated, i.e.*

$$\mathbb{E}[XY] = \mathbb{E}[X]\mathbb{E}[Y] \qquad \textit{equivalently,} \qquad \mathrm{Cov}(X, Y) = 0. \tag{4.24}$$

Proof. Applying (4.23)

$$\begin{aligned}\mathbb{E}[\,|XY|\,] &= \iint_{\mathbb{R}^2} |xy|\, \mathbb{P}_{X,Y}(dxdy) = \int_{\mathbb{R}} \left(\int_{\mathbb{R}} |xy|\, \mathbb{P}_Y(dy) \right) \mathbb{P}_X(dx) \\ &= \left(\int_{\mathbb{R}} |x|\, \mathbb{P}_X(dx) \right) \left(\int_{\mathbb{R}} |y|\, \mathbb{P}_Y(dy) \right) = \mathbb{E}[\,|X|\,]\mathbb{E}[\,|Y|\,] < +\infty.\end{aligned}$$

Thus the function $\varphi : (x, y) \mapsto xy$ is summable with respect to $\mathbb{P}_{X,Y}$ and (4.23) can be applied to φ to get

$$\begin{aligned}\mathbb{E}[XY] &= \iint_{\mathbb{R}^2} xy\, \mathbb{P}_{X,Y}(dxdy) = \int_{\mathbb{R}} \left(\int_{\mathbb{R}} xy\, \mathbb{P}_Y(dy) \right) \mathbb{P}_X(dx) \\ &= \left(\int_{\mathbb{R}} x\, \mathbb{P}_X(dx) \right) \left(\int_{\mathbb{R}} y\, \mathbb{P}_Y(dy) \right) = \mathbb{E}[X]\mathbb{E}[Y].\end{aligned}$$

Remark 4.38 Notice that proving that two variables are uncorrelated is a much easier task than proving that they are independent. In fact, X and Y are uncorrelated if and only if $\mathbb{E}[XY] = \mathbb{E}[X]\mathbb{E}[Y]$, which can be checked with only one equality. In contrast, X and Y are independent means that (4.22) holds *for every* couple of events A and B, which amounts to a huge, possibly infinite, set of equalities.

4.3.3 Independence of many random variables

The notion of independence extends to many variables, even sequences of random variables.

Definition 4.39 *Let X_n, $n \in \mathbb{N}$, be random variables on $(\Omega, \mathcal{E}, \mathbb{P})$.*

(i) We say that the random variables X_n are pairwise independent *if for every $i, j \in \mathbb{N}$, $i \neq j$, the random variables X_i and X_j are independent, see Definition 4.34.*

(ii) *n random variables $X_1, \dots, X_n$ are said to be* independent *if for any $k \geq 2$ and for any choice $X_{r_1}, \dots, X_{r_k}$ of k of them and for any $A_1, \dots, A_k \in \mathcal{B}(\mathbb{R})$*

$$\begin{aligned} &\mathbb{P}(X_{r_1} \in A_1, X_{r_2} \in A_2, \dots, X_{r_k} \in A_k) \\ &\qquad = \mathbb{P}(X_{r_1} \in A_1)\mathbb{P}(X_{r_2} \in A_2) \cdots \mathbb{P}(X_{r_k} \in A_k). \end{aligned} \tag{4.25}$$

(iii) *The random variables of the sequence $\{X_n\}$ are said to be* independent *if for every $k \geq 2$ the variables $X_1, \dots, X_k$ are independent.*

Equation (4.25) says that the joint distribution of $X_{r_1}, \dots, X_{r_k}$ in $\mathcal{B}(\mathbb{R}^k)$ is the product measure of the distributions of $X_{r_1}, \dots, X_{r_k}$, i.e.

$$\mathbb{P}_{X_{r_1},\dots,X_{r_k}} = \mathbb{P}_{X_{r_1}} \times \cdots \times \mathbb{P}_{X_{r_k}} \text{ on } \mathcal{B}(\mathbb{R}^k).$$

Thus independence of $X_1, \dots, X_n$ is equivalent to the following: choose k random variables between $X_1, \dots, X_n$ and group them to form two vector valued random variables say $Y := (X_{r_1}, \dots, X_{r_h})$ and $Z = (X_{r_{h+1}}, \dots, X_{r_k})$, then $X_1, \dots, X_n$ are independent if and only if Y and Z are independent,

$$\mathbb{P}_{Y,Z} = \mathbb{P}_Y \times \mathbb{P}_Z \qquad \text{on } \mathcal{B}(\mathbb{R}^k)$$

whatever the variables used to build Y and Z are.

Although three or more independent variables are trivially pairwise independent, the converse is false as the following example shows.

Example 4.40 Let A_1, A_2, A_3 be the events in Example 4.32. Then the corresponding random variables $X_i = \mathbb{1}_{A_i}$ $i = 1, 2, 3$ are pairwise independent but not independent as a whole.

Remark 4.41 Beware that the notion of independence is somewhat delicate. For instance, suppose that X, Y, Z, T are independent random variables. Then (X, Y) and (Z, T) are independent by definition and Proposition 4.36 yields that $X + Y$ and $Z + T$ are independent variables, too. However, in general, $X + Y$ and $Z + Y$ are not independent, as shown by Example 4.42.

Example 4.42 Throw three dice and let X, Y, Z be the outcome of each dice. Then $X + Y$ and $Z + Y$ are not independent. In fact, let $A := \{X + Y = 6\}$ and $B = \{Y + Z = 6\}$. Then

$$\begin{aligned} \mathbb{P}(X + Y = 6) &= \sum_{j=1}^{5} \mathbb{P}(X = j,\ Y = 6 - j) \\ &= \sum_{j=1}^{5} \mathbb{P}(X = j)\,\mathbb{P}(Y = 6 - j) = \frac{5}{6^2} \end{aligned}$$

and, similarly, $\mathbb{P}(Z+Y=6)=\frac{5}{6^2}$. But

$$\mathbb{P}(X+Y=6, Z+Y=6)=\sum_{j=1}^{5}\mathbb{P}(Y=j,\ X=6-j,\ Z=6-j)$$

$$=\sum_{j=1}^{5}\mathbb{P}(Y=j)\,\mathbb{P}(X=6-j)\,\mathbb{P}(Z=6-j)=\frac{5}{6^3}\neq\frac{25}{6^4}.$$

4.3.4 Sum of independent random variables

Theorem 4.43 *Let X and Y be two independent random variables with values in $\mathbb{Z}$. Let $\{p_k\}$ and $\{q_k\}$ be the mass densities of $\mathbb{P}_X$ and $\mathbb{P}_Y$, respectively. Then $X+Y$ is a random variable with values in $\mathbb{Z}$ and the mass density $\{r_k\}$ of $\mathbb{P}_{X+Y}$ is given by*

$$r_k=\sum_{j=-\infty}^{+\infty}p_j q_{k-j}\qquad\forall k\in\mathbb{Z}.$$

Proof. For any $k\in\mathbb{Z}$ one has

$$\mathbb{P}_{X+Y}(\{k\})=\mathbb{P}(X+Y=k)=\mathbb{P}\Big(\bigcup_j\Big\{x\in\Omega\,\Big|\,X=j, Y=k-j\Big\}\Big)$$

$$=\sum_{j=-\infty}^{+\infty}\mathbb{P}\,(X=j, Y=k-j)=\sum_{j=-\infty}^{+\infty}\mathbb{P}(X=j)\mathbb{P}(Y=k-j)$$

$$=\sum_{j=-\infty}^{+\infty}p_j q_{k-j}.$$

A similar result holds for the sum of two independent random variables following an absolutely continuous distribution. We have the following.

Theorem 4.44 *Let X and Y be two independent random variables with $\mathbb{P}_X(dt)=f(t)\,dt$ and $\mathbb{P}_Y(dt)=g(t)\,dt$, where f and g are two summable functions. Then $X+Y$ follows the absolutely continuous distribution, $\mathbb{P}_{X+Y}(dt)=\tau(t)\,dt$ with*

$$\tau(t)=\int_{-\infty}^{+\infty}f(s)g(t-s)\,ds.$$

Proof. Let $\varphi:\mathbb{R}\to\mathbb{R}$ be a $\mathcal{B}(\mathbb{R})$-measurable non-negative function. Then

$$\int_{\mathbb{R}}\varphi(t)\,\mathbb{P}_{X+Y}(dt)=\iint_{\mathbb{R}^2}\varphi(x+y)\,\mathbb{P}_{X,Y}(dxdy)\tag{i}$$

$$= \int_{\mathbb{R}} \left(\int_{\mathbb{R}} \varphi(x+y)\, \mathbb{P}_Y(dy) \right) \mathbb{P}_X(dx) \qquad \text{(ii)}$$

$$= \int_{-\infty}^{+\infty} f(x) \left(\int_{-\infty}^{+\infty} \varphi(x+y) g(y)\, dy \right) dx \qquad \text{(iii)}$$

$$= \int_{-\infty}^{+\infty} f(x) \left(\int_{-\infty}^{+\infty} \varphi(t) g(t-x)\, dt \right) dx \qquad \text{(iv)}$$

$$= \int_{-\infty}^{+\infty} \varphi(t) \left(\int_{-\infty}^{+\infty} f(x) g(t-x)\, dx \right) dt \qquad \text{(v)}$$

$$= \int_{-\infty}^{+\infty} \varphi(t) \tau(t)\, dt.$$

Here we have taken advantage of (4.2) in (i), of the independence of X and Y in (ii), and of the Fubini theorem in (iii). Moreover, in (iv) we have changed variable $y \mapsto t := x + y$ and in (v) we have used the Fubini theorem in order to change the order of integration.

4.3.5 Exercises

Exercise 4.45 (Sum of two uniformly distributed random variables) *Let X and Y be two random variables, uniformly distributed on the same interval $[a, b]$. Compute the distribution of $X + Y$, and compare with the distribution of $2X$.*

Solution. With the notation of Theorem 4.44 we have

$$f(t) = g(t) = \begin{cases} \dfrac{1}{b-a} & \text{if } t \in (a, b), \\ 0 & \text{otherwise.} \end{cases}$$

The product $f(s)g(t-s)$ is not zero (and equal to $\frac{1}{(b-a)^2}$) if and only if s and $t - s \in (a, b)$, i.e. if and only if

$$s \in (a, t-a), \ t \in (2a, a+b], \ s \in (t-b, b), \ t \in (a+b, 2b).$$

Thus

$$\tau(t) = \begin{cases} \dfrac{t-2a}{(b-a)^2} & \text{if } t \in (2a, a+b], \\ \dfrac{2b-t}{(b-a)^2} & \text{if } t \in (a+b, 2b), \\ 0 & \text{otherwise.} \end{cases}$$

The density distribution $\tau(t)$ is called the *triangular distribution* on the interval $(2a, 2b)$: the graph of τ is shown in Figure 4.3. The distribution of $2X$ is uniform on the interval $(2a, 2b)$ with density $\frac{1}{2(b-a)}$.

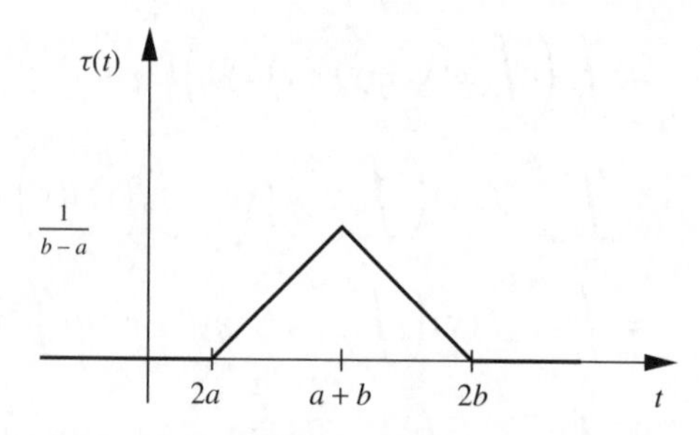

Figure 4.3 *The sum of two independent random variables uniformly distributed on (a, b) follows the triangular distribution on the interval $(2a, 2b)$.*

Exercise 4.46 (Sum of independent Poisson random variables) *Assume X and Y are independent random variables with $\mathbb{P}_X = P(\lambda)$ and $\mathbb{P}_Y = P(\mu)$. Prove the following:*

(i) $\mathbb{P}_{X+Y} = P(\lambda + \mu)$.

(ii) For any $k, n \in \mathbb{N}$

$$\mathbb{P}\Big(X = k \,\Big|\, X + Y = n\Big) = B(n, p)(\{k\})$$

where $p := \frac{\lambda}{\lambda+\mu}$.

Solution. (i) Let $k \in \mathbb{Z}$; then

$$\mathbb{P}_{X+Y}(\{k\}) = \sum_{j=-\infty}^{+\infty} P(\lambda)(\{j\})P(\mu)(\{k-j\}) = \sum_{j=0}^{k} P(\lambda)(\{j\})P(\mu)(\{k-j\})$$

$$= \sum_{j=0}^{k} \frac{\lambda^j}{j!}e^{-\lambda}\frac{\mu^{k-j}}{(k-j)!}e^{-\mu} = e^{-(\lambda+\mu)}\frac{1}{k!}\sum_{j=0}^{k}\binom{k}{j}\lambda^j\mu^{k-j}$$

$$= e^{-(\lambda+\mu)}\frac{1}{k!}(\lambda+\mu)^k = P(\lambda+\mu)(\{k\}).$$

(ii) For any $n \in \mathbb{N}$ and $k \in \{0, \dots, n\}$ one computes

$$\mathbb{P}\Big(X = k \,\Big|\, X + Y = n\Big) = \frac{\mathbb{P}(X{=}k, Y{=}n-k)}{\mathbb{P}(X+Y=n)} = \frac{\mathbb{P}(X=k)\mathbb{P}(Y{=}n-k)}{\mathbb{P}(X+Y{=}n)}$$

$$= \frac{e^{-\lambda}\lambda^k}{k!}\frac{e^{-\mu}\mu^{n-k}}{(n-k)!}\frac{n!}{e^{-\lambda-\mu}(\lambda+\mu)^n}$$

$$= \binom{n}{k}\left(\frac{\lambda}{\lambda+\mu}\right)^k\left(\frac{\mu}{\lambda+\mu}\right)^{n-k}$$

$$= \binom{n}{k}p^k(1-p)^{n-k} = B(n, p)(\{k\}).$$

Exercise 4.47 (Sum of independent Gaussian random variables) *Assume that X and Y are two independent random variables following two Gaussian distributions, $\mathbb{P}_X = N(\ell, \sigma)$ and $\mathbb{P}_Y = N(m, \delta)$. Show that $X + Y$ is a Gaussian random variable of parameters $\ell + m$ and $\sqrt{\sigma^2 + \delta^2}$, $\mathbb{P}_{X+Y} = N(\ell + m, \sqrt{\sigma^2 + \delta^2})$.*

Solution. By considering $X - \ell$ and $Y - m$ instead of X and Y, it suffices to prove the claim when $\ell = m = 0$. By Theorem 4.44, $\mathbb{P}_{X+Y}(dt) = \tau(t)\, dt$ where

$$\tau(t) = \frac{1}{2\pi\sigma\delta} \int_{-\infty}^{+\infty} e^{-\frac{s^2}{2\sigma^2}} e^{-\frac{(t-s)^2}{2\delta^2}}\, ds.$$

One computes

$$\frac{s^2}{\sigma^2} + \frac{(t-s)^2}{\delta^2} = \alpha^2 s^2 - 2\alpha\beta st + \beta^2 t^2 + \frac{t^2}{\delta^2} - \beta^2 t^2 = (\alpha s - \beta t)^2 + \gamma^2 t^2$$

where

$$\alpha := \sqrt{\frac{1}{\sigma^2} + \frac{1}{\delta^2}}, \qquad \beta := \frac{1}{\delta^2 \alpha}, \qquad \gamma^2 := \frac{1}{\sigma^2 + \delta^2}.$$

Thus

$$\begin{aligned}
\tau(t) &= \frac{1}{2\pi\sigma\delta} e^{-\frac{\gamma^2 t^2}{2}} \int_{-\infty}^{+\infty} e^{-\frac{(\alpha s - \beta t)^2}{2}}\, ds = \frac{1}{2\pi\sigma\delta} e^{-\frac{\gamma^2 t^2}{2}} \frac{\sqrt{2\pi}}{\alpha} \\
&= \frac{1}{\sqrt{2\pi}\sqrt{\sigma^2 + \delta^2}} e^{-\frac{t^2}{2(\sigma^2+\delta^2)}}.
\end{aligned}$$

Exercise 4.48 (Minimum of exponential random variables) *Assume that X and Y are two independent random variables with $\mathbb{P}_X = \exp(\lambda)$ and $\mathbb{P}_Y = \exp(\mu)$. Show that the random variable $Z := \min(X, Y)$ follows the exponential distribution of parameter $\lambda + \mu$, $\mathbb{P}_Z = \exp(\lambda + \mu)$.*

Solution. Let $\varphi : \mathbb{R} \to \mathbb{R}$ be a $\mathcal{B}(\mathbb{R})$-measurable non-negative function. Then

$$\begin{aligned}
\int_{\mathbb{R}} \varphi(t)\, \mathbb{P}_{\min(X,Y)}(dt) &= \iint_{\mathbb{R}^2} \varphi(\min(x, y))\, \mathbb{P}_{X,Y}(dxdy) \\
&= \int_{\mathbb{R}} \left(\int_{\mathbb{R}} \varphi(\min(x, y))\, \mathbb{P}_X(dx) \right) \mathbb{P}_Y(dy) \\
&= \int_0^{+\infty} \lambda e^{-\lambda t} \left(\int_0^{+\infty} \varphi(\min(t, s)) \mu e^{-\mu s}\, ds \right) dt \\
&= \lambda\mu \int_0^{+\infty} e^{-\lambda t} \varphi(t) \left(\int_t^{+\infty} e^{-\mu s} ds \right) dt \\
&\quad + \lambda\mu \int_0^{+\infty} e^{-\mu s} \varphi(s) \left(\int_s^{+\infty} e^{-\lambda t} dt \right) ds
\end{aligned}$$

$$= \lambda \int_0^{+\infty} \varphi(t) e^{-(\lambda+\mu)t}\, dt + \mu \int_0^{+\infty} \varphi(s) e^{-(\lambda+\mu)s}\, ds$$
$$= \int_0^{+\infty} \varphi(t)(\lambda+\mu) e^{-(\lambda+\mu)t}\, dt$$
$$= \int_{\mathbb{R}} \varphi(t) \exp(\lambda+\mu)(dt).$$

Exercise 4.49 (Sum of Gamma distributed random variables) *Let X and Y be two independent random variables with $\mathbb{P}_X = \Gamma(\alpha, \lambda)$ and $\mathbb{P}_Y = \Gamma(\beta, \lambda)$, where α, β and λ are positive constants. Prove that $\mathbb{P}_{X+Y} = \Gamma(\alpha+\beta, \lambda)$.*

Solution. We recall that the Euler Beta function

$$B(p, q) := \int_0^{\infty} x^{p-1}(1-x)^{q-1}\, dx, \qquad p, q > 0$$

can be computed in terms of the Euler Gamma function Γ, in fact,

$$B(p, q) = \frac{\Gamma(p)\,\Gamma(q)}{\Gamma(p+q)}.$$

Let $f_{\alpha,\lambda}(t)$ and $f_{\beta,\lambda}(t)$ be the densities of $\mathbb{P}_X$ and $\mathbb{P}_Y$, respectively, see (3.48). Since X and Y are independent, Theorem 4.44 yields $\mathbb{P}_{X+Y}(dt) = r(t)\, dt$ where $r(t)$ is given by

$$r(t) = \int_{\mathbb{R}} f_{\alpha,\lambda}(x) f_{\beta,\lambda}(t-x)\, dx.$$

Since $f_{\alpha,\lambda}(s)$ and $f_{\beta,\lambda}(s)$ are zero if $s \le 0$, we infer that $r(t) = 0 = f_{\alpha+\beta,\lambda}(t)$ for $t \le 0$. For $t > 0$ we have

$$r(t) = \int_0^{\infty} \frac{\lambda^\alpha x^{\alpha-1} e^{-\lambda x}}{\Gamma(\alpha)} \frac{\lambda^\beta (t-x)^{\beta-1} e^{-\lambda(t-x)}}{\Gamma(\beta)}\, dx$$
$$= \frac{\lambda^{\alpha+\beta} e^{-\lambda t}}{\Gamma(\alpha)\,\Gamma(\beta)} \int_0^{\infty} x^{\alpha-1}(t-x)^{\beta-1}\, dx.$$

After substituting $y := x/t$, the last integral is equal to

$$\int_0^{\infty} t^{\alpha-1} y^{\alpha-1} t^{\beta-1}(1-y)^{\beta-1} t\, dy = t^{\alpha+\beta-1} B(\alpha, \beta) = t^{\alpha+\beta-1} \frac{\Gamma(\alpha)\,\Gamma(\beta)}{\Gamma(a+\beta)},$$

so that

$$r(t) = \frac{\lambda^{\alpha+\beta} t^{\alpha+\beta-1} e^{-\lambda t}}{\Gamma(\alpha+\beta)} = f_{\alpha+\beta,\lambda}(t).$$

Exercise 4.50 *Consider the communication network in Figure 4.4. The capacities of the links, expressed in Mb/s, are $C_{1,2} = C_{1,3} = 5$, $C_{2,4} = C_{3,4} = 10$ and*

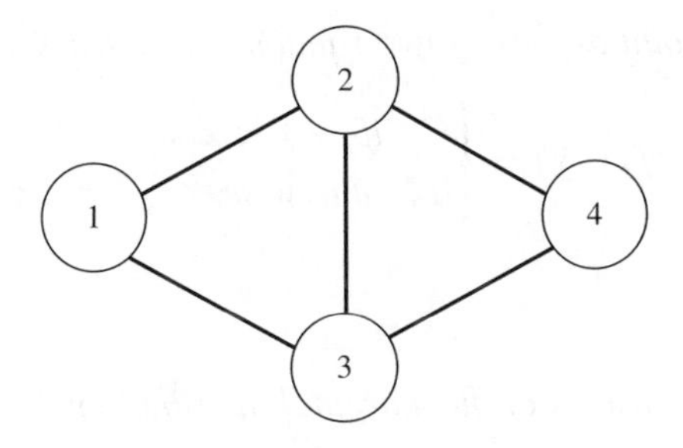

Figure 4.4 The network involved in Exercise 4.50.

$C_{2,3} = 8$, *all working in the same direction. Let* F_{ij} *be the event 'The link* C_{ij} *is broken'. Assume that the events* F_{ij} *are independent and that* $\mathbb{P}(F_{ij}) = 0.2$ *for each link. For each couple of connected nodes* (i, j) *compute the capacity of the network in Mb/s when the link* C_{ij} *is broken.*

Exercise 4.51 *Flip two fair coins. Let* $A =$*'head on the first coin',* $B =$*'the same outcome on both coins' and* $C =$*'head on the second coin'. Show that:*

- A, B *and* C *are pairwise independent, but are not independent.*
- C *and* $A \cap B$ *are not independent.*

Exercise 4.52 *Let X and Y be two independent random variables, and assume that* $\mathbb{P}_X = B(2, 1/2)$ *and* $\mathbb{P}_Y = B(2, 1/3)$*, respectively. Compute* $\mathbb{P}(XY = 2)$. $\left[\frac{1}{6}\right]$

Exercise 4.53 *Let X and Y be two independent random variables and assume that X is uniformly distributed on the interval* $[0, 1]$ *while* $\mathbb{P}_Y = \exp(\lambda)$*;*

(i) Compute $\mathbb{E}\left[\, Y^2/(1 + X) \,\right]$.

(ii) Compute $\mathbb{P}(X \leq Y)$.

(iii) Compute the joint distribution of X^2 *and* XY.

Exercise 4.54 *Assume that X and Y are two independent random variables on* $(\Omega, \mathcal{E}, \mathbb{P})$*. Assume* $X(\Omega) = Y(\Omega) = \{-1, 1\}$*, and that* $\mathbb{P}(X = 1) = \frac{1}{2}$ *and* $\mathbb{P}(Y = 1) = \frac{1}{3}$*. Compute* $\mathbb{P}(XY = 1)$. $\left[\frac{1}{2}\right]$

Exercise 4.55 *Assume X and Y are independent random variables. Suppose that X and Y follow the geometric distribution of parameters* $p = 1/2$ *and* $q = 1/4$*, respectively. Compute* $\mathbb{P}(X = Y)$. $\left[\frac{1}{5}\right]$

Exercise 4.56 *The joint density of two random variables X and Y is*

$$f(x, y) = \begin{cases} C & \text{if } -1 < x < y < 1, \\ 0 & \text{otherwise.} \end{cases}$$

(i) Compute C. $\left[\frac{1}{2}\right]$

(ii) Compute the densities of the marginal distributions $\mathbb{P}_X$ *and* $\mathbb{P}_Y$.

(iii) Are X and Y independent? [*No*]

(iv) For each $a \in \mathbb{R}$ *compute* $\mathbb{P}(X + Y > a)$.

$$\left[\begin{cases} 0 & \text{if } a \geq 1, \\ \dfrac{(2-a)^2}{8} & \text{if } a \in [0, 1[, \\ 1 - \dfrac{(a+2)^2}{8} & \text{if } a \in]-2, 0[, \\ 1 & \text{if } a \leq -2. \end{cases}\right]$$

Answer the same questions for the joint density

$$f(x, y) = \begin{cases} C & \text{if } x^2 + y^2 \leq R^2, \\ 0 & \text{otherwise,} \end{cases}$$

where $R > 0$ *is given.*

Exercise 4.57 *Let X and Y be two independent random variables with absolutely continuous distributions,* $\mathbb{P}_X(dt) = f_X(t)\,dt$, $\mathbb{P}_Y(dt) = f_Y(t)\,dt$ *where*

$$f_X(x) = \begin{cases} 6x(1-x) & x \in [0, 1], \\ 0 & \text{otherwise,} \end{cases} \quad \text{and} \quad f_Y(y) = \begin{cases} 2y & y \in [0, 1], \\ 0 & \text{otherwise,} \end{cases}$$

respectively.

(i) Check that $\int_{\mathbb{R}} f_X(s)\,ds = \int_{\mathbb{R}} f_Y(s)\,ds = 1$.

(ii) Compute the density of the random variable $Z = X^2 Y$.

$$\left[x^2 - 8x^{3/2} + 6x \text{ if } x \in [0, 1], 0 \text{ otherwise}\right].$$

Exercise 4.58 *Let X and Y be independent random variables following the standard Gaussian distribution* $N(0, 1)$;

(i) Compute $\operatorname{Cov}(3X + 2Y, X + 5Y + 10)$.

(ii) Compute $\mathbb{P}(X + 4Y \geq 2)$.

(iii) Compute $\mathbb{P}((X - Y)^2 > 9)$.

Exercise 4.59 *Assume X and Y are independent random variables following two exponential distributions. Compute the distribution of X/Y.*

Exercise 4.60 *Assume X and Y are independent random variables following the exponential distribution* $\exp(\lambda)$. *Compute the distribution of $|X - Y|$.*

Exercise 4.61 *Assume that both the independent random variables X and Y follow the distribution $\Gamma(\alpha, \lambda)$. Compute the joint law of $U := X + Y$ and $V := \dfrac{X}{X+Y}$.*

Solution. Let φ be a non-negative $\mathcal{B}(\mathbb{R}^2)$ function and let $f_{\alpha,\lambda}$ and $f_{\beta,\lambda}$ be the densities of $\mathbb{P}_X$ and $\mathbb{P}_Y$, respectively. We have

$$\iint_{\mathbb{R}^2} \varphi(u, v)\, \mathbb{P}_{U,V}(dudv) = \iint_{\mathbb{R}^2} \varphi(x + y, \frac{x}{x+y})\, \mathbb{P}_{X,Y}(dxdy)$$
$$= \iint_{\mathbb{R}^2} \varphi(x + y, \frac{x}{x+y}) f_{\alpha,\lambda}(x) f_{\beta,\lambda}, (y)\, dxdy.$$

Consider the map $g(x, y) := (x + y, \frac{x}{x+y})$. g is a one-to-one map between $\mathbb{R}^2 \setminus \{x + y = 0\}$ onto $\{(r, s) \in \mathbb{R}^2 \,|\, r = 0\}$ with inverse $h(r, s) := (rx, r(1 - s))$. By changing variable the last integral amounts to

$$= \iint_{\mathbb{R}^2} \varphi(r, s) f_{\alpha,\lambda}(rs) f_{\beta,\lambda}(r(1 - s))\, |J(Dh(r, s))|\, drds$$

so that

$$\mathbb{P}(U \le u, V \le v) = \iint_{r \le u,\ s \le v} f_{\alpha,\lambda}(rs) f_{\beta,\lambda}(r(1 - s))\, |J(Dh(r, s))|\, drds.$$

Computing the previous double integral yields

$$\mathbb{P}(U \le u, V \le v)$$
$$= \int_0^u dr \int_0^{\min(1,v)} r \frac{\lambda^\alpha (rs)^{\alpha-1} e^{-\lambda rs}}{\Gamma(\alpha)} \frac{\lambda^\beta (r(1 - s))^{\beta-1} e^{-\lambda r(1-s)}}{\Gamma(\beta)}\, ds =$$
$$= \frac{\Gamma(\alpha + \beta)}{\Gamma(\alpha)\Gamma(\beta)} \int_0^u f_{\alpha+\beta,\lambda}(r) dr \int_0^{\min(1,v)} s^{\alpha-1} (1 - s)^{\beta-1}\, ds.$$

Thus, the random variable U follows a Γ distribution of parameters $\alpha + \beta$ and λ (a fact which was already proved in Exercise 4.49) and the distribution of V is supported in the interval $[0, 1]$, with density $\dfrac{\Gamma(\alpha + \beta)}{\Gamma(\alpha)\Gamma(\beta)} v^{\alpha-1}(1 - v)^{\beta-1}$.

Exercise 4.62 *Assume $X_1, \ldots, X_n$ are independent random variables following the standard Gaussian distribution $N(0, 1)$, and let $Y := \sum_{k=1}^n X_k^2$. Show that $\mathbb{P}_Y = \chi^2(n)$. We recall that $\chi^2(n) = \Gamma(\frac{n}{2}, \frac{1}{2})$.*

Exercise 4.63 *Mark and John have decided to meet at the Pub 'Born to be Stochastic' between noon and 1pm without waiting for more than 15 min for each other. Assume Mark and John's arrival times at the Pub are independent random variables that are uniformly distributed in the time interval from noon to 1pm. Compute the probability of Mark and John meeting at the Pub.*

Solution. Denote with M and J the random variables describing the arrival time for Mark and John, respectively. Using the minute as the unit of measurement, both M and J are uniformly distributed in the interval $(0, 60)$. Mark and John meet if and only if $|G - M| \leq 15$. Since J and M are independent, the joint distribution of (J, M) is uniformly distributed on the square $[0, 60]^2$, that is, it is absolutely continuous with density $\frac{1}{3600}\mathbb{1}_{[0,60]^2}(x, y)$. Denoting by R the set

$$R := \left\{(s, t) \in [0, 60]^2 \,\middle|\, |s - t| \leq 15\right\},$$

we conclude that

$$\mathbb{P}(|G - M| \leq 15) = \iint_R \frac{1}{3600}\, dsdt = \frac{7}{16}.$$

Exercise 4.64 *Flip a fair coin and throw a fair dice together many times. Compute the probability of the event 'We obtain a head on the coin before obtaining a six on the dice'. Compute the probability of the event 'The success of the first process occurs before the first success of the second process'.*

Solution. We are considering two independent Bernoulli processes whose probability of success is $p = 1/2$ and $q = 1/6$, respectively. Let H be the random variable 'Flipping of the coin at which the first head occurs' and let S be the random variable 'Throw of the dice at which the first six occurs'. Thus, H and S are geometric random variables of parameter $p = 1/2$ and $q = 1/6$, respectively:

$$\mathbb{P}(H = k) = p(1 - p)^{k-1}, \quad \mathbb{P}(S = k) = q(1 - q)^{k-1}, \qquad k = 1, 2, \ldots.$$

We want to compute $\mathbb{P}(H < S)$. Since

$$\{H < S\} = \bigcup_{k=2}^{\infty} \{S = k,\ H < k\}$$

and H and S are independent random variables, we get

$$\mathbb{P}(H < S) = \sum_{k=2}^{\infty} \mathbb{P}(S = k, H < k) = \sum_{k=2}^{+\infty} \mathbb{P}(S = k)\mathbb{P}(H < k).$$

On the other hand,

$$\mathbb{P}(H < k) = \sum_{j=1}^{k-1} \mathbb{P}(H = j) = \sum_{j=1}^{k-1} p(1 - p)^{j-1}$$

$$= \sum_{i=0}^{k-2} p(1-p)^i = 1-(1-p)^{k-1},$$

hence we infer

$$\mathbb{P}(H < S) = \sum_{k=2}^{+\infty} q(1-q)^{k-1}\left(1-(1-p)^{k-1}\right) = \frac{p(1-q)}{p+q-pq}.$$

In particular, in the case of a fair coin and a fair dice we have $\mathbb{P}(H < S) = 5/7$.

Exercise 4.65 *Let $W_1, \dots, W_n$ be identically distributed independent random variables following the exponential distribution* $\exp(\lambda)$. *Compute the law of $S_n := \sum_{k=0}^{n} W_k$.*

Solution. By assumption, for each $k = 1, \dots, n$ we have $\mathbb{P}_{W_k} = \rho(s)\, ds$ with

$$\rho(s) = \begin{cases} 0 & \text{if } s < 0, \\ \lambda e^{-\lambda s} & \text{if } s \geq 0. \end{cases}$$

By Theorem 4.44, $\mathbb{P}_{W_1+W_2}(ds) = \rho_2(s)\, ds$ where

$$\rho_2(s) = \begin{cases} 0 & \text{if } s < 0, \\ \int_0^s \lambda^2 e^{-\lambda t} e^{-\lambda(s-t)}\, dt = \lambda e^{-\lambda s}(\lambda s) & \text{if } s \geq 0. \end{cases}$$

By induction, one proves that $\mathbb{P}_{S_n}(ds) = \rho_n(s)\, ds$ where

$$\rho_n(s) = \begin{cases} 0 & \text{if } s < 0, \\ \lambda e^{-\lambda s} \dfrac{(\lambda s)^{n-1}}{(n-1)!} = \dfrac{\lambda^n s^{n-1} e^{-\lambda s}}{\Gamma(n)} & \text{if } s \geq 0. \end{cases}$$

i.e. S_n follows the Gamma distribution $\Gamma(n, \lambda)$, also known as the Erlang distribution of parameters n and λ.

Clearly, $\mathbb{P}(S_n \leq t) = 0$ if $t \leq 0$. Assume now $t > 0$. Define $I_0 = 0$ and, for each $k \geq 1$ let $I_k := \mathbb{P}(S_k > t)$. Integrating by parts, we get the recursion formula

$$I_{k+1} = \int_t^{\infty} \rho_{k+1}(s)\, ds = \int_t^{+\infty} \lambda e^{-\lambda s} \frac{(\lambda s)^k}{k!}\, ds$$

$$= \int_{\lambda t}^{+\infty} e^{-\tau} \frac{\tau^k}{k!}\, d\tau = e^{-\lambda t} \frac{(\lambda t)^k}{k!} + I_k$$

for every $k \geq 1$. This yields $I_n = \sum_{k=0}^{n-1} e^{-\lambda t} \frac{(\lambda t)^n}{n!}$ and therefore

$$\mathbb{P}(W_1 + W_2 + \cdots + W_n \leq t) = 1 - I_n = 1 - \sum_{k=0}^{n-1} e^{-\lambda t} \frac{(\lambda t)^k}{k!}.$$

4.4 Sequences of independent random variables

Let $\{X_n\}$ be a sequence of random variables defined on the same probability space $(\Omega, \mathcal{E}, \mathbb{P})$ and with the same finite expected value: $\mathbb{E}[X_n] = E \in \mathbb{R}\ \forall n$. We are concerned with the convergence

$$\frac{X_1 + X_2 + \cdots + X_n}{n} \to E \tag{4.26}$$

under appropriate independence relations of the sequence $\{X_n\}$.

Consider, for instance, a sequence $\{X_n\}$ of independent Bernoulli trials of parameter $1/2$, so that $\mathbb{E}[X_n] = 1/2$ and $\mathrm{Var}\,[X_n] = 1/4$. Thus $S_n := X_1 + X_2 + \cdots + X_n$ yields the number of successes in n trials and S_n/n is the *rate of success* in n trials. The limit as $n \to \infty$ of $S_n(x)/n$ may even not exist and, in general, it should depend on x, that is, on the particular sequence of trials: if success always occurs, then the rate of success is constantly 1, while, if failure always occurs, then such a rate is zero for every n. Nevertheless, experience suggests that *almost certainly*, that is *for almost all sequences of trials*, such a rate approaches $1/2$ as $n \to \infty$. The mathematical formulation of this intuition is the so-called *strong law of large numbers*.

4.4.1 Weak law of large numbers

Before dealing with the law of large numbers, we start with an estimate of the average of a fixed number of random variables.

Theorem 4.66 **($\mathcal{L}^2$ and weak estimates)** *Let $X_1, \ldots, X_n$ be real-valued random variables on a probability space $(\Omega, \mathcal{E}, \mathbb{P})$, and let $S_n(x) := \sum_{k=1}^n X_k(x)$. If the $\{X_n\}$ are pairwise uncorrelated, with the same expected value $E := \mathbb{E}[X_k]\ \forall k$ and with finite variance, then*

$$\int_\Omega \left|\frac{S_n(x)}{n} - E\right|^2 \mathbb{P}(dx) \le \frac{C^2}{n} \tag{4.27}$$

where $C^2 := \max_k \mathrm{Var}[X_k]$; in particular,

$$\mathbb{P}\Big(|\frac{S_n}{n} - E| \ge \delta\Big) \le \frac{C^2}{\delta^2 n} \qquad \forall \delta > 0. \tag{4.28}$$

Proof. Observe that

$$\mathbb{E}[S_n] = \sum_{k=1}^n \mathbb{E}[X_k] = nE,$$

hence, by (4.17),

$$\int_\Omega \left|\frac{S_n(x)}{n} - E\right|^2 \mathbb{P}(dx) = \mathrm{Var}\left[\frac{S_n}{n}\right] = \frac{1}{n^2}\sum_{k=1}^n \mathrm{Var}[X_k] \le \frac{C^2}{n}.$$

Therefore, the Chebyshev inequality yields

$$\mathbb{P}\Big(|\frac{S_n}{n} - E| \ge \delta\Big) \le \frac{1}{\delta^2}\int_\Omega |\frac{S_n(x)}{n} - E|^2\,\mathbb{P}(dx) \le \frac{C^2}{\delta^2\,n}.$$

Inequality (4.28) says that the event 'The distance of $S_n(x)/n$ from its mean E is greater that δ' has probability smaller than $C^2/(\delta^2 n)$. If $\mathbb{P}$ is the uniform probability on Ω, then (4.28) reads as 'The distance between $S_n(x)/n$ and its mean E is larger than δ in no-more than $100\,C^2/(\delta^2 n)$ per cent of cases'. Notice that we have obtained an estimate on the percentage of cases, a fact which agrees with having only a probabilistic knowledge of the phenomenon. In particular, increasing n yields a better relative estimate, but, likely, the absolute number of cases may increase.

Clearly, (4.28) is not interesting if $C^2/(\delta^2 n)$ is larger than or equal to 1.

Theorem 4.66 is not the only result of this kind; experience teaches than if n is large enough, then the left-hand term in (4.28) is much smaller than the right-hand one. The following theorem, also known as the *law of large deviations* provides a hint.

Theorem 4.67 (Cernoff) *Let $X_1, \dots, X_n$ be equidistributed independent random variables such that $\mathbb{E}\big[\,e^{\theta_0|X_1|}\,\big] < +\infty$ for some $\theta_0 > 0$. Let $E := \mathbb{E}\big[\,X_1\,\big]$ and $S_n := \sum_{k=1}^n X_k$. Then for any $\delta > 0$ there exists a positive constant $C = C(\delta) > 0$ such that*

$$\mathbb{P}\Big(|\frac{S_n}{n} - E| > \delta\Big) \le 2\,e^{-C(\delta)n}.$$

$C(\delta)$ can be explicitly computed:

$$C(\delta) := \sup_{0<\theta<\theta_0} (\theta\delta - \log M(\theta)), \qquad \textit{where} \qquad M(\theta) := \mathbb{E}\big[\,e^{\theta|X_1|}\,\big].$$

Theorem 4.66 yields immediately a version of the *weak law of large numbers*, or WLN, for short.

Theorem 4.68 (Weak law of large numbers) *Let $\{X_n\}$ be a sequence of random variables on a probability space $(\Omega, \mathcal{E}, \mathbb{P})$. Assume that the X_n's are pairwise uncorrelated with the same expected value $\mathbb{E}\big[\,X_n\,\big] = E$ and with equibounded variances, $\mathrm{Var}\big[\,X_n\,\big] \le C^2$ $\forall n$. For every $n = 1, 2, \dots,$ set $S_n := \sum_{k=1}^n X_k$. Then*

$$\int_\Omega \Big|\frac{S_n(x)}{n} - E\Big|^2\,\mathbb{P}(dx) \to 0 \qquad \textit{as } n \to \infty$$

and

$$\mathbb{P}\Big(|\frac{S_n}{n} - E| > \delta\Big) \to 0 \qquad \textit{as } n \to \infty. \tag{4.29}$$

Convergence in probability (4.29) also holds in other situations. Since the almost sure convergence discussed below implies convergence in probability, the

Etemadi theorem, Theorem 4.81, improves a classical result by Khinchin, as follows.

Theorem 4.69 *Let $\{X_n\}$ be a sequence of summable, pairwise independent and equidistributed random variables on a probability space $(\Omega, \mathcal{E}, \mathbb{P})$. Set $S_n := \sum_{k=1}^{n} X_k$. Then for every $\delta > 0$*

$$\mathbb{P}\Big(|\frac{S_n}{n} - E| > \delta\Big) \to 0 \qquad \textit{as } n \to \infty.$$

4.4.2 Borel–Cantelli lemma

We are now going to discuss an interesting property of the intersection of infinitely many events.

Assume $(\Omega, \mathcal{E}, \mathbb{P})$ is a probability space and let $\{A_n\} \subset \mathcal{E}$ be a sequence of events. Define

$$\limsup_n A_n := \bigcap_{m=1}^{\infty} \bigcup_{n=m}^{\infty} A_n$$

i.e. the set of those $x \in \Omega$ belonging to infinitely many A_n's.

Proposition 4.70 (Borel–Cantelli lemma) *Let $\{A_n\}$ be a sequence of events on a probability space $(\Omega, \mathcal{E}, \mathbb{P})$. If $\sum_{n=1}^{\infty} \mathbb{P}(A_n) < +\infty$, then*

$$\mathbb{P}(\limsup_n A_n) = 0.$$

Proof. Since $\mathbb{P}$ is denumerably subadditive and since the series $\sum_{n=1}^{\infty} \mathbb{P}(A_n)$ converges, we infer

$$\mathbb{P}(\limsup_n A_n) \le \mathbb{P}\Big(\bigcup_{n=m}^{\infty} A_n\Big) \le \sum_{n=m}^{\infty} \mathbb{P}(A_n) \to 0 \qquad \text{as } m \to \infty.$$

If the events $\{A_n\}$ are independent, then the statement of the Borel–Cantelli lemma is indeed an equivalence. In fact, the following holds.

Proposition 4.71 *Assume that $\{A_n\}$ is a sequence of independent events on a probability space $(\Omega, \mathcal{E}, \mathbb{P})$. If the series $\sum_{n=1}^{\infty} \mathbb{P}(A_n)$ diverges, then $\mathbb{P}(\limsup_n A_n) = 1$.*

Proof. Let $m, q \in \mathbb{N}$, $m \le q$. By the independence assumption, using also the inequality $1 - x \le e^{-x}$ $\forall x \in [0, 1]$, we get

$$\mathbb{P}\Big(\bigcap_{n=m}^{q} A_n^c\Big) = \prod_{n=m}^{q} \mathbb{P}(A_n^c) = \prod_{n=m}^{q} (1 - \mathbb{P}(A_n))$$

$$\le \prod_{n=m}^{q} e^{-\mathbb{P}(A_n)} = \exp\Big(-\sum_{n=m}^{q} \mathbb{P}(A_n)\Big).$$

As q goes to infinity we get

$$\mathbb{P}\Big(\bigcap_{n=m}^{\infty} A_n^c\Big) \to 0 \qquad \text{for every } m,$$

thus,

$$\mathbb{P}\Big((\limsup_n A_n)^c\Big) = \lim_{m\to\infty} \mathbb{P}\Big(\bigcap_{n=m}^{\infty} A_n^c\Big) = 0.$$

Combining Propositions 4.70 and 4.71 we get the following.

Corollary 4.72 (Kolmogorov 0-1 law) *Let $\{A_n\}$ be a sequence of independent events in a probability space $(\Omega, \mathcal{E}, \mathbb{P})$. Then*

$$\mathbb{P}(\limsup_n A_n) = 0 \qquad \textit{if and only if} \qquad \sum_{n=1}^{\infty} \mathbb{P}(A_n) < \infty$$

and

$$\mathbb{P}(\limsup_n A_n) = 1 \qquad \textit{if and only if} \qquad \sum_{n=1}^{\infty} \mathbb{P}(A_n) = +\infty.$$

Actually, the claims in Proposition 4.71 and Corollary 4.72 still hold for pairwise independent events $\{A_n\}$.

Theorem 4.73 *Let $\{A_n\}$ be a sequence of pairwise independent events in a probability space $(\Omega, \mathcal{E}, \mathbb{P})$. If $\sum_{n=1}^{\infty} \mathbb{P}(A_n) = +\infty$, then*

$$\lim_{n\to\infty} \frac{\sum_{k=1}^{n} \mathbb{1}_{A_k}(x)}{\sum_{k=1}^{n} \mathbb{P}(A_k)} = 1 \qquad \mathbb{P}\text{-}a.e.;$$

in particular, for $\mathbb{P}$-almost every $x \in \Omega$ the series $\sum_{k=1}^{\infty} \mathbb{1}_{A_k}(x)$ diverges, i.e. x belongs to infinitely many A_n's for $\mathbb{P}$-a.e. x, that is, $\mathbb{P}(\limsup_n A_n) = 1$.

We omit the proof of Theorem 4.73, which can be found e.g. in [4].

Remark 4.74 Notice that the assumption of independence cannot be dropped. As an example, throw a fair dice and let A be the event '6 on the top'. Define $A_n :=$ A for every n. Thus $\sum_{n=1}^{\infty} \mathbb{P}(A_n) = 1/6 + 1/6 + \cdots = +\infty$ while $\limsup_n A_n =$ A so that $\mathbb{P}(A) = 1/6$.

4.4.3 Convergences of random variables

Three different kinds of convergence of random variables appear in discussing the law of large numbers.

Definition 4.75 *Let X and X_n, $n \in \mathbb{N}$, be random variables in a probability space $(\Omega, \mathcal{E}, \mathbb{P})$.*

(i) *The sequence* $\{X_n\}$ *is said to* converge in $\mathcal{L}^2$ *to* X *if*

$$\int_\Omega |X_n - X|^2 \, \mathbb{P}(dx) \to 0 \qquad \text{as } n \to \infty.$$

(ii) *The sequence* $\{X_n\}$ *is said to* converge in probability *to* X *if for every* $\delta > 0$

$$\mathbb{P}\Big(|X_n - X| > \delta\Big) \to 0 \qquad \text{as } n \to \infty.$$

(iii) *The sequence* $\{X_n\}$ *is said to* converge $\mathbb{P}$-almost surely *or* $\mathbb{P}$-almost everywhere *if the probability of the event*

$$E := \Big\{x \in \Omega \,\Big|\, X_n(x) \to \pm\infty \text{ or } X_n(x) \not\to X(x)\Big\}$$

$$= \Big\{x \in \Omega \,\Big|\, \limsup_{n \to \infty} |X_n(x) - X(x)| > 0\Big\}$$

is null, $\mathbb{P}(E) = 0$. *With a shortened notation we write* $X_n \to X$ $\mathbb{P}$*-a.s., or* $X_n \to X$ $\mathbb{P}$*-a.e., or*

$$\mathbb{P}(X_n \to X) = 1$$

when $\{X_n\}$ *converges to* X $\mathbb{P}$*-almost surely.*

Of course, the above three types of limit are unique $\mathbb{P}$-a.e., provided they exist. The three types of convergence are related in several ways although they are not equivalent. In particular, it can be shown, see Appendix B, that

(i) If $X_n \to X$ in $\mathcal{L}^2$, then $X_n \to X$ in probability.

(ii) If $X_n \to X$ $\mathbb{P}$-a.e., then $X_n \to X$ in probability.

The following example shows that convergence in probability and convergence $\mathbb{P}$-a.e. are not equivalent.

Example 4.76 Divide $[0, 1]$ in one part, two parts, three parts, etc. Enumerate sequentially the intervals thus obtained, i.e. $I_0 = [0, 1]$, $I_1 = [0, 1/2]$, $I_2 = [1/2, 1]$, $I_3 = [0, 1/3]\dots$. We therefore obtain a sequence $\{I_n\}$ of intervals. Consider the uniform probability on $[0, 1]$. It is then easy to check that the sequence $\{\mathbb{1}_{I_n}(x)\}$ converges in probability to the null random variable but does not converge pointwisely at any x, in particular, $\{\mathbb{1}_{I_n}(x)\}$ does not converge almost surely.

The following proposition clarifies the meaning of the almost sure convergence.

Proposition 4.77 *Let* X, X_n, $n \in \mathbb{N}$ *be random variables on the probability space* $(\Omega, \mathcal{E}, \mathbb{P})$. *The following are equivalent:*

(i) $X_n \to X$ $\mathbb{P}$*-almost surely,* $\mathbb{P}(X_n \to X) = 1$.

(ii) For every $\delta > 0$ *then with zero probability* $|X_n(x) - X(x)| > \delta$ *for infinitely many indexes* n.

(iii) For every $\delta > 0$ *then* $\mathbb{P}$*-almost surely* $|X_n(x) - X| \leq \delta$ *for* n *large enough (depending on* x*).*

More formally, if for $\delta > 0$ and $n \in \mathbb{N}$,

$$A_{n,\delta} := \Big\{ x \in \Omega \,\Big|\, |X_n - X| > \delta \Big\}$$

and

$$E_\delta := \liminf_n A_{n,\delta},$$

then (ii) means that $\mathbb{P}(E_\delta) = 0$ for every $\delta > 0$.

Proof of Proposition 4.77.
(i) $\Leftrightarrow$ (ii). Let

$$E = \Big\{ x \in \Omega \,\Big|\, \limsup_{n\to\infty} |X_n(x) - X(x)| > 0 \Big\}$$

and let $\{\delta_k\}$ be a sequence monotonically decreasing to 0. Trivially, $\{E_{\delta_k}\}$ is an increasing sequence of events and $E = \cup_k E_{\delta_k}$. This shows that E is an event and that $\mathbb{P}(E_\delta) \leq \mathbb{P}(E)\ \forall \delta > 0$. If (i) holds, then $\mathbb{P}(E_\delta) \leq \mathbb{P}(E) = 1 - \mathbb{P}(X_n \to X) = 0$, i.e. (ii).

Conversely, if (ii) holds, i.e. $\mathbb{P}(E_\delta) = 0$ for every $\delta > 0$, from the continuity property of the measure, we get

$$\mathbb{P}(E) = \lim_k \mathbb{P}(E_{\delta_k}) = 0,$$

hence $\mathbb{P}(X_n \to X) = 1 - \mathbb{P}(E) = 1$.
(ii) $\Leftrightarrow$ (iii). In fact, (ii) is equivalent to $\mathbb{P}(E_\delta^c) = 1$ for every $\delta > 0$, hence (ii) is equivalent to (iii) since

$$E_\delta^c = \Big\{ x \in \Omega \,\Big|\, |X_n(x) - X(x)| \leq \delta \text{ definitely} \Big\}.$$

Let X, X_n, $n \geq 1$, be random variables. For every $\delta > 0$ and $n \in \mathbb{N}$ set

$$A_{n,\delta} = \Big\{ x \in \Omega \,\Big|\, |X_n(x) - X(x)| > \delta \Big\}$$

and $E_\delta := \limsup_{n\to\infty} A_{n,\delta}$. If $\sum_{n=1}^{\infty} \mathbb{P}(A_{n,\delta}) < +\infty$, then by the Borel–Cantelli lemma, $\mathbb{P}(E_\delta) = 0$. Taking into account Proposition 4.77, we then state the following.

Proposition 4.78 *If for every* $\delta > 0$ *the series* $\sum_{n=1}^{\infty} \mathbb{P}(A_{n,\delta})$ *converges, then* $X_n \to X$ $\mathbb{P}$*-almost surely.*

4.4.4 Strong law of large numbers

Let $\{X_n\}$ be a sequence of random variables and let $S_n := \sum_{j=1}^{n} X_j$. We now discuss $\mathbb{P}$-almost sure convergence of the averages S_n/n of the X_n's.

Lemma 4.79 *Let $\{X_n\}$ be a sequence of random variables in the probability space $(\Omega, \mathcal{E}, \mathbb{P})$ such that $\mathbb{E}\left[X_n\right] = E$ $\forall n$. If $\sum_{n=1}^{\infty} \mathrm{Var}\left[X_n\right] < +\infty$, then $X_n \to E$ $\mathbb{P}$-almost surely.*

Proof. Since, by assumption, the series $\sum_{n=1}^{\infty} \int_\Omega |X_n(x) - E|^2 \,\mathbb{P}(dx)$ converges, Lebesgue dominated convergence theorem for series yields that the series $\sum_{k=1}^{\infty} |X_k(x) - E|$ converges in $\mathcal{L}^2$ and $\mathbb{P}$-almost surely. In particular, $X_n \to E$ $\mathbb{P}$-almost surely.

Another proof of Lemma 4.79. For any positive δ and any $n \in \mathbb{N}$, let $A_{n,\delta} = \{x \in \Omega \,|\, |X_n(x) - E| > \delta\}$. By the Chebyshev inequality

$$\mathbb{P}(A_{n,\delta}) \le \frac{1}{\delta^2} \int_\Omega |X_n(x) - E|^2 \,\mathbb{P}(dx).$$

Thus, from the assumptions, the series $\sum_{n=1}^{\infty} \mathbb{P}(A_{n,\delta})$ converges. The claim then follows from Proposition 4.78 since δ is an arbitrary positive number.

We now state a version of the *strong law of large numbers*.

Theorem 4.80 (Rajchman) *Let $\{X_n\}$ be a sequence of pairwise uncorrelated random variables on the probability space $(\Omega, \mathcal{E}, \mathbb{P})$. Assume $\mathbb{E}\left[X_n\right] = E$ and $\mathrm{Var}\left[X_n\right] \le C^2$ $\forall n$, and for $n = 1, 2, \ldots$, let $S_n := \sum_{k=1}^{n} X_k$. Then $S_n/n \to E$ in $\mathcal{L}^2$, in probability and $\mathbb{P}$-almost surely.*

Proof. The $\mathcal{L}^2$ convergence and the convergence in probability are already proved in Theorem 4.68. It remains to prove $\mathbb{P}$-almost sure convergence.

Replacing X_n by $X_n - E$, we may assume $\mathbb{E}\left[X_n\right] = 0$. Set, for convenience, $S_0 = 0$. Since the random variables X_n are pairwise uncorrelated, for any i, j, $0 \le i < j$, we have

$$\mathrm{Var}\,[S_j - S_i] = \mathrm{Var}\left[\sum_{k=i+1}^{j} X_k\right] = \sum_{k=i+1}^{j} \mathrm{Var}\,[X_k] \le C^2(j - i). \tag{4.30}$$

Applying (4.30) with $j = n^2$ and $i = 0$ we get

$$\mathrm{Var}\,[S_{n^2}] \le C^2 n^2,$$

hence

$$\sum_{n=1}^{\infty} \mathrm{Var}\left[\frac{S_{n^2}}{n^2}\right] \le C^2 \sum_{n=1}^{\infty} \frac{1}{n^2} < +\infty,$$

and consequently, since $\mathbb{E}\left[S_{n^2} \right] = 0$, we get

$$\frac{S_{n^2}}{n^2} \to 0 \qquad \mathbb{P}\text{-almost surely}$$

by Lemma 4.79.

For any $p \in \mathbb{N}$, let $n = n(p)$ be such that $n^2 \le p < (n+1)^2$. Clearly,

$$\frac{S_{n^2}}{p} = \frac{S_{n^2}}{n^2}\frac{n^2}{p} \to 0 \qquad \mathbb{P}\text{-almost surely.} \tag{4.31}$$

On the other hand, since $(n+1)^2 - n^2 = 2n + 1 \le 2\sqrt{p} + 1$, by (4.30) we get

$$\sum_{p=1}^{\infty} \operatorname{Var}\left[\frac{S_p}{p} - \frac{S_{n^2}}{p}\right] \le C^2 \sum_{p=1}^{\infty} \frac{p - n^2}{p^2} \le C^2 \sum_{p=1}^{\infty} \frac{2\sqrt{p} + 1}{p^2} < +\infty,$$

and, since $\mathbb{E}\left[\frac{S_p}{p} - \frac{S_{n^2}}{p} \right] = 0$, Lemma 4.79 implies that

$$\frac{S_p}{p} - \frac{S_{n^2}}{p} \to 0 \qquad \mathbb{P}\text{-almost surely.} \tag{4.32}$$

Since

$$\frac{S_p}{p} = \left(\frac{S_p}{p} - \frac{S_{n^2}}{p}\right) + \frac{S_{n^2}}{p},$$

from (4.31) and (4.32) we finally conclude that S_p/p converges to zero $\mathbb{P}$-almost surely.

Theorem 4.80 is one of the possible results known as the *strong law of large numbers*. It can be generalized in various directions, weakening certain assumptions and strengthening others. We refer the reader to the specialized literature, see e.g. [4–6]. Here we quote the following theorem, where no assumption on the variances is required.

Theorem 4.81 (Etemadi) *Let $\{X_n\}$ be a sequence of pairwise independent, equally distributed and summable random variables on the probability space $(\Omega, \mathcal{E}, \mathbb{P})$. Let $E = \mathbb{E}\left[X_1 \right] \in \mathbb{R}$ and for $n \ge 1$ let $S_n = \sum_{k=1}^{n} X_k$. Then*

$$\frac{S_n(x)}{n} \to E \qquad \mathbb{P}\text{-almost surely.}$$

We remark that the summability assumption can be weakened: integrability suffices, allowing $E = +\infty$ or $E = -\infty$, see Exercise 4.11.

Let us first state a few lemmas.

Lemma 4.82 $\displaystyle\sum_{n>x} \frac{1}{n^2} \le \frac{2}{x}$ *for any $x > 0$.*

Proof. In fact, let $m = \lfloor x \rfloor + 1$. We then have

$$\sum_{n=m}^{\infty} \frac{1}{n^2} \le \frac{1}{m^2} + \int_m^{\infty} \frac{1}{t^2}\, dt = \frac{1}{m^2} + \frac{1}{m} \le \frac{2}{x}.$$

Lemma 4.83 *Let $\alpha > 1$ and $k_n := \lfloor \alpha^n \rfloor$. Then* $\displaystyle\sum_{n, k_n \ge j} \frac{1}{k_n^2} \le \frac{4\alpha^2}{\alpha^2 - 1} \frac{1}{j^2}.$

Proof. We first show that $k_n \ge \alpha^n/2$. In fact, if $\alpha^n \le 2$, then $k_n \ge 1 \ge \alpha^n/2$; if $\alpha^n > 2$ then $k_n \ge \alpha^n - 1 > \alpha^n/2$.

Let $n_0 := \left\lfloor \frac{\log j}{\log \alpha} \right\rfloor$. If $k_n \ge j$, then $\alpha^n \ge j$, so that $n \ge \frac{\log j}{\log \alpha} \ge n_0$. Thus

$$\sum_{n,\, k_n \ge j} \frac{1}{k_n^2} \le 4 \sum_{n=n_0}^{\infty} \frac{1}{\alpha^{2n}} = \frac{4}{\alpha^{2n_0}} \frac{1}{1 - \alpha^{-2}} \le \frac{4\alpha^2}{\alpha^2 - 1} \frac{1}{j^2}.$$

Lemma 4.84 *Let $\{X_n\}$ be a sequence of identically distributed real valued random variables. For any $n \in \mathbb{N}$, set*

$$Y_n(x) := X_n(x) \mathbb{1}_{\{|X_n| < n\}}(x) = \begin{cases} X_n(x) & \textit{if } |X_n(x)| \le n, \\ 0 & \textit{otherwise} \end{cases}$$

and let $S_n := \sum_{j=1}^n X_j$ and $T_n := \sum_{j=1}^n Y_j$. If $T_n(x)/n \to \mu$ $\mathbb{P}$-almost surely, then $S_n(x)/n \to \mu$ almost surely.

Proof. Let $A_n := \{x \in \Omega \mid X_n(x) \ne Y_n(x)\}$. Applying the Cavalieri formula (3.14) we get

$$\sum_{n=1}^{\infty} \mathbb{P}(A_n) = \sum_{n=1}^{\infty} \mathbb{P}(|X_n| > n) = \sum_{n=1}^{\infty} \mathbb{P}(|X_1| > n) = \mathbb{E}\big[\,|X_1|\,\big] < +\infty.$$

Thus, we infer from the Borel–Cantelli lemma that for almost every $x \in \Omega$ there exists $\overline{n} = \overline{n(x)}$ such that $X_n(x) = Y_n(x)$ for every $n \ge \overline{n}$, hence

$$\frac{S_n(x) - T_n(x)}{n} = \frac{S_{n_0}(x) - T_{n_0}(x)}{n} \to 0 \qquad \text{as } n \to \infty \qquad \mathbb{P}\text{-almost surely.}$$

Since by assumption $T_n(x)/n \to \mu$ almost surely, also $S_n(x)/n \to \mu$ $\mathbb{P}$-almost surely.

Lemma 4.85 *Let $\{X_n\}$ be a sequence of summable, identically distributed random variables. For any $n \ge 1$, let $Y_n := X_n \mathbb{1}_{\{|X_n| \le n\}}$. Then*

$$\sum_{j=1}^{\infty} \frac{\mathbb{E}\big[\,Y_j^2\,\big]}{j^2} \le 4\mathbb{E}\big[\,|X_1|\,\big].$$

Proof. The Cavalieri formula yields

$$\mathbb{E}\left[\,Y_j^2\,\right] = \int_0^\infty \mathbb{P}(Y_j^2 > t)\,dt = \int_0^\infty 2s\,\mathbb{P}(|Y_j| > s)\,ds = \int_0^j 2s\,\mathbb{P}(|Y_j| > s)\,ds$$

$$\le \int_0^j 2s\,\mathbb{P}(|X_j| > s)\,ds = \int_0^j 2s\,\mathbb{P}(|X_1| > s)\,ds$$

$$= \int_0^\infty 2s\,\mathbb{1}_{[0,j]}(s)\mathbb{P}(|X_1| > s)\,ds.$$

Then, using the Beppo Levi theorem, Lemma 4.82 and Cavalieri formula, we conclude

$$\sum_{j=1}^\infty \frac{\mathbb{E}\left[\,Y_j^2\,\right]}{j^2} = \int_0^\infty 2s\Big(\sum_{j=1}^\infty \frac{1}{j_2}\mathbb{1}_{[0,j]}(s)\Big)\mathbb{P}(|X_1| > s)\,ds$$

$$= \int_0^\infty 2s\sum_{j>s} \frac{1}{j^2}\mathbb{P}(|X_1| > s)\,ds$$

$$\le 4\int_0^\infty \mathbb{P}(|X_1| > s)\,ds = 4\mathbb{E}\left[\,|X_1|\,\right].$$

Proof of the Etemadi theorem. Both $\{X_n^+\}$ and $\{X_n^-\}$ are sequences of summable, identically distributed and pairwise independent random variables, thus we may restrict ourselves to the case $X_n \ge 0\ \forall n$. The random variables $Y_n(x) = X_n(x)\mathbb{1}_{\{|X_n| \le n\}}$, $x \in \Omega$, are bounded, non-negative and pairwise independent. Let $T_k(x) := \sum_{j=1}^k Y_j(x)$. Thanks to Lemma 4.84 it is enough to show that for $\mathbb{P}$-almost all $x \in \Omega$

$$\frac{T_k(x)}{k} \to \mathbb{E}\left[\,X_1\,\right] \qquad \text{as } k \to \infty.$$

Let $\alpha > 1$ and $k_n := \lfloor \alpha^n \rfloor$. Since the random variables $\{Y_n\}$ are pairwise independent, they are also uncorrelated, thus, using Chebyshev inequality we get

$$\mathbb{P}\Big(|T_{k_n} - \mathbb{E}\left[\,T_{k_n}\,\right]| > k_n\delta\Big) \le \frac{\text{Var}\,[T_{k_n}]}{\delta^2 k_n^2} = \frac{1}{\delta^2 k_n^2}\sum_{j=1}^{k_n} \text{Var}\,[Y_j] \qquad \forall \delta > 0.$$

Thus, by Lemmas 4.83 and 4.84 we infer

$$\sum_{n=1}^\infty \mathbb{P}\Big(|T_{k_n} - \mathbb{E}\left[\,T_{k_n}\,\right]| > k_n\delta\Big)$$

$$\le \frac{1}{\delta^2}\sum_{n=1}^\infty \frac{1}{k_n^2}\sum_{j=1}^{k_n} \text{Var}\,[Y_j] = \frac{1}{\delta^2}\sum_{j=1}^\infty \text{Var}\,[Y_j]\Big(\sum_{n,\,k_n \ge j} \frac{1}{k_n^2}\Big)$$

$$\leq \frac{1}{\delta^2}\frac{4\alpha^2}{\alpha^2-1}\sum_{j=1}^{\infty}\frac{\text{Var}\,[Y_j]}{j^2} \leq \frac{16\alpha^2}{a^2-1}\frac{1}{\delta^2}\mathbb{E}\left[\,|X_1|\,\right] < +\infty$$

for every $\delta > 0$. Therefore, see Proposition 4.78,

$$\frac{T_{k_n}(x) - \mathbb{E}\left[\,T_{k_n}\,\right]}{k_n} \to 0 \qquad \mathbb{P}\text{-almost surely.} \tag{4.33}$$

Since the random variables are non-negative, Beppo Levi's theorem yields

$$\mathbb{E}\left[\,Y_n\,\right] = \mathbb{E}\left[\,X_n\mathbb{1}_{\{X_n\leq n\}}\,\right] = \mathbb{E}\left[\,X_1\mathbb{1}_{\{X_1\leq n\}}\,\right] \to \mathbb{E}\left[\,X_1\,\right].$$

Hence by the Cesaro theorem,

$$\frac{\mathbb{E}\left[\,T_{k_n}\,\right]}{k_n} \to \mathbb{E}\left[\,X_1\,\right]. \tag{4.34}$$

Thus, (4.33) and (4.34) yield

$$\frac{T_{k_n}(x)}{k_n} \to \mathbb{E}\left[\,X_1\,\right] \qquad \mathbb{P}\text{-almost surely.} \tag{4.35}$$

For any $k \geq 1$, let $n = n(k)$ such that $k_n < k \leq k_{n+1}$. Clearly n goes to infinity as k goes to infinity. Since $Y_k \geq 0$ for any k, we have

$$\frac{k_n}{n}\frac{T_{k_n}(x)}{k_n} \leq \frac{T_k(x)}{k} \leq \frac{T_{k_{n+1}}(x)}{k_{n+1}}\frac{k_{n+1}}{n}.$$

From $k_n \leq \alpha^n \leq k_n + 1 \leq k \leq k_{n+1} \leq \alpha^{n+1}$, we have

$$\frac{\alpha^n - 1}{\alpha^{n+1}} \leq \frac{k_n}{n}, \qquad \frac{k_{n+1}}{n} \leq \frac{\alpha^{n+1}}{\alpha^n} = \alpha,$$

hence the inequalities

$$\frac{\alpha^n-1}{\alpha^n}\frac{T_{k_n}(x)}{k_n} \leq \frac{T_k(x)}{k} \leq \frac{T_{k_{n+1}}(x)}{k_{n+1}}\alpha. \tag{4.36}$$

Letting $k \to \infty$ in (4.36), we then get from (4.35)

$$\frac{1}{\alpha}\mathbb{E}\left[\,X_1\,\right] \leq \liminf_{k\to\infty}\frac{T_k(x)}{k} \leq \limsup_{k\to\infty}\frac{T_k(x)}{k} \leq \alpha\,\mathbb{E}\left[\,X_1\,\right] \tag{4.37}$$

$\mathbb{P}$-almost surely. Since $\alpha > 1$ is arbitrary, we conclude that

$$\lim_{k\to\infty}\frac{T_k(x)}{x} = \mathbb{E}\left[\,X_1\,\right] \qquad \mathbb{P}\text{-almost surely,}$$

see Exercise 4.110. The claim is proven.

Finally, we quote a further estimate on the averages of a sequence of random variables, known as the *law of the iterated logarithm* which yields a more precise asymptotic behaviour as $n \to \infty$.

Theorem 4.86 (Hartman–Wintner) *Let* $\{X_n\}$ *be a sequence of independent identically distributed random variables with zero expected value and variance,* $E = \mathbb{E}[X_n] = 0$ *and* $\mathrm{Var}[X_n] = C^2$ *(thus* $S_n/n \to 0$ $\mathbb{P}$*-almost surely), and, for* $n \geq 1$*, let* $S_n = \sum_{k=1}^n X_k$*. Then*

$$\limsup_{n\to\infty} \frac{S_n(x)}{C\sqrt{2\,n\log\log n}} = 1 \qquad \mathbb{P}\text{-almost surely.}$$

In particular, for any $\delta > 0$ $\mathbb{P}$-almost surely we have

$$|S_n(x)| \leq C\sqrt{2}(1+\delta)\sqrt{n\log\log n} \qquad \text{definitely.}$$

4.4.4.1 Existence of a sequence of independent random variables

A problem one should consider is the one of the existence of a sequence of independent identically distributed random variables on the same probability space $(\Omega, \mathcal{E}, \mathbb{P})$. Let us construct a sequence of this kind going back to the Bernoulli infinite process: fix a probability space $(\Omega, \mathcal{E}, \mathcal{P})$. Denote by Ω^∞ the set of all sequences taking values in Ω. As for the infinite Bernoulli scheme, one proves, using the Method I of construction of measures, see Section 2.2.5 and Appendix B, the existence of a probability measure $(\mathcal{E}^\infty, \mathbb{P}^\infty)$ on Ω^∞ such that for any $n \in \mathbb{N}$ and for any $E_1, \dots, E_n \in \mathcal{E}$ the set $E \subset \Omega^\infty$ defined by

$$E = \Big\{x = \{x_n\} \in \Omega^\infty \,\Big|\, x_i \in E_i \ \forall i = 1, \dots, n\Big\}$$

is an event, $E \in \mathcal{E}^\infty$ and

$$\mathbb{P}^\infty(E) = \prod_{i=1}^n \mathbb{P}(E_i).$$

One can now easily check the following.

Proposition 4.87 *Let* $X : \Omega \to \mathbb{R}$ *be a random variable on* $(\Omega, \mathcal{E}, \mathbb{P})$*. Then the random variables* X_k *defined on* $(\Omega^\infty, \mathcal{E}^\infty, \mathbb{P}^\infty)$ *for* $k = 1, 2, \dots$ *by*

$$X_k(x) := X(x_k) \qquad \textit{if} \qquad x = \{x_k\}$$

are independent random variables equidistributed as X, i.e. $(\mathbb{P}^\infty)_{X_k} = \mathbb{P}_X \quad \forall k$.

4.4.5 A few applications of the law of large numbers

4.4.5.1 Bernstein polynomials

The following approximation theorem is a classical application of the weak law of large numbers. Let $f : [0, 1] \to \mathbb{R}$ be a function. The polynomials

$$p_n(x) := \sum_{j=0}^{n} \binom{n}{j} f\Big(\frac{j}{n}\Big) x^j (1-x)^{n-j}$$

are called *Bernstein polynomials* of f. Clearly the degree of $p_n(x)$ cannot exceed n.

Theorem 4.88 (Bernstein) *Let* $f \in C^0([0, 1])$. *Then* $p_n(x) \to f(x)$ *uniformly in the interval* $[0, 1]$.

In particular, the Bernstein theorem proves that the space of polynomials is dense in $C^0([0, 1])$ with respect to the uniform convergence; a property which the space of partial sums of power series does not satisfy.

Proof. Let $\{X_n\}$ be a sequence of independent Bernoulli trials of parameter x. Thus $\mathbb{E}\big[X_n\big] = x$ and $\mathrm{Var}\,[X_n] = x(1-x) \le 1/4$. Recall the weak estimate

$$\mathbb{P}\Big(|\frac{S_n}{n} - x| > \delta\Big) \le \frac{1}{4\,\delta^2\, n} \qquad \forall \delta > 0 \tag{4.38}$$

where $S_n := \sum_{k=1}^{n} X_k$.

The random variable S_n counts the number of successes in n trials, hence S_n follows the binomial distribution of parameters n and x, so that for every $j \in \mathbb{N}$

$$\mathbb{P}\Big(\frac{S_n}{n} = \frac{j}{n}\Big) = B(n, x)(\{j\}) = \binom{n}{j} x^j (1-x)^{n-j}.$$

Thus, if $f \in C^0([0, 1])$, we have

$$\mathbb{E}\big[\,f(S_n/n)\,\big] = \sum_{j=0}^{n} f\Big(\frac{j}{n}\Big)\mathbb{P}(S_n = j) = p_n(x).$$

Let $\varepsilon > 0$. f is continuous, hence it is bounded, i.e. there exists a constant $M > 0$ such that $||f||_\infty \le M$. Moreover, f is uniformly continuous, so that there exists $\delta > 0$ such that $|f(x) - f(y)| < \varepsilon$ whenever $|x - y| \le \delta$. Thus, by (4.38),

$$\begin{aligned}
|p_n(x) - f(x)| &= \Big|\mathbb{E}\big[\,f(S_n/n)\,\big] - \mathbb{E}[\,1\,]f(x)\Big| = \Big|\mathbb{E}\big[\,f(S_n/n) - f(x)\,\big]\Big| \\
&\le \mathbb{E}\big[\,|f(S_n/n) - f(x)|\,\big] \\
&= \int_{|\frac{S_n(t)}{n} - x| > \delta} \Big|f\Big(\frac{S_n(t)}{n}\Big) - f(x)\Big|\,\mathbb{P}(dt)
\end{aligned}$$

$$+\int_{|\frac{S_n(t)}{n}-x|\leq\delta}\left|f\Big(\frac{S_n(t)}{n}\Big)-f(x)\right|\mathbb{P}(dt)$$
$$\leq\frac{2M}{4\delta^2 n}+\varepsilon.$$

Consequently, choosing $\overline{n} := M/(2\delta^2\varepsilon)$ we conclude that $|p_n(x) - f(x)| \leq 2\varepsilon$ whenever $n \geq \overline{n}$ and $x \in [0, 1]$. Since ε is arbitrary, $p_n(x) \to f(x)$ uniformly in $[0, 1]$.

4.4.5.2 The Monte Carlo method

Let $Q = [0, 1]^N$ and $f \in C^0(Q)$. We want to estimate

$$\int_Q f(x)\,dx.$$

We may consider a method which is analogous to the one-dimensional Simpson method. Divide each edge of Q in k parts, then one has to evaluate f in k^N points, i.e. in a very large number of points (consider for instance $k = 100$, $N = 4$). During the Second World War, Enrico Fermi (1901–1954), John von Neumann (1903–1957) and Stanislaw Ulam (1909–1984) used a probabilistic method, called the Monte Carlo method.

Example 4.89 Take a list of $2n$ phone numbers, for instance from a couple of pages of the telephone directory and, with the last two digits of each number, form n couples of integers $(i, j) \in \{0, \dots, 99\}^2$. Let $T(n)$ be the number of couples (i, j) such that $i^2 + j^2 \leq (99)^2$. One can see that, as n increases, the ratio $T(n)/n$ gets nearer and nearer to $0.769 \simeq \frac{\pi}{4}(0.99)^2$.

Let $f : \mathbb{R}^N \to \mathbb{R}$ be a bounded measurable function and set $M := ||f||_\infty$. Assume $(\mathcal{B}(\mathbb{R}^N), \mu)$ is a probability measure on $\mathbb{R}^N$. We want to compute

$$\overline{f} := \int_{\mathbb{R}^N} f(t)\,\mu(dt).$$

The idea is to 'randomly choose' points X_j in $\mathbb{R}^N$ and to approximate $\int f(t)\,\mu(dt)$ with $\frac{1}{n}\sum_{j=1}^n f(X_j)$.

We can model the random choice of points as a sequence of random variables $\{X_n\}$ that follow the distribution μ, $\mathbb{P}_X = \mu$, with a suitable degree of independence. For instance, one can consider a random variable on a probability space $(\Omega, \mathcal{E}, \mathbb{P})$ with values in Q and following the distribution $\mathbb{P}_X = \mu$, and choose the corresponding sequence $\{X_n\}$ of independent and identically distributed random variables on the Bernoulli scheme $(\Omega^\infty, \mathcal{E}^\infty, \mathbb{P}^\infty)$ following the distribution $\mathbb{P}_X$. Then, for every positive integer n we get

$$\mathbb{E}\big[\,f(X_n)\big] = \mathbb{E}\big[\,f(X)\big] = \int_{\mathbb{R}^N} f(t)\,\mu(dt) = \overline{f},$$

$$\mathbb{E}\left[\,|f(X_n)|\,\right] = \mathbb{E}\left[\,|f(X)|\,\right] = \int_{\mathbb{R}^N} |f(t)|\,\mu(dt) \le M,$$

$$\mathrm{Var}\,[f(X_n)] = \mathrm{Var}\,[f(X)] = \int_{\mathbb{R}^N} |f(t) - \overline{f}|^2\,\mu(dt) \le 4M^2.$$

The random variables $Y_n := f(X_n)$ are independent, see Proposition 4.36, identically distributed and their variance is not larger than $4M^2$. Applying the strong law of large numbers

$$\frac{1}{n}\sum_{j=1}^{n} f(X_j(x)) \;\to\; \overline{f}$$

$\mathbb{P}$-almost surely on Ω^∞, in $\mathcal{L}^2$ and in probability. In particular, for any positive $\delta > 0$ and any positive integer n

$$\mathbb{P}^\infty\left(\left|\frac{1}{n}\sum_{j=1}^{n} f(X_j) - \overline{f}\right| > \delta\right) \le \frac{4M^2}{\delta^2 n}. \tag{4.39}$$

Consider, e.g. the case $M = 1$ and $n = 10^6$. Then the probability that the distance between $\frac{1}{n}\sum_{j=1}^{n} f(X_j)$ and $\overline{f}$ is larger than 0.02 is less than 0.01. In particular, if μ is uniform, in 99% of cases (i.e. of random choices of n points) the error is smaller than 0.02.

Notice that, in order to discuss the Monte Carlo method, it is enough to suppose that the random variables $\{X_n\}$ are such that $\mathbb{P}_{X_n} = \mu$ for each n and that the random variables $Y_n = f(X_n)$ are pairwise uncorrelated, an easier condition since true independence is very hard to prove in practice. If we insist on having an independence condition on the X_n's, the pairwise independence of the X_n's suffices.

Example 4.90 Consider $E \subset \mathbb{R}^N$, so that

$$\mu(E) = \int \mathbb{1}_E(t)\,\mu(dt).$$

Let X be a random variable on $(\Omega, \mathcal{E}, \mathbb{P})$ following the distribution μ and let $\{X_n\}$ be the sequence of independent and equidistributed random variables with $\mathbb{P}_{X_n} = \mathbb{P}_X$ on the Bernoulli scheme $(\Omega^\infty, \mathcal{E}^\infty, \mathbb{P}^\infty)$. For any random point X_k, we say that we have a success if $X_k \in E$, failure otherwise. The number of successes in n trials $T_{n,E}$ is given by

$$T_{n,E} := \sum_{j=1}^{n} f(X_j), \qquad \text{where} \qquad f(t) := \mathbb{1}_E(t).$$

Since the variables $f(X_j)$ are pairwise independent,

$$\frac{1}{n} T_{n,E} \;\to \mathbb{P}(E)$$

$\mathbb{P}^\infty$-almost surely, in $\mathcal{L}^2$ and in probability with respect to the measure $\mathbb{P}^\infty$.

4.4.5.3 Empirical distribution function

Let $X : \Omega \to \mathbb{R}$ be a random variable on the probability space $(\Omega, \mathcal{E}, \mathbb{P})$. We want to approximate the law $F_X(t) := \mathbb{P}(X \le t)$.

Let us apply Monte Carlo method: fix $t \in \mathbb{R}$ and set

$$E_t := \left\{x \in \Omega \,\middle|\, X(x) \le t\right\}.$$

Consider a sequence of independent identically distributed random variables $\{X_n\}$ following the distribution $\mathbb{P}_X$ of X. For each $n \in \mathbb{N}$, let $T_{n,t}(x)$ be the number of random variables X_j such that $j \le n$ and $X_j(x) \le t$ (i.e. count how many times, in n trials, a number, randomly chosen according the distribution $\mathbb{P}_X$, is not larger than t). Of course,

$$T_{n,t}(x) = \sum_{j=1}^{n} \mathbb{1}_{E_t}(x)$$

and Monte Carlo method yields

$$\frac{1}{n} T_{n,t}(x) \;\to\; \mathbb{E}\left[\,\mathbb{1}_{E_t}\,\right] = \mathbb{P}(E_t) = F_X(t)$$

where the convergence is in probability, in $\mathcal{L}^2$ and $\mathbb{P}^\infty$-almost surely.

The family of random variables $\{F_n(x,t)\}_n$ defined for $x \in \Omega^\infty$ by

$$F_n(x,t) := \frac{1}{n} T_{n,t}(x)$$

is called the *empirical distribution function* of X. One may think of these random variables as distributed on the vertical line through $(t, 0)$ on the plane where the graph $(t, F_X(t))$ of $F_X(t)$ lives.

As we have seen, Monte Carlo method yields

$$F_n(x,t) \;\to\; F_X(t) \qquad \text{as } n \to \infty \tag{4.40}$$

$\mathbb{P}^\infty$-almost surely, in $\mathcal{L}^2$ and in probability at every t. With the same procedure one can prove that we also have

$$F_n(x,t^-) \;\to\; F_X(t^-) \qquad \text{as } n \to \infty \tag{4.41}$$

$\mathbb{P}^\infty$-almost surely, in $\mathcal{L}^2$ and in probability at every t. One may also prove that the convergence in (4.40) is uniform with respect to $t \in \mathbb{R}$ and thus get the following result.

Theorem 4.91 (Glivenko–Cantelli) *With the notations above,*

$$\sup_{t \in \mathbb{R}} |F_n(t,x) - F_X(t)| \to 0 \qquad \mathbb{P}^\infty\textit{-almost surely.}$$

Theorem 4.91 easily follows from (4.40), (4.41) and the following lemma.

Lemma 4.92 *Let f and $f_n : \mathbb{R} \to \mathbb{R}$, $\forall n \geq 1$ be monotone nondecreasing and right-continuous functions such that $f(-\infty) = f_n(-\infty) = 0$ and $f(+\infty) = f_n(+\infty) = 1$. Assume that*

$$f_n(t) \to f(t) \qquad \text{and} \qquad f_n(t^-) \to f(t^-) \tag{4.42}$$

for any $t \in \mathbb{R}$. Then

$$\sup_{t \to \mathbb{R}} |f_n(t) - f(t)| \to 0 \qquad \text{as } n \to \infty.$$

Proof. Let $k \geq 1$. For any $j = 0, 1, \ldots, k$, set $E_j := \{t \mid f(t) \geq j/k\}$ and let $t_j := \inf E_j$. Clearly, $t_0 = -\infty$, $t_k \leq +\infty$ and $t_{j-1} \leq t_j$ for every j. Since f is right-continuous, $|f(t) - f(s)| \leq \frac{1}{k}$ if $t_{j-1} \leq t, s < t_j$; in particular, $|f(t) - f(t_j^-)|$ and $|f(t) - f(t_{j-1}^+)| \leq \frac{1}{k}$. Let $t_{j-1} \leq t < t_j$, then

$$\begin{aligned}
f_n(t) \leq f_n(t_j^-) &= f(t_j^-) - (f_n(t_j^-) - f(t_j^-)) \\
&\leq f(t) + |f_n(t_j^-) - f(t_j^-)| + \frac{1}{k} \\
f_n(t) \geq f_n(t_{j-1}^+) &= f(t_{j-1}^+) - (f_n(t_{j-1}^+) - f(t_{j-1}^+)) \\
&\geq f(t) - \frac{1}{k} - |f_n(t_{j-1}) - f(t_{j-1})|
\end{aligned}$$

hence

$$|f_n(t) - f(t)| \leq R_n + \frac{2}{k} \qquad \forall t \in \mathbb{R} \tag{4.43}$$

where

$$R_n := \max_{j=1,\ldots,k} (|f_n(t_{j-1}) - f(t_{j-1})| + |f_n(t_j^-) - f(t_j^-)|).$$

By assumption (4.42) and, since by assumption $f_n(-\infty) = f(-\infty)$, $f_n(+\infty) = f(+\infty)$, we infer that R_n converges to 0 as $n \to \infty$. Thus, by (4.43) $|f_n(t) - f(t)| \leq 3/k$ uniformly on $\mathbb{R}$ for large enough n's. Since k is arbitrary, the claim is proven.

4.4.5.4 Entropy

Consider a source of information producing a random sequence of symbols from a finite alphabet $E = \{1, \ldots, q\}$. Assume that each symbol $i \in E$ has a given probability $p(i)$. We can model the situation as follows: consider a random variable $X : \Omega \to E$ defined on a probability space $(\Omega, \mathcal{E}, \mathbb{P})$ with mass density $p = (p(i))$ and let $\{X_n\}$ be the sequence of independent random variables following the same distribution defined on the Bernoulli scheme $(\Omega^\infty, \mathcal{E}^\infty, \mathbb{P}^\infty)$. For each $n \geq 1$, let $M_n = (X_1, \ldots, X_n)$ be a sequence of n symbols in E. Then, the probability that the source of information has produced the sequence M_n is the product of the probabilities of the symbols in M_n,

$$\mathbb{P}(M_n) = \prod_{i=1}^{n} p(X_i). \tag{4.44}$$

Consider, for instance, the case of a fair dice, $E = \{1, \dots, 6\}$, where each symbol is equiprobable, i.e. $p(i) = 1/6$ for every i. Then the probability of each message is $1/6^n$.

In general, when different symbols have different probabilities, then $\mathbb{P}(M_n)$ depends on the symbols appearing in M_n: if, e.g. $M_n = (1, \dots, 1)$ then $\mathbb{P}(M_n) = p(1)^n$ while if $M_n = (2, \dots, 2)$, then $\mathbb{P}(M_n) = p(2)^n$ and so on. However, we claim that, when n is large enough, $\mathbb{P}$-*almost surely* $\mathbb{P}(M_n)$ has the same behaviour. In fact, (4.44) can be written as

$$\frac{1}{n} \log \mathbb{P}(M_n) = \frac{1}{n} \sum_{k=1}^{n} \log(p(X_k)).$$

On the other hand, the random variables $Y_n := -\log(p \circ X_n)$ are independent, see Proposition 4.36, and follow the same distribution, that is the distribution of $Y := \log(p(X))$. Thus they have also finite expected value and variance. In particular,

$$\mathbb{E}\left[\, Y_n \,\right] = \mathbb{E}\left[\, Y \,\right] = \int \log(p(x))\, \mathbb{P}_X(dx) = \sum_{i=1}^{q} \log(p(i))\, p(i).$$

The strong law of large numbers applies, and we infer that

$$\frac{1}{n} \log \mathbb{P}(M_n) \ \to\ \mathbb{E}\left[\, Y \,\right] \qquad \text{as } n \to \infty$$

$\mathbb{P}$-almost surely, in $\mathcal{L}^2$ and in probability. The number

$$H(p) := -\sum_{y \in E} p(y) \log p(y),$$

is called the *entropy* of the probability distribution of the alphabet E.

The *entropy function*

$$H(p) := -\sum_{i=1}^{q} p_i \log p_i$$

defined on the set

$$\Delta := \left\{ p = (p_1, \dots, p_n) \in \mathbb{R}^n \;\middle|\; \sum_{i=1}^{n} p_i = 1,\ p_i \geq 0\ \forall i \right\},$$

plays a crucial role in information theory. Here we just point out that $H(p)$ achieves its maximum on Δ in just one point, i.e. $(1/n, \dots, 1/n)$. Thus the entropy of a set E of n-symbols is maximized if and only if the symbols in E are equiprobable.

4.4.5.5 Waiting time

Assume that the intervals of time occurring between two breakdowns in a certain piece of machinery have all the same length E. Of course, during an interval of time of length nE, exactly n breakdowns occur, so that the number of breakdowns per unit of time is

$$N = \frac{n}{nE} = \frac{1}{E}.$$

A similar relation occurs when the times between two subsequent breakdowns are independent random variables with the same expected value; namely, we have the following.

Proposition 4.93 *Let $\{T_n\}$ be a sequence of non-negative random variables on $(\Omega, \mathcal{E}, \mathbb{P})$. Moreover, assume the T_n's are:*

- *summable, with the same expected value, pairwise uncorrelated and with equibounded variances; or*
- *integrable, identically distributed and pairwise independent.*

Set $E := \mathbb{E}\left[X_n\right]$ and let $S_n := \sum_{k=1}^{n} T_k$, and, for each $t \in \mathbb{R}$, let

$$N_t(x) := \sup\left\{n \,\middle|\, S_n(x) \le t\right\}.$$

Then

$$\lim_{t\to\infty} \frac{N_t(x)}{t} = \frac{1}{E} \qquad \mathbb{P}\text{-almost surely.}$$

Example 4.94 Let $\{T_n\}$ be the length of time between the $(n-1)$th and the nth event, so that S_n is the length of time one has to wait to have n events. N_t is the number of events occurring until time t. We say that $\{N_t\}$ is the *counting process* associated with the time random variables $\{T_n\}$. Finally, N_t/t is the average number of events occurring in $[0, t]$.

Proof of Proposition 4.93. For each $x \in \Omega$, $N_t := N_t(x)$ is an integer and, by definition, $S_{N_t} \le t \le S_{N_t+1}$. The random variables $T_k(x)$ are summable, so that $T_k < +\infty$ $\mathbb{P}$-almost surely $\forall k$. As t goes to $+\infty$,

$$\sum_{k=1}^{N_t(x)+1} T_k(x) = S_{N_t+1}(x) \to +\infty$$

so that $N_t(x) \to \infty$ $\mathbb{P}$-almost surely. Since $S_{N_t}(x) \le t \le S_{N_t+1}(x)$ we have

$$\frac{S_{N_t}(x)}{N_t(x)} \le \frac{t}{N_t(x)} \le \frac{S_{N_t+1}(x)}{N_t(x)+1}\frac{N_t(x)+1}{N_t(x)}. \tag{4.45}$$

By the appropriate version of the strong law of large numbers, either the Rajchman theorem, Theorem 4.80, or the Etemadi theorem, Theorem 4.81, or its extension to integrable random variables, Exercise 4.111, we have

$$\frac{S_n(x)}{n} \to E \qquad \mathbb{P}\text{-almost surely.}$$

Since $\frac{N_t(x)+1}{N_t(x)} \to 1$ $\mathbb{P}$-almost surely, (4.45) yields

$$\frac{t}{N_t(x)} \to E \qquad \mathbb{P}\text{-almost surely.}$$

4.4.6 Central limit theorem

4.4.6.1 Convergence in law and weak convergence

Let $\{X_n\}$ be a sequence of random variables on a probability space $(\Omega, \mathcal{E}, \mathbb{P})$. Assume $\{X_n\}$ converges $\mathbb{P}$-almost surely to a random variable X. What can be said about the laws F_{X_n} and the distributions $\mathbb{P}_{X_n}$?

Example 4.95 Let X be a random variable and let $F_X$be its law. For each $n \in \mathbb{N}$, set $X_n := X + \frac{1}{n}$, so that X_n converges uniformly to X. When we consider the laws we have

$$F_{X_n}(t) = \mathbb{P}(X_n \le t) = \mathbb{P}\Big(X \le t - \frac{1}{n}\Big)$$

hence $F_{X_n}(t)$ converges to $F_X(t^-)$ which agrees with $F_X(t)$ if and only if F_X is continuous at t.

Example 4.96 Consider the example provided in Exercise 4.105. There $\{X_n\}$ is a sequence of Bernoulli trials of probability $p = 1/2$, $\{X_n\}$ converges to $1/2$ $\mathbb{P}$-almost surely, but, for each t, $F_{X_n}(t)$ converges pointwisely to the function

$$\phi(t) := \begin{cases} 0 & \text{if } t < 1/2, \\ 1/2 & \text{if } t = 1/2, \\ 1 & \text{if } t > 1/2. \end{cases}$$

The function $\phi(t)$ is not right-continuous, so it cannot be the law of a random variable.

These examples justify the following definition.

Definition 4.97 *Let $\{X_n\}$ be a sequence of random variables. We say that $\{X_n\}$* converges in law *to a random variable X if $F_{X_n}(t)$ converges to $F_X(t)$ at each point t where F_X is continuous.*

It can be shown, see Proposition B.73, that if $\{X_n\}$ converges to X in probability (in particular if $\{X_n\}$ converges to X $\mathbb{P}$-almost surely), then $\{X_n\}$ converges to X in law.

The notion of convergence in law is strongly related to the notion of *weak convergence* of measures.

Definition 4.98 *Let $(\mathcal{B}(\mathbb{R}), \mu)$ and $(\mathcal{B}(\mathbb{R}), \mu_n)$, $n \geq 1$, be probability measures on $\mathbb{R}$. We say that $\{\mu_n\}$* weakly converges *to μ and one writes $\mu_n \rightharpoonup \mu$, if*

$$\int_{\mathbb{R}} \varphi(t)\, \mu_n(dt) \to \int_{\mathbb{R}} \varphi(t)\, \mu(dt)$$

for every continuous bounded function $\varphi : \mathbb{R} \to \mathbb{R}$.

It can be shown, see Theorem B.71, that $\mu_n \rightharpoonup \mu$ if and only if $F_n(t) := \mu_n([-\infty, t]) \to F(t) := \mu([-\infty, t])$ at every $t \in \mathbb{R}$ such that $F(t)$ is continuous. In particular, *if X and X_n, $n \geq 1$, are random variables then $\{X_n\}$ converges to X in law if and only if $\{\mathbb{P}_{X_n}\}$ weakly converges to $\mathbb{P}_X$.*

4.4.6.2 Central limit theorem

Recall that the *normal distribution* $N(0, 1)$ is the absolutely continuous probability measure on $\mathbb{R}$, whose density and law are respectively given by:

$$f(x) := \frac{1}{\sqrt{2\pi}} e^{-\frac{x^2}{2}}, \qquad \Phi(t) := \frac{1}{\sqrt{2\pi}} \int_{-\infty}^{t} e^{-\frac{x^2}{2}}\, dx.$$

The following convergence result can be proven.

Theorem 4.99 Central limit theorem *Let $\{X_n\}$ be a sequence of independent, identically distributed random variables such that $\mathbb{E}\left[X_n \right] = E$ and $\mathrm{Var}\left[X_n \right] = \sigma^2$. For each $n \geq 1$, set $S_n := \sum_{k=1}^{n} X_k$. Then for any $t \in \mathbb{R}$*

$$\mathbb{P}\Big(\frac{S_n - nE}{\sigma \sqrt{n}} \leq t \Big) \to \Phi(t) \qquad \text{as } n \to \infty. \tag{4.46}$$

The central limit theorem has a long history which dates to the works of Abraham de Moivre (1667–1754) and Pierre-Simon Laplace (1749–1827) on the limits of Bernoulli trials. The proof of Theorem 4.99 requires the introduction of ideas and methods that are beyond the scope of this book. In this volume we limit ourselves to a few simple remarks and refer the interested reader to, e.g. [4–10] where several versions of the theorem, that also applies to sequences of random variables following different distributions, are considered.

With the same assumptions of the central limit theorem, the weak law (4.29) yields

$$\mathbb{P}\Big(\frac{S_n}{n} - E \leq t \Big) \to \begin{cases} 0 & \text{if } t < 0, \\ 1 & \text{if } t > 0. \end{cases}$$

On the other hand, (4.46) yields

$$\mathbb{P}\Big(\frac{S_n}{n} - E \leq \frac{\sigma t}{\sqrt{n}} \Big) = \mathbb{P}\Big(\frac{S_n - nE}{\sigma \sqrt{n}} \leq t \Big) \to \Phi(t) \neq 0, \tag{4.47}$$

In words, the central limit theorem provides the right rescaling $(1/\sqrt{n})$ in order to achieve a nontrivial limit (different from 0 and 1). The convergence stated in (4.46) is equivalent to the weak convergence of the distribution of $(S_n - nE)/(\sigma\sqrt{n})$ to $N(0, 1)$, so that

$$\mathbb{P}\Big(\frac{\sigma a}{\sqrt{n}} < \frac{S_n}{n} - E \leq \frac{\sigma b}{\sqrt{n}}\Big) \to \Phi(b) - \Phi(a) \qquad \text{as } n \to \infty.$$

Trivially, we cannot rewrite (4.47) as

$$\mathbb{P}\Big(\frac{S_n}{n} - E \leq \sigma b\Big) \longrightarrow \Phi(b\sqrt{n}),$$

and, moreover, (4.47) does not straightforwardly imply that

$$\Big|\mathbb{P}\Big(\frac{S_n}{n} - E \leq \sigma b\Big) - \Phi(b\sqrt{n})\Big| \longrightarrow 0. \tag{4.48}$$

In order to get (4.48) one needs to know that the convergence in (4.46) is *uniform with respect to* $t \in \mathbb{R}$. Lemma 4.92 provides this piece of information: hence, (4.46) can be improved to

$$\sup_{t\in\mathbb{R}} \Big|\mathbb{P}\Big(\frac{S_n - nE}{\sigma\sqrt{n}} \leq t\Big) - \Phi(t)\Big| \longrightarrow 0 \qquad \text{as } n \to \infty,$$

or, equivalently, to

$$\sup_{t\in\mathbb{R}} \Big|\mathbb{P}(S_n \leq t) - \Phi\Big(\frac{t - nE}{\sigma\sqrt{n}}\Big)\Big| \longrightarrow 0 \qquad \text{as } n \to \infty.$$

Another piece of information that is missing is the speed of such convergence. The following theorem provides a useful estimate under an additional hypothesis.

Theorem 4.100 (Berry–Esseen) *Let* $\{X_n\}$ *be a sequence of independent, indentically distributed random variables such that* $\mathbb{E}\big[X_n\big] = 0$, $\mathrm{Var}\big[X_n\big] = \sigma^2 > 0$ *and* $\gamma := \mathbb{E}\big[|X_n|^3\big] < \infty$. *Let* $S_n = \sum_{k=1}^n X_k$. *Then*

$$\Big|\mathbb{P}\Big(\frac{S_n}{\sqrt{n}} \leq \sigma t\Big) - \Phi(t)\Big| \leq \frac{C}{\sqrt{n}}, \tag{4.49}$$

where $C := \frac{0.8\gamma}{\sigma^3}$.

Estimate (4.49) is uniform with respect to $t \in \mathbb{R}$. Therefore, with the same notations, we also have

$$\Big|\mathbb{P}(S_n \leq t) - \Phi\Big(\frac{t - nE}{\sigma\sqrt{n}}\Big)\Big| \leq \frac{C}{\sqrt{n}}. \tag{4.50}$$

Estimate (4.50) allows us to approximate $F_{S_n}(t)$ with a suitable renormalization of the standard normal law,

$$F_{S_n}(t) \simeq \Phi\Big(\frac{t - nE}{\sigma\sqrt{n}}\Big) \tag{4.51}$$

with a known bound on the error.

In many applications, where $\mathbb{P}_{X_n}$ is not *too badly* distributed around its expected value, both experience and numerics suggest that the approximation in (4.51) is good enough for $n \geq 30$, so that approximation (4.51) is often used as if it were an equality.

Remark 4.101 One may ask if the sequence $\{S_n/\sqrt{n}\}$ converges $\mathbb{P}$-almost surely to a random variable Z following the normal law $N(0, 1)$. The answer is negative. In fact, applying the iterated logarithm law, Theorem 4.86,

$$\limsup_{n\to\infty} \frac{S_n}{C\sqrt{2n\log\log n}} = 1 \qquad \mathbb{P}\text{-almost surely,}$$

where $C^2 = \text{Var}\,[X_n]$, so that

$$\limsup_{n\to+\infty} \frac{S_n}{\sqrt{n}} = +\infty \qquad \mathbb{P}\text{-almost surely.}$$

As a consequence, if $S_n/\sqrt{n} \to Z$ almost surely, then $Z = +\infty$ $\mathbb{P}$-almost surely, a contradiction.

The proof of the Berry–Esseen theorem, Theorem 4.100, as well as the proof of central limit theorem requires the introduction of ideas and methods that are beyond our scope. The interested reader may consult e.g. [4–7, 10].

Example 4.102 (Bernoulli trials and central limit theorem) Let X_n, $n \geq 1$, be independent Bernoulli trials of parameter p. Then $\mathbb{E}\big[X_n\big] = p$, $\sigma^2 := \text{Var}\,[X_n] = p(1-p)$ and

$$\mathbb{E}\big[\,|X_1 - p|^3\,\big] = p^3(1-p) + (1-p)^3 p = p(1-p)(1-2p).$$

For $n \geq 1$, the random variable $S_n := \sum_{k=1}^n X_k$ follows the Bernoulli distribution $B(n, p)$,

$$\mathbb{P}(S_n = k) = B(n, p)(\{k\}) = \binom{n}{k} p^k (1-p)^{n-k} \qquad \forall k = 0, \ldots, n,$$

hence

$$F_{S_n}(t) = \sum_{k\leq t} B(n, p)(\{k\}).$$

Thus, the Berry–Esseen formula reads

$$\Big|F_{S_n}(t) - \Phi\Big(\frac{t - np}{\sqrt{np(1-p)}}\Big)\Big| \leq \frac{C}{\sqrt{n}}$$

where $C = 0.8\,\gamma\sigma^{-3} = 0.8p^{-1/2}(1-p)^{-3/2}$.

Experience teaches that the left-hand term in the previous estimate is in many cases much smaller than the right-hand one. An empirical rule for a sequence of Bernoulli trials of parameter p is that, if both np and $n(1-p)$ are larger than 5, then the approximation of $B(n, p)$ with the normal distribution works well in applications. A better approximation is achieved by a translation of the normal law of 0.5 to the left: in fact, the law of the Bernoulli distribution is constant on intervals of length 1 and is right-continuous. This procedure is called the *continuity correction* and it must be done any time we approximate a discrete law with the normal law. Thus a better approximation of $F_{B(n,p)}$ is

$$F_{B(n,p)}(t) \simeq \Phi\Big(\frac{t - np + 0.5}{\sqrt{np(1-p)}}\Big). \tag{4.52}$$

Another possibility is to normalize the Bernoulli distribution before 'comparing' it with the normal law. In fact,

$$\mathbb{P}\Big(\frac{S_n - np}{\sqrt{n}} \le \sigma\, t\Big) = \sum_{k-np\le\sigma\sqrt{n}t} B(n, p)(\{k\}) = \sum_{x_k\le t} B(n, p)(\{k\})$$

where $x_k := \frac{k-np}{\sigma\sqrt{n}}$. Let $\rho(t)$ be the piecewise constant function

$$\rho(t) = \sum_{k=0}^{n} y_k \mathbb{1}_{[x_k,x_{k+1}[}(t), \qquad y_k := \sigma\sqrt{n}B(n, p)(\{k\}).$$

Since $x_{k+1} - x_k = \frac{1}{\sigma\sqrt{n}}$, we get

$$\mathbb{P}\Big(\frac{S_n - np}{\sqrt{n}} \le \sigma\ t\Big) = \int_{-\infty}^{t} \rho(t)\, dt.$$

The Berry–Esseen theorem yields the estimate

$$\Big|\int_{-\infty}^{t} \rho(s)\, ds - \Phi(t)\Big| \le \frac{C}{\sqrt{n}}, \qquad C \le \frac{0.8\,\gamma}{\sigma^3},$$

see Figure 4.5.

4.4.7 Exercises

Exercise 4.103 *Let* $\{X_n\}$ *be a sequence of independent random variables with the same expected value* E, $\mathbb{E}\big[X_n\big] = E\ \forall n$, *and with equibounded variances,* $\mathrm{Var}\big[X_n\big] \le C^2\ \forall n$. *For* $n = 1, 2, \ldots$, *let* $Y_n := \frac{1}{n}\sum_{k=1}^{n} X_k$ *and let* $F_{Y_n}(t)$ *be its law. Prove that*

$$F_{Y_n}(t) \to \begin{cases} 0 & \text{if } t < E, \\ 1 & \text{if } t > E, \end{cases}$$

that is, $\{Y_n\}$ *converges in law to the constant random variable* $Y(x) := E\ \forall x \in \Omega$.

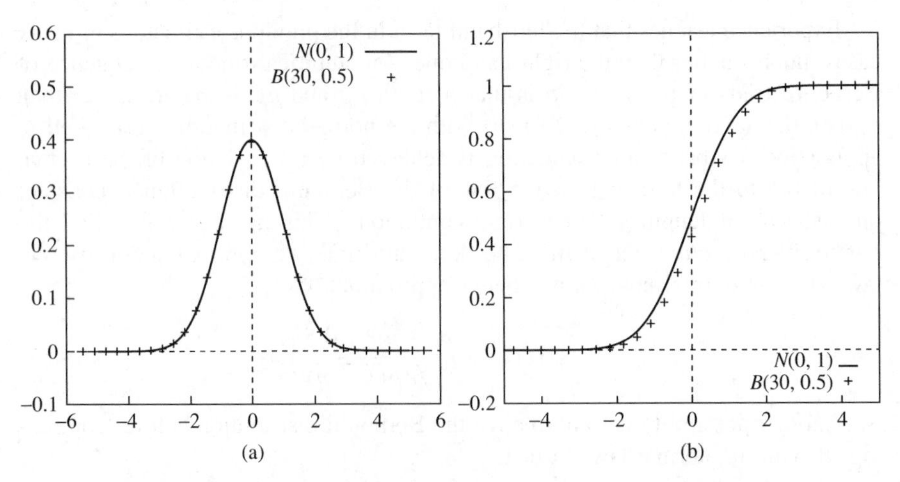

Figure 4.5 *A comparison between (a) the densities and (b) the laws of a rescaling of $B(30, 1/2)$ and $N(0, 1)$.*

Exercise 4.104 *Let $\{X_n\}$ be a sequence of independent Bernoulli trials of parameter $1/2$. For $n \geq 1$, set $Y_n := \frac{1}{n}\sum_{k=1}^{n} X_k$ and let $F_{Y_n}(t)$ be the law of Y_n. Show that*

$$\mathbb{P}\Big(|Y_n - \frac{1}{2}| = 0\Big) \to 0.$$

Notice that the weak law of large numbers yields instead

$$\mathbb{P}\Big(|Y_n - \frac{1}{2}| \leq \delta\Big) \to 1 \qquad \forall \delta > 0.$$

Solution. Let $S_n = \sum_{j=1}^{n} X_j$. We have

$$\begin{aligned}\mathbb{P}\Big(Y_n = \frac{1}{2}\Big) &= \mathbb{P}\Big(S_n = \frac{n}{2}\Big) = B(n, 1/2)(\{n/2\}) \\ &= \begin{cases} 0 & \text{if } n \text{ is odd,} \\ \frac{1}{2^n}\binom{n}{n/2} & \text{if } n \text{ is even} \end{cases} = O\Big(\frac{1}{\sqrt{n}}\Big)\end{aligned}$$

by the Stirling formula.

Exercise 4.105 *Let $\{X_n\}$ be a sequence of independent Bernoulli trials of parameter $1/2$. For $n \geq 1$, set $Y_n := \frac{1}{n}\sum_{k=1}^{n} X_k$ and let $F_{Y_n}(t)$ be the law of Y_n. Show that*

$$F_{Y_n}(t) \to \begin{cases} 0 & \textit{if } t < 1/2, \\ 1/2 & \textit{if } t = 1/2, \\ 1 & \textit{if } t > 1/2. \end{cases}$$

Notice that the limit function is not right-continuous.

Solution. The weak law of large numbers yields $F_{Y_n}(t) \to 1$ if $t > 1/2$ and $F_{Y_n}(t) \to 0$ if $t < 1/2$, see Exercise 4.103. Assume now $t = 1/2$ and let $S_n := \sum_{j=1}^n X_j$. For any $q \in \mathbb{N}$ we have

$$\mathbb{P}\Big(Y_n = q\Big) = \mathbb{P}\Big(S_n = qn\Big) = B(n, 1/2)(\{nq\})$$
$$= \begin{cases} \frac{1}{2^n}\binom{n}{nq} & \text{if } nq \in \{0, \dots, n\}, \\ 0 & \text{otherwise.} \end{cases}$$

Therefore

$$\mathbb{P}\Big(Y_n \leq \frac{1}{2}\Big) = \frac{1}{2^n}\sum_{k=0}^{\lfloor n/2 \rfloor}\binom{n}{k} = \begin{cases} \frac{1}{2} & \text{if } n \text{ is odd,} \\ \frac{1}{2} - \frac{1}{2}\binom{n}{n/2} & \text{if } n \text{ is even.} \end{cases}$$

Hence $\mathbb{P}(Y_n \leq 1/2) = \frac{1}{2} + O(\frac{1}{\sqrt{n}})$ by the Stirling formula.

Exercise 4.106 *Let X, X_n, $n \in \mathbb{N}$, be random variables on a probability space $(\Omega, \mathcal{E}, \mathbb{P})$. Show that* $\limsup_n X_n(x) \leq X(x)$ *$\mathbb{P}$-almost surely if and only if for every $\delta > 0$, $\mathbb{P}$-almost surely $X_n(x) \leq X(x) + \delta$ definitely. We mean that for every $\delta > 0$ we have $\mathbb{P}(F_\delta) = 1$ where*

$$F_\delta := \Big\{x \in \Omega \,\Big|\, X_n(x) \leq X(x) + \delta \text{ definitely}\Big\}.$$

Exercise 4.107 *Let $X_1, \dots, X_n$ be n independent, identically distributed random variables on a probability space $(\Omega, \mathcal{E}, \mathbb{P})$ and set $E := \mathbb{E}\big[X_n\big]$ and $\mu := \mathbb{P}_{X_n}$. Show that*

$$\int_\Omega \Big|\frac{1}{n}\sum_{j=1}^n X_j - E\Big|^2 \mathbb{P}(dx)$$
$$= \int_{\mathbb{R}^n} \Big|\frac{1}{n}\sum_{j=1}^n t_j - E\Big|^2 \mu(dt_1)\mu(dt_2)\cdots\mu(dt_n)$$

and

$$\mathbb{P}\Big(\Big|\frac{1}{n}\sum_{j=1}^n X_j - E\Big| > \delta\Big)$$
$$= \mu \times \cdots \times \mu\Big(\Big\{t = (t_j) \in \mathbb{R}^n \,\Big|\, \Big|\frac{1}{n}\sum_{j=1}^n t_j - E\Big| > \delta\Big\}\Big).$$

Exercise 4.108 *Let $\big\{X_n\big\}$ be a sequence of random variables such that X_n is independent of $X_1, \dots, X_{n-1}$ for every $n \geq 2$. Show that the random variables X_n are independent.*

Exercise 4.109 *Let $\{X_n\}$ and $\{Y_n\}$ converge to X and Y $\mathbb{P}$-almost surely, respectively. Show that $\{X_n + Y_n\}$ converges to $X + Y$ $\mathbb{P}$-almost surely.*

Exercise 4.110 *Let X, X_n, $n \geq 1$ be random variables defined on a probability space $(\Omega, \mathcal{E}, \mathbb{P})$ and assume that for any $\alpha > 1$,*

$$\frac{1}{\alpha}X(x) \leq \liminf_{n\to\infty} X_n(x) \leq \limsup_{n\to\infty} X_n(x) \leq \alpha X(x) \qquad \mathbb{P}\text{-almost surely.} \tag{4.53}$$

Show that $X_n \to X$ $\mathbb{P}$-almost surely.

Solution. By assumption for any $\alpha > 1$ there exists a set $N_\alpha \in \mathcal{E}$ with $\mathbb{P}(N_\alpha) = 0$ such that inequalities (4.53) hold for each $x \in \Omega \setminus N_\alpha$. Let $\{\alpha_n\}$ be a monotone decreasing sequence converging to 1 and let $N = \cup_n N_{\alpha_n}$ so that $\mathbb{P}(N) = 0$. For any $x \in \Omega \setminus N$ and any $n \in \mathbb{N}$ we have

$$\frac{1}{\alpha_n}X(x) \leq \liminf_{k\to\infty} X_k(x) \leq \limsup_{k\to\infty} X_k \leq \alpha_n X(x).$$

As n goes to infinity we get

$$X(x) \leq \liminf_{k\to\infty} X_k(x) \leq \limsup_{k\to\infty} X_k \leq X(x),$$

i.e.

$$\lim_{k\to\infty} X_k(x) = X(x) \qquad \forall x \in \Omega \setminus N.$$

Exercise 4.111 *Prove the following extension of the Etemadi version of the strong law of large numbers, Theorem 4.81.*

Theorem *Let $\{X_n\}$ be integrable, pairwise independent and identically distributed random variables. For every $n \geq 1$, set $S_n(x) := \sum_{k=1}^n X_k(x)$. Then*

$$\mathbb{E}\left[\frac{S_n}{n}\right] \to \mathbb{E}\left[X_1\right].$$

Solution. Thanks to the Etemadi theorem, it remains to prove the claim for sequences $\{X_n\}$ of random variables such that $X_n \geq 0$ $\forall n$ and $\mathbb{E}\left[X_n\right] = +\infty$ $\forall n$.

For any $M > 0$, let $Y_{n,M}(x) := X_n(x)\mathbb{1}_{\{0\leq X_n \leq M\}}(x)$. The random variables $\{Y_{n,M}\}$ are pairwise independent, equidistributed and summable. Therefore, if $S_{n,M}(x) := \sum_{k=1}^n X_{n,M}(x)$, the Etemadi theorem yields

$$\mathbb{E}\left[\frac{S_{n,M}(x)}{n}\right] \to \mathbb{E}\left[Y_{1,M}\right]$$

so that

$$\liminf_{n\to\infty} \mathbb{E}\left[\frac{S_n}{n}\right] \geq \mathbb{E}\left[Y_{1,M}\right] \qquad \forall M$$

since $S_n(x) \geq S_{n,M}(x)$. Since M is arbitrary, the Beppo Levi theorem implies $\mathbb{E}[Y_{1,M}] \to \mathbb{E}[X_1] = \infty$, hence the claim is proven.

Exercise 4.112 *Throw 100 fair dice and compute, approximately the probability that the sum of the values of the dice is in* $[325, 375]$. $[\simeq 0.86]$

Solution. Let X_n, $n = 1, 2, \ldots, 100$, be the result of the nth dice. The random variables X_n are independent, identically distributed with expected value $\mu = 7/2$ and variance $\sigma^2 = 35/12$.

We want to compute $\mathbb{P}\left(\sum_{i=1}^{100} X_i \in [325, 375]\right)$. By the central limit theorem with the continuity correction, see (4.52), we get

$$\mathbb{P}\left(\sum_{i=1}^{100} X_i \in]324, 375]\right) = \mathbb{P}\left(324.5 \leq \sum_{i=1}^{100} X_i \leq 375.5\right)$$

$$= \mathbb{P}\left(\frac{324.5 - 100\frac{7}{2}}{10\sqrt{\frac{35}{12}}} \leq \frac{\sum_{i=1}^{100} X_i - 100\frac{7}{2}}{10\sqrt{\frac{35}{12}}} \leq \frac{375.5 - 100\frac{7}{2}}{10\sqrt{\frac{35}{12}}}\right)$$

$$\simeq 2\Phi(1.49 - 1) \simeq 0.864.$$

5

Discrete time Markov chains

In this chapter we give a short survey of the properties of a class of simple *stochastic processes*, namely, the *discrete time homogeneous Markov chains* with discrete states, that widely appear both in pure and applied mathematics, and have many applications in science and technology, see e.g. [11–15].

After introducing stochastic matrices in 5.1, and discrete time stochastic processes and their *transition matrices*, we deal with discrete time Markov chains and the Markov properties in Section 5.2. Section 5.3 is devoted to computing a few interesting parameters related to a Markov chain. Convergence of the powers of the transition matrices and the characterization of the limit in terms of *return times* are investigated in Sections 5.5 and 5.7. Markov chains with a finite number of states are peculiar. The *canonical form* of a finite stochastic matrix is presented in Section 5.4; in Section 5.5 we discuss the convergence of powers of a finite regular transition matrix, the relation between the limit and the return times and, finally, summarize the relations between the parameters introduced in Section 5.3 and the transition matrix. In Section 5.6 we deal with the *ergodic property* of Markov chains with an irreducible transition matrix. This eventually leads to a probabilistic method for the computation of the expected value of a random variable, known as *Markov chain Monte Carlo*, which plays a prominent role in numerical and statistical applications. Section 5.7 contains the convergence results for arbitrary transition matrices; therein the notion of *periodicity* is introduced and the renewal theorem is proven.

5.1 Stochastic matrices

Before discussing stochastic processes and Markov chains, it is worth introducing both stochastic vectors and matrices.

A First Course in Probability and Markov Chains, First Edition. Giuseppe Modica and Laura Poggiolini.

5.1.1 Definitions

Let us start by fixing some notations. Let $\mathbb{R}^{N*}$ be the linear space of row vectors having N real components. To each matrix $\mathbf{P} \in M_{N,N}(\mathbb{R})$ we can associate the linear map $P : \mathbb{R}^{N*} \to \mathbb{R}^{N*}$ defined by $x \mapsto x\mathbf{P}$. The row vectors $e^1 := (1, 0, \dots, 0)$, ..., $e^n := (0, \dots, 0, 1)$ are a basis for $\mathbb{R}^{N*}$, called the *standard basis* of $\mathbb{R}^{N*}$. For each $i = 1, \dots, N$ the row vector $P(e^i) := e^i\mathbf{P}$ is the ith row of the matrix $\mathbf{P}$.

A vector $x = (x_i) \in \mathbb{R}^{N*}$ is a *stochastic vector* if $x_i \geq 0\ \forall i$ and $\sum_{i=1}^N x_i = 1$. The set of stochastic vectors in $\mathbb{R}^{N*}$ is denoted by $\mathcal{T}$. Finally, a matrix $\mathbf{P} = (\mathbf{P}^i_j) \in M_{N,N}$ is called *stochastic* if each of its rows is a stochastic vector

$$\mathbf{P}^i_j \geq 0, \qquad \sum_{j=1}^N \mathbf{P}^i_j = 1 \qquad \forall i, j.$$

Proposition 5.1 *Let* $\mathbf{P} \in M_{N,N}(\mathbb{R})$ *and let* $P(x) := x\mathbf{P}$. *Then* $\mathbf{P}$ *is stochastic if and only if* $P(\mathcal{T}) \subset \mathcal{T}$.

Proof. Assume $\mathbf{P}$ is stochastic and let $x = (x_1, \dots, x_N) \in \mathcal{T}$. Thus $(x\mathbf{P})_j = \sum_{i=1}^N x_i \mathbf{P}^i_j$ is non-negative and

$$\sum_{j=1}^N (x\mathbf{P})_j = \sum_{j=1}^N \sum_{i=1}^N x_i \mathbf{P}^i_j = \sum_{i=1}^N x_i \sum_{j=1}^N \mathbf{P}^i_j = \sum_{i=1}^N x_i = 1.$$

This proves that $P(x) \in \mathcal{T}$.

Conversely, assume $P(\mathcal{T}) \subset \mathcal{T}$. Since for each $i = 1, \dots, N$, the ith row of $\mathbf{P}$ is $e^i\mathbf{P} = P(e^i) \in \mathcal{T}$, we conclude that $\mathbf{P}$ is stochastic.

With some care, the same holds when the set of indexes is denumerable. Let S be a denumerable set. We call a *vector indexed by* S a sequence of real numbers and denote it as $x = \{x_j\}_{j \in S}$ or $x = (x_j)_{j \in S}$. We denote by ℓ^1 the set of all vectors $x = (x_j)$ indexed by S such that $\sum_{j \in S} |x_j| < \infty$. A vector $x = (x_j)$ indexed by S is called a *stochastic vector* if

$$x_j \geq 0 \ \ \forall j \in S \qquad \text{and} \qquad \sum_{j \in S} x_j = 1.$$

The set of stochastic vectors will be denoted by $\mathcal{T}$. Trivially $\mathcal{T} \subset \ell^1$.

We shall also deal with double sequences $\mathbf{P} = (\mathbf{P}^i_j)_{i,j \in S}$ that we call *matrices* as in the case when S is finite. The set of relevant matrices are those for which

$$||\mathbf{P}|| := \sup_{i \in S} \sum_{j \in S} |\mathbf{P}^i_j| < +\infty.$$

A *stochastic matrix* is a matrix $\mathbf{P} = (\mathbf{P}^i_j)$ such that

$$\mathbf{P}^i_j \geq 0, \qquad \sum_{j \in S} \mathbf{P}^i_j = 1 \qquad \forall i, j \in S.$$

Notice that $||\mathbf{P}|| = 1$ and $\mathbf{P}^i_j \leq 1$ $\forall i, j \in S$ if $\mathbf{P}$ is stochastic.

If $x = (x_i) \in \ell^1$ and $\mathbf{P}$ is a matrix such that $||\mathbf{P}|| < \infty$, we notice that $(x\mathbf{P})_j := \sum_{i \in S} x_i \mathbf{P}^i_j$ is well defined, being the sum of an absolute convergent series. Moreover,

$$\sum_{j \in S} |(x\mathbf{P})_j| \leq \sum_{j \in S} \sum_{i \in S} |x_i||\mathbf{P}^i_j| = \sum_{i \in S} |x_i| \sum_{j \in S} |\mathbf{P}^i_j| \leq ||\mathbf{P}|| \sum_{i \in S} |x_i|,$$

i.e. $x\mathbf{P} \in \ell^1$. Therefore every matrix $\mathbf{P}$ such that $||\mathbf{P}|| < \infty$ defines a map $P(x) := x\mathbf{P}$ from ℓ^1 into itself.

Observe also that if $\mathbf{P}$ and $\mathbf{Q}$ are two matrices indexed by S such that $||\mathbf{P}||, ||\mathbf{Q}|| < \infty$, then their *product* $\mathbf{PQ}$ is well defined by $(\mathbf{PQ})^i_j := \sum_{k \in S} \mathbf{P}^i_k \mathbf{Q}^k_j$, being the sum of an absolutely convergent series, and, moreover, $||\mathbf{PQ}|| \leq ||\mathbf{Q}|| \, ||\mathbf{P}|| < \infty$. Denoting by $P, Q : \ell^1 \to \ell^1$ the maps defined by $P(x) := x\mathbf{P}$, $Q(x) := x\mathbf{Q}$, respectively, then $Q(P(x)) = (x\mathbf{P})\mathbf{Q} = x(\mathbf{PQ})$.

Proposition 5.2 *Let $\mathbf{P}$ be a matrix indexed by S such that $||\mathbf{P}|| < \infty$ and let $P : \ell^1 \to \ell^1$, $P(x) := x\mathbf{P}$. Then $\mathbf{P}$ is a stochastic matrix if and only if $P(\mathcal{T}) \subset \mathcal{T}$.*

The proof goes as the proof of proposition 5.1.

An easy consequence of proposition 5.1 and 5.2 is that products of stochastic matrices are stochastic matrices hence, in particular, any power of a stochastic matrix is a stochastic matrix. Notice also that a convex combination of stochastic matrices, $\mathbf{P} = \sum_{k=1}^N t_i \mathbf{P}_i$, $t_i \geq 0$, $\sum_{i=1}^n t_i = 1$, $\mathbf{P}_i$ stochastic, is a stochastic matrix since

$$\sum_{j \in S} \mathbf{P}^i_j = \sum_{j \in S} \sum_{k=1}^n t_k (\mathbf{P}_k)^i_j = \sum_{k=1}^n t_k \sum_{j \in S} (\mathbf{P}_k)^i_j = \sum_{k=1}^n t_k = 1.$$

In other words, the set of stochastic matrices is a convex set of non-negative matrices.

5.1.2 Oriented graphs

An *oriented graph* is a pair (S, E) where S is a finite or denumerable set of points, called *nodes*, and E is a set of ordered pairs of nodes $E = \{(i, j) \mid i, j \in S\}$, called *arcs*. We say that the arc (i, j) *goes from i to j*, or, equivalently, that it joins i to j. An *oriented path* from i to j is a finite union of arcs $(i, i_1)(i_1, i_2) \cdots (i_k, j)$.

An oriented graph (S, E) can be represented by a matrix $\mathbf{A}$ indexed by S, called the *incidence matrix* of the graph and defined as

$$\mathbf{A}^i_j = \begin{cases} 1 & \text{if there is an arc from } i \text{ to } j, \\ 0 & \text{otherwise.} \end{cases}$$

For each $i, j \in S$ we have

$$(\mathbf{A}^2)^i_j = \sum_{k \in S} \mathbf{A}^i_k \mathbf{A}^k_j > 0$$

if and only if there exists $k \in S$ such that $\mathbf{A}^i_k = \mathbf{A}^k_j = 1$, i.e. if and only if there exists at least a path from i to j made of exactly two arcs. Similarly, for each positive integer n

$$(\mathbf{A}^n)^i_j = \sum_{i_1, i_2, \ldots, i_{n-1} \in S} \mathbf{A}^i_{i_1} \mathbf{A}^{i_1}_{i_2} \mathbf{A}^{i_2}_{i_3} \cdots \mathbf{A}^{i_{n-1}}_j$$

is positive if and only if there exists at least one path from i to j made of exactly n arcs.

The same reasoning applies to *weighted oriented graphs*, i.e. to those oriented graphs such that a weight $p_{ij} > 0$ is attached to each arc (i, j) of the graph. In fact, we can repeat the previous argument replacing the matrix $\mathbf{A}$ with the matrix

$$(\mathbf{A}')^i_j = \begin{cases} p_{ij} & \text{if there is an arc from } i \text{ to } j, \\ 0 & \text{otherwise.} \end{cases}$$

This way, a weighted oriended graph is represented by a matrix.

Conversely, to each non-negative matrix $\mathbf{P}$ indexed by S one associates a *weighted oriented graph* as follows: let $\mathbf{P} = (p_{ij})_{i,j \in S}$. Whenever $p_{ij} > 0$ draw a directed arc from i to j and associate the weight p_{ij} to such arc.

For example, assume $S = \{1, 2, 3\}$. Then the stochastic matrix

$$\mathbf{P} = \begin{pmatrix} 2/9 & 0 & 7/9 \\ 1/2 & 0 & 1/2 \\ 1/7 & 6/7 & 0 \end{pmatrix} \tag{5.1}$$

can be represented as shown in Figure 5.1.

Definition 5.3 *Let $\mathbf{P}$ be a stochastic matrix indexed by S and let $i, j \in S$. If there exists $n \geq 0$ such that $p^{(n)}_{ij} := (\mathbf{P}^n)^i_j > 0$, we say that i* leads *to j, or that j is* accessible *from i, and denote it as $i \to j$. If $i \to j$ and $j \to i$, we say that i* and

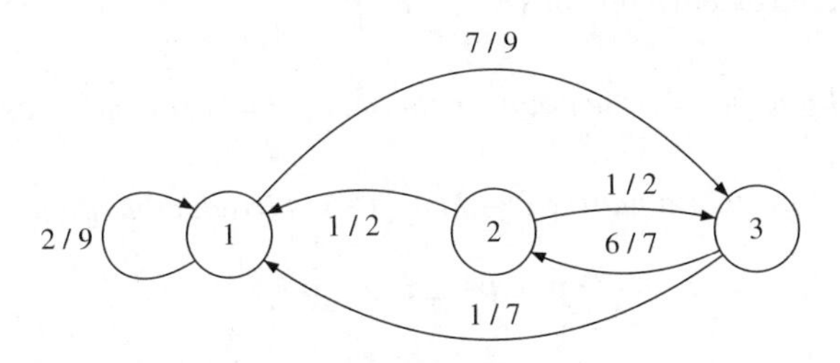

Figure 5.1 A graph representing the matrix $\mathbf{P}$ *defined in (5.1).*

j are comunicating *and write* $i \leftrightarrow j$. *If* $i \leftrightarrow j$ *for every* $i, j \in S$ *then the matrix* $\mathbf{P}$ *is called* irreducible.

Since $\mathbf{P}^0 = \text{Id}$, clearly $i \to i$ for any $i \in S$. Moreover, if $i \to j$ and $j \to k$, then $i \to k$. Furthermore, the relation $i \leftrightarrow j$ is an equivalence relation on the set of indexes S, and therefore S can be split into equivalence classes of comunicating indexes.

It is clear that i leads to j if and only if either $i = j$ or $i \neq j$ and there exists a path from i to j.

If the set of nodes S is finite, $S = \{1, \ldots, N\}$, a path from i to j exists if and only if there exists at least one path made by no more than N arcs [actually $(N-1)$ if $i \neq j$]. Thus a path from i to j, $i \neq j$, exists if and only if

$$\mathbf{C}^i_j := (\mathbf{P} + \mathbf{P}^2 + \cdots + \mathbf{P}^{N-1})^i_j > 0.$$

In particular, for each node $i \in S$, the set

$$\left\{ j \in S \,\middle|\, \mathbf{C}^i_j > 0 \right\}$$

is the set of the nodes that are reachable by at least an oriented path starting from i.

Whenever $S = \{1, \ldots, N\}$, $i \to j$ if and only if either $i = j$ or $i \neq j$ and there exists a path from i to j made by at most $N - 1$ arcs. Thus i leads to j if and only if

$$\mathbf{B}^i_j := (\text{Id} + \mathbf{P} + \mathbf{P}^2 + \mathbf{P}^{N-1})^i_j > 0. \tag{5.2}$$

In particular, the set of indexes (nodes) accessible from i is the set of indexes (nodes) j such that $\mathbf{B}^i_j > 0$,

$$\left\{ j \in \{1, \ldots, N\} \,\middle|\, i \to j \right\} = \left\{ j \,\middle|\, \mathbf{B}^i_j > 0 \right\}. \tag{5.3}$$

5.1.3 Exercises

Exercise 5.4 *Show that the set* $\mathcal{T} \subset \mathbb{R}^{N*}$ *of stochastic vectors is closed and convex. Show that the standard basis of* $\mathbb{R}^{N*}$ *is the set of extreme points of* $\mathcal{T}$, *i.e. that* $\mathcal{T}$ *is the convex envelope of* $\{e^1, \ldots e^N\}$.

Exercise 5.5 *Find the extreme points of the set of stochastic matrices in* $M_{N,N}(\mathbb{R})$.

Exercise 5.6 *Show that a matrix* $\mathbf{P} \in M_{N,N}(\mathbb{R})$ *is irreducible if and only if all the elements of*

$$\mathbf{P} + \mathbf{P}^2 + \cdots + \mathbf{P}^N$$

are positive.

5.2 Markov chains

In this section, after a short introduction to stochastic processes and their transition matrices, we introduce *Markov chains*, a restricted class of stochastic processes $\{X_n\}$ with finite or denumerable state-space. Their main property is that the joint distributions of the involved variables are fixed by the transition matrices of the process and by the distribution of the initial variable X_0.

5.2.1 Stochastic processes

Let S be a finite or denumerable set. Without any loss of generality we can assume $S = \mathbb{Z}$ or $S = \mathbb{N}$ or, when S is finite, $S = \{1, \dots, N\}$. A sequence of random variables on a probability space $(\Omega, \mathcal{E}, \mathbb{P})$ $\{X_n\}$, $X_n : \Omega \to S$ is called a *(discrete time) stochastic process* with values in S. The elements of S are called *states* and S is the *state-space* of the process. One can look at $\{X_n\}$ in several ways:

(i) The definition: a sequence $\{X_n\}$ of random variables with values in S.

(ii) A sequence of states $\{X_n(x)\} \subset S$ indexed by the parameter $x \in \Omega$, that is, the sequence of states taken by the point $x \in \Omega$.

(iii) A map $X : \mathbb{N} \times \Omega \to S$ on the product $\Omega \times \mathbb{N}$.

We are interested in describing the probabilistic laws related to the evolution of the process and in computing the probabilities of various events detected by the random variables $\{X_n\}$. Notice that, formally, we also need to have a probability space structure on $\Omega \times \mathbb{N}$ in such a way that $X : \mathbb{N} \times \Omega \to S$, $X(n, x) := X_n(x)$ is a random variable. Here, we just point out that, in general, describing the *evolution of a stochastic process*, i.e. computing for each $n \in \mathbb{N}$ the joint distribution of $(X_0, X_1, \dots, X_n)$ is a difficult task. In fact, in general, the complexity of this computation grows exponentially with respect to n since $(X_0, \dots, X_n)$ is a random variable with values in $S^n := S \times \cdots \times S$. A huge complexity reduction in computing the joint distributions appears for a subclass of stochastic processes, the *Markov chains*, see Section 5.3.

Description (ii) provides a helpful point of view on the process: at the 'first step' $n = 0$, we start from a random point $X_0(x) \in S$ then, for each positive integer n, we move from $X_n(x)$ to $X_{n+1}(x)$. Each $x \in \Omega$ fixes the sequence of states $\{X_n(x)\} \subset S$, i.e. the *path* on S defined by x. If $j := X_n(x)$ we say that the path (defined by) x *visits* j at step n. Then the set $\{x \in \Omega \mid X_n(x) = j\}$ is the set of parameters such that the corresponding path visits j at step n. By abuse of language (in general, two points $x, y \in \Omega$, $x \neq y$, can fix the same path), one refers to $\{x \in \Omega \mid X_n(x) = j\}$ as the set of paths that visit j at the nth step. Consequently $\mathbb{P}(X_n = j)$ is the probability that a path visits j at the nth step and $\mathbb{P}(X_n = j \mid X_0 = i)$ is the probability that a path starting from i visits j at step n.

5.2.2 Transition matrices

Let $\{X_n\}$ be a stochastic process with a finite or denumerable state-space S. For each $n \geq 0$, let $\mathbf{P}(n) := (\mathbf{P}(n)^i_j)_{i,j \in S}$ be the matrix

$$\mathbf{P}(n+1)^i_j := \begin{cases} \mathbb{P}\Big(X_{n+1} = j \,\Big|\, X_n = i\Big) & \text{if } \mathbb{P}(X_n = i) > 0, \\ \delta^i_j & \text{if } \mathbb{P}(X_n = i) = 0. \end{cases}$$

We call the stochastic matrix $\mathbf{P}(n)$ the *matrix of transition probabilities* or the *transition matrix* of the process $\{X_n\}$ from the nth step to the $(n+1)$th step.

For every $n \geq 0$, the matrix $\mathbf{P}(n+1)$ links the distribution of the random variables X_n and X_{n+1}. In fact, for every $n \geq 0$, let $\pi(n) := (\pi(n)_i)_{i \in S}$ be the mass density of $\mathbb{P}_{X_n}$, that is, $\pi(n)_i := \mathbb{P}(X_n = i)\ \forall i \in S$. Then

$$\pi(n+1)_j = \sum_{i \in S} \mathbb{P}\Big(X_{n+1} = j \,\Big|\, X_n = i\Big)\mathbb{P}(X_n = i) = \sum_{i \in S} \pi(n)_i \mathbf{P}(n+1)^i_j$$

i.e.

$$\pi(n+1) = \pi(n)\mathbf{P}(n+1). \tag{5.4}$$

Also notice that iterating (5.4), we get for each $n, k \geq 0$

$$\pi(n+k) = \pi(k)\mathbf{P}(k+1)\mathbf{P}(k+2)\cdots\mathbf{P}(k+n). \tag{5.5}$$

5.2.3 Homogeneous processes

Definition 5.7 *A stochastic process $\{X_n\}$ with a finite or denumerable state-space S is said to be* homogeneous *if the transition matrices of the random variables $\{X_n\}$ are independent of n, i.e. if there exists a stochastic matrix $\mathbf{P} = (\mathbf{P}^i{}_j)$, $i, j \in S$, such that for any $n \in \mathbb{N}$ and for any $(i, j) \in S$ such that $\mathbb{P}(X_n = i) > 0$, we have*

$$\mathbb{P}\Big(X_{n+1} = j \,\Big|\, X_n = i\Big) = \mathbf{P}^i_j.$$

Thus, if $\{X_n\}$ is a homogeneous stochastic process with transition matrix $\mathbf{P}$ with a finite or denumerable state-space S, then for every $n, k \geq 0$, the mass densities $\pi(n)$ and $\pi(n+k)$ of X_n and X_{n+k}, respectively, are related by

$$\pi(n+k) = \pi(k)\mathbf{P}^n.$$

5.2.4 Markov chains

5.2.4.1 Markov property

Let us start with a definition.

Definition 5.8 (Markov chain) *A stochastic process $\{X_n\}$ with a finite or denumerable state-space S is said to have the* Markov property *if for any positive* *integers k, n, $k \leq n$, and for any choice of states $i_0, \dots, i_{n+1}$ in S we have*

$$\begin{aligned}\mathbb{P}\Big(X_{n+1} = i_{n+1} \,\Big|\, X_n = i_n, \dots, X_{k+1} = i_{k+1}, X_k = i_k\Big) \\ = \mathbb{P}\Big(X_{n+1} = i_{n+1} \,\Big|\, X_n = i_n\Big)\end{aligned} \tag{5.6}$$

whenever the involved conditional probabilities are defined, i.e. whenever the conditioning events have positive probabilities. If $\{X_n\}$ has the Markov property, we say that $\{X_n\}$ is a Markov chain *with state-space S.*

We now state in a more appealing way the Markov property (5.6). Recall that an event is said to be detected by $X_0, \dots X_{n-1}$ if it is detected by their joint distribution, i.e. if it is the disjoint union of events of the following kind

$$\Big\{x \in \Omega \,\Big|\, X_{n-1}(x) = i_{n-1}, \dots, X_1(x) = i_1, X_0(x) = i_0\Big\}.$$

We then state the following.

Theorem 5.9 *Let $\{X_n\}$ be a stochastic process with a finite or denumerable state-space S. Then $\{X_n\}$ is a Markov chain if and only if for any triplet of integers r, n, k with $0 \leq r < n < n + k$ and for any triplet of events G, F, E detected by $X_0, \dots, X_r$, $X_{r+1}, \dots, X_n$ and $X_{n+1}, \dots, X_{n+k}$, respectively, we have*

$$\mathbb{P}\Big(E \,\Big|\, F \cap G\Big) = \mathbb{P}\Big(E \,\Big|\, F\Big). \tag{5.7}$$

Moreover, if $\mathbb{P}(F \cap G) > 0$, then

$$\mathbb{P}\Big(E \cap F \,\Big|\, G\Big) = \mathbb{P}\Big(E \,\Big|\, F\Big)\mathbb{P}\Big(F \,\Big|\, G\Big). \tag{5.8}$$

Equality (5.7) says that from a statistical point of view, the future state of the chain depends only on the present state, memoryless of the past history. Equality (5.8) is similar to the formula for the probability of the intersection of independent events and will play a crucial role in what follows.

Proof of theorem 5.9. Choose a sequence $\{i_n\}$ of points in S, and for $n \geq 0$ set

$$A_n := \Big\{x \in \Omega \,\Big|\, X_n(x) = i_n\Big\}.$$

The Markov property, (5.6) and the law of total probability yield

$$\mathbb{P}\Big(A_{n+1} \,\Big|\, A_n \cap B\Big) = \mathbb{P}\Big(A_{n+1} \,\Big|\, A_n\Big) \tag{5.9}$$

for any $n \geq 0$ and for any event B detected by the random variables $X_0, \dots, X_{n-1}$. Clearly, (5.9) is meaningful if and only if $\mathbb{P}(A_n \cap B) > 0$. We now claim that we also have

$$\mathbb{P}\Big(A_{n+1+k} \cap \dots \cap A_{n+1} \,\Big|\, A_n \cap B\Big) = \prod_{j=n}^{k+n} \mathbb{P}\Big(A_{j+1} \,\Big|\, A_j\Big) \tag{5.10}$$

if $\mathbb{P}(A_{n+k} \cap \cdots \cap A_n \cap B) > 0$. In fact, applying (5.9) and recalling that $\mathbb{P}(E \cap F \mid G) = \mathbb{P}(E \mid F \cap G)\mathbb{P}(F \mid G)$ we get

$$\mathbb{P}\Big(A_{n+1+k} \cap \cdots \cap A_{n+1} \,\Big|\, A_n \cap B\Big)$$

$$= \mathbb{P}\Big(A_{n+1+k} \,\Big|\, A_{n+k} \cap \cdots \cap A_{n+1} \cap A_n \cap B\Big)$$

$$\mathbb{P}\Big(A_{n+k} \cap \cdots \cap A_{n+1} \,\Big|\, A_n \cap B\Big)$$

$$= \mathbb{P}\Big(A_{n+1+k} \,\Big|\, A_{n+k}\Big)\mathbb{P}\Big(A_{n+k} \cap \cdots \cap A_{n+1} \,\Big|\, A_n \cap B\Big).$$

An induction argument thus gives (5.10).

Equality (5.7) is an immediate consequence of (5.10) and of the law of total probability. Moreover, (5.8) follows from (5.7) since

$$\mathbb{P}(E \cap F \mid G) = \mathbb{P}(E \mid F \cap G)\mathbb{P}(F \cap G \mid G)$$

$$= \mathbb{P}(E \mid F \cap G)\mathbb{P}(F \mid G) = \mathbb{P}(E \mid F)\mathbb{P}(F \mid G).$$

We point out that equalities in (5.10) and the Markov property are in fact special cases of (5.7). Finally, we mention another consequence of (5.7). Let $\{X_n\}$, r, k, n, and E, G be as in Theorem 5.9 and let $\{F_i\}_{i \in I}$ be a family of disjoint events detected by $X_{r+1}, \dots, X_n$ and $F = \cup_{i \in I} F_i$. Then using the law of total probability, we have

$$\mathbb{P}\Big(E \cap (\cup_i F_i) \,\Big|\, G\Big) = \sum_{i \in I} \mathbb{P}(E \mid F_i)\mathbb{P}(F_i \mid G). \tag{5.11}$$

Let $\{X_n\}$ be a stochastic process with a finite or denumerable state-space S. As we have seen, the transition matrices $\{\mathbf{P}(n)\}$ of the process allow to compute for any $n \geq 0$ the mass density $\pi(n+1)$ of X_{n+1} in terms of the mass density $\pi(n)$ of X_n by $\pi(n+1) = \pi(n)\mathbf{P}(n+1)$. Thus by an induction argument the mass density of X_{n+1} can be computed in terms of the mass density of X_0,

$$\pi(n+1) = \pi(0)\mathbf{P}(1)\mathbf{P}(2)\cdots\mathbf{P}(n+1).$$

However, in general, the sequence $\{\mathbf{P}(n)\}$ does not contain enough information to compute the joint distribution of two of the X_n's: for instance, in general, we cannot compute from $\mathbf{P}$ the probability $\mathbb{P}(X_n = j \mid X_0 = i)$ for $n \geq 2$ and fixed i and j. Examples can be given of different stochastic processes having the same initial state, the same transition matrices but different joint distributions.

A dramatic reduction of complexity occurs for Markov chains: *if* $\{X_n\}$ *is a Markov chain, then for every* $n \geq 1$ *the joint distribution of* $X_0, \dots, X_n$ *can be computed in terms of the transition matrices* $\mathbf{P}(1), \dots, \mathbf{P}(n)$ *and of the mass density of* X_0. In fact, for any choice of $i_0, \dots, i_n \in S$, set

$A_j := \{x \in \Omega \mid X_j(x) = i_j\}$. From (5.10) or (5.7) we then infer

$$\begin{aligned}\mathbb{P}\Big(X_{n+k} = i_{n+k}, \dots, X_{n+1} = i_{n+1} \Big| X_n = i_n\Big) \\ = \mathbb{P}\Big(A_{n+k} \cap \dots \cap A_{n+1} \Big| A_n\Big) = \prod_{j=n}^{n+k-1} \mathbb{P}\Big(A_{j+1} \Big| A_j\Big) \\ = \mathbf{P}(n)^{i_n}_{i_{n+1}} \mathbf{P}(n+1)^{i_{n+1}}_{i_{n+2}} \cdots \mathbf{P}(n+k)^{i_{n+k-1}}_{i_{n+k}}\end{aligned} \tag{5.12}$$

whenever $\mathbf{P}(A_{n+k} \cap \dots \cap A_n) > 0$. In particular,

$$\begin{aligned}\mathbb{P}\Big(X_k = i_k, \dots, X_1 = i_1, X_0 = i_0\Big) \\ = \pi(0)_{i_0} \mathbf{P}(1)^{i_0}_{i_1} \mathbf{P}(2)^{i_1}_{i_2} \cdots \mathbf{P}(k)^{i_{k-1}}_{i_k}.\end{aligned} \tag{5.13}$$

5.2.4.2 Homogeneity

A Markov chain $\{X_n\}$ with a finite of denumerable state-space is *homogeneous* if $\mathbf{P}(n) = \mathbf{P}\ \forall n$. Therefore, (5.13) reads

$$\begin{aligned}\mathbb{P}\Big(X_{k+n} = i_k, \cdots, X_{n+1} = i_1 \Big| X_n = i_0\Big) \\ = \prod_{j=0}^{k-1} \mathbf{P}^{i_j}_{i_{j+1}} = \mathbf{P}^{i_0}_{i_1} \mathbf{P}^{i_1}_{i_2} \cdots \mathbf{P}^{i_{k-1}}_{i_k}.\end{aligned} \tag{5.14}$$

As a consequence, since the right-hand side of (5.14) is independent of n, we then deduce for any $n, k \geq 0$ and for any $i_0, \dots, i_n \in S$ the equality

$$\begin{aligned}\mathbb{P}\Big(X_{n+k} = i_k, \cdots, X_{n+1} = i_1 \Big| X_n = i_0\Big) \\ = \mathbb{P}\Big(X_k = i_k, \cdots, X_1 = i_1 \Big| X_0 = i_0\Big).\end{aligned} \tag{5.15}$$

This is the *renewal property of homogeneous Markov chains*: if $\mathbb{P}(X_n = i_0) = \mathbb{P}(X_0 = i_0)$, then the joint distribution of the k variables following X_n is the same as the joint distribution of the first k variables. In words, the probabilistic knowledge of X_n, that is of $\mathbb{P}(X_n = i_0)$, *restarts* the chain: the stochastic process $\{X_{n+k}\}_k$ is a Markov chain with the same transition matrix $\mathbf{P}$ of $\{X_k\}_k$.

5.2.4.3 Strong Markov property

Definition 5.10 *Let $\{X_n\}$ be a Markov chain on $(\Omega, \mathcal{E}, \mathbb{P})$ with a finite or denumerable state-space S. A random variable $T : \Omega \to \mathbb{N} \cup \{+\infty\}$ is called a* stopping time *for $\{X_n\}$ if for any non-negative integer n, the set*

$$\Big\{x \in \Omega \Big| T(x) = n\Big\} \tag{5.16}$$

is detected by the random variables $X_0, X_1, \dots, X_n$.

Trivially, each of the following are stopping times:

- all constants;
- the *first passage time*, i.e. the step at which the first visit in $A \subset S$ occurs;
- the *kth passage time*, i.e. the step at which the k visit in $A \subset S$ occurs;
- the first step at which we leave $A \subset S$,

Observe that the step at which we leave a state j definitively, called the *last exit time*, is not a stopping time.

Theorem 5.11 (Strong Markov property) *Let $\{X_n\}$ be a homogeneous Markov chain with values in a finite or denumerable set S defined on a probability space $(\Omega, \mathcal{E}, \mathbb{P})$. Let T be a stopping time for $\{X_n\}$. Then, for any $i, j \in S$ and any $n \geq 1$*

$$\mathbb{P}\Big(X_{T+n} = j \,\Big|\, T < +\infty, X_T = i\Big) = \mathbb{P}\Big(X_n = j \,\Big|\, X_0 = i\Big).$$

That is, the sequence $\{Y_n\}$, $Y_n(x) := X_{T(x)+n}(x)$ restricted to the event $\{x \in \Omega \,|\, T < +\infty\} \subset \Omega$ is a homogeneous Markov chain whose transition matrix is the transition matrix of $\{X_n\}$. By abuse of language, one says that the chain 'restarts' at each stopping time.

Proof. By the Markov property and homogeneity, for each integer h we get

$$\begin{aligned}
\mathbb{P}\Big(X_{T+n} = j, T = h, X_T = i\Big) &= \mathbb{P}\Big(X_{h+n} = j, T = h, X_h = i\Big) \\
&= \mathbb{P}\Big(X_{h+n} = j \,\Big|\, X_h = i, T = h\Big)\mathbb{P}\Big(T = h, X_h = i\Big) \\
&= \mathbb{P}\Big(X_{h+n} = j \,\Big|\, X_h = i\Big)\mathbb{P}\Big(T = h, X_h = i\Big) \\
&= \mathbb{P}\Big(X_n = j \,\Big|\, X_0 = i\Big)\mathbb{P}\Big(T = h, X_h = i\Big).
\end{aligned}$$

Thus, summing with respect to h, we conclude

$$\begin{aligned}
&\mathbb{P}\Big(X_{T+n} = j, T < +\infty, X_T = i\Big) \\
&\qquad = \mathbb{P}\Big(X_n = j \,\Big|\, X_0 = i\Big)\sum_{h=0}^{\infty}\mathbb{P}\Big(T = h, X_T = i\Big) \\
&\qquad = \mathbb{P}\Big(X_n = j \,\Big|\, X_0 = i\Big)\mathbb{P}\Big(T < +\infty, X_T = i\Big).
\end{aligned}$$

5.2.5 Canonical Markov chains

Example 5.12 A typical example which may help intuition is that of *random walks*. A person is at a random position k, $k \in \mathbb{Z}$, and at each step moves either

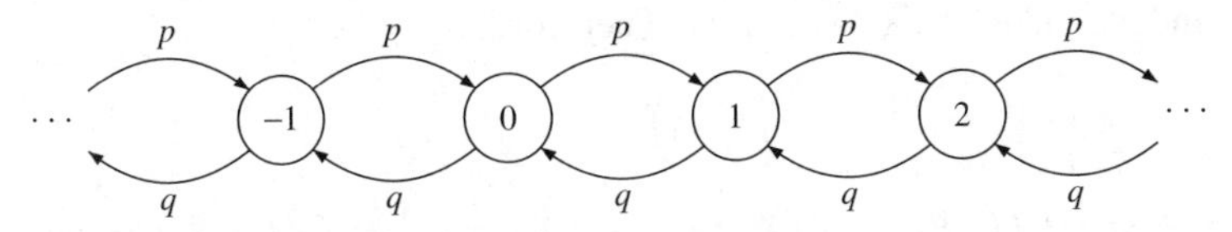

Figure 5.2 A graph representing the transition matrix of an unbounded random walk.

to the position $k-1$ or to the position $k+1$ according to a Bernoulli trial of parameter p, for example by flipping a coin. Let X_n be the position occupied at the nth step and let $\{\zeta_n\}$ be a sequence of *independent* random variables that follow $B(1, p)$ and such that ζ_n decides if the person moves backward or foward at step n. Then

$$X_{n+1} = X_n + 2\zeta_n - 1$$

so that $\{X_n\}$ is a Markov chain, see Theorem 5.13, whose transition matrix is

$$\mathbf{P} = \begin{pmatrix} \dots & \dots & \dots & \dots & \dots & \dots & \dots \\ \dots & p & 0 & 0 & 0 & 0 & \dots \\ \dots & 0 & p & 0 & 0 & 0 & \dots \\ \dots & q & 0 & p & 0 & 0 & \dots \\ \dots & 0 & q & 0 & p & 0 & \dots \\ \dots & 0 & 0 & q & 0 & p & \dots \\ \dots & 0 & 0 & 0 & q & 0 & \dots \\ \dots & 0 & 0 & 0 & 0 & q & \dots \\ \dots & \dots & \dots & \dots & \dots & \dots & \dots \end{pmatrix}$$

Matrix $\mathbf{P}$ is schematically represented in Figure 5.2.

Example 5.12 is indeed a standard way to construct Markov chains. The following holds.

Theorem 5.13 *Let S be a finite or denumerable set, and let $X_0 : \Omega \to S$ be a random variable on $(\Omega, \mathcal{E}, \mathbb{P})$. Let $\{\xi_n\}$, $\xi_n : \Omega \to \mathbb{R}^N$ be a sequence of independent, identically distributed random variables on $(\Omega, \mathcal{E}, \mathbb{P})$ that are also independent of X_0 and let $f : S \times \mathbb{R}^N \to S$ be a Borel function. Then the sequence $\{X_n\}$, $X_n : \Omega \to S$ defined by*

$$X_{n+1}(x) := f(X_n(x), \xi_n(x)), \qquad \forall n \geq 0, \tag{5.17}$$

is a homogeneous Markov chain with state-space S and transition matrix

$$\mathbf{P}^i_j := \mathbb{P}(f(i, \xi_k) = j) \qquad \forall i, j \in S.$$

Proof. From the definition, it is clear that for any integer k, the random variable X_k is a function of X_0 and $\xi_1, \dots, \xi_{k-1}$. Consequently, for any integer n, the random variable ξ_n, which is by definition independent of $(X_0, \xi_1, \dots, \xi_{n-1})$, is also independent of $(X_0, \dots, X_n)$. Therefore,

$$\begin{aligned}\mathbb{P}\Big(X_{n+1} = j \,\Big|\, X_n = i, \dots, X_0 = i_0\Big) \\ = \mathbb{P}\Big(f(i, \xi_n) = j \,\Big|\, X_n = i, \dots, X_0 = i_0\Big) = \mathbb{P}(f(i, \xi_n) = j);\end{aligned} \tag{5.18}$$

the last equality holds since the random variable $(X_0, \dots, X_n)$ is independent of ξ_n, hence of $f(i, \xi_n)$. Since the right-hand side of (5.18) is independent of the values of $X_{n-1}, \dots, X_0$, we conclude that

$$\mathbb{P}\Big(X_{n+1} = j \,\Big|\, X_n = i, \dots, X_0 = i_0\Big) = \mathbb{P}\Big(X_{n+1} = j \,\Big|\, X_n = i\Big),$$

i.e. $\{X_n\}$ is a Markov chain. The transition matrix is then $\mathbf{P} = (\mathbf{P}^i{}_j)$, $\mathbf{P}^i{}_j := \mathbb{P}(f(i, \xi_n) = j)$ and is independent of n since the random variables ξ_n are identically distributed.

Theorem 5.14 *Let S be a finite or denumerable set, and let $\mathbf{P}$ be a stochastic matrix. Let $X_0 : \Omega \to S$ be a random variable on $(\Omega, \mathcal{E}, \mathbb{P})$ and let $\{\xi_n\}$, $\xi_n : \Omega \to [0, 1]$ be a sequence of independent, uniformly distributed random variables that are also independent of X_0. Define*

$$f(i, s) := \min\left\{ j \,\middle|\, \sum_{h=1}^{j} \mathbf{P}^i_h \geq s \right\} \qquad \forall i > 0, s \in \mathbb{R}.$$

Then the sequence $\{X_n\}$, $X_n : \Omega \to S$ defined by

$$X_{n+1}(x) = f(X_n(x), \xi_n(x)), \quad x \in \Omega, n \geq 0,$$

is a homogeneous Markov chain with state-space S and transition matrix $\mathbf{P}$.

Proof. By Theorem 5.13, the sequence $\{X_n\}$ is a Markov chain. Moreover, $f(i, \xi_n(x)) = j$ if and only if

$$\sum_{h=1}^{j-1} \mathbf{P}^i_h < \xi_n(x) \leq \sum_{h=1}^{j} \mathbf{P}^i_h$$

hence

$$\mathbb{P}(f(i, \xi_n(x)) = j) = \sum_{h=1}^{j} \mathbf{P}^i_h - \sum_{h=1}^{j-1} \mathbf{P}^i_h = \mathbf{P}^i{}_j.$$

The previous theorem constructs in fact a homogeneous Markov chain $\{X_n\}$ with a given transition matrix and a given initial data X_0. In particular, Theorem 5.14 reduces the problem of the existence of a Markov chain with a given transition matrix $\mathbf{P}$ on a probability space $(\Omega, \mathcal{E}, \mathbb{P})$ and a given initial data X_0 to the existence of a sequence of independent random variables that are uniformly distributed on [0, 1], see Sections 2.2.5 and 4.4.4.

5.2.6 Exercises

Exercise 5.15 *Let S be a finite or denumerable set and assume X and Y are random variables taking values in S. Show that X and Y are independent if and only if $\forall(i, j) \in S \times S$ if $\mathbb{P}(X = i) > 0$ then the probability $\mathbb{P}(Y = j \mid X = i)$ does not depend on i.*

Exercise 5.16 *Let $\{\mathbf{P}(n)\}$ be the sequence of transition matrices of a Markov chain $\{X_n\}$ with values in a finite or denumerable set S. For $i, j \in S$ and $n, k \geq 0$ compute $\mathbb{P}\Big(X_{n+k} = j \,\Big|\, X_k = i\Big)$.*

Solution. Applying (5.11) we get

$$\begin{aligned}
&\mathbb{P}\Big(X_{k+2} = j \,\Big|\, X_k = i\Big) \\
&\qquad = \sum_{h \in S} \mathbb{P}\Big(X_{k+2} = j \,\Big|\, X_{k+1} = h\Big)\mathbb{P}\Big(X_{k+1} = h \,\Big|\, X_k = i\Big) \\
&\qquad = (\mathbf{P}(k+1)\mathbf{P}(k+2))^i_j.
\end{aligned}$$

Thus, with an induction argument, we have

$$\mathbb{P}\Big(X_{n+k} = j \,\Big|\, X_k = i\Big) = \Big(\mathbf{P}(k+1)\cdots\mathbf{P}(k+n)\Big)^i_j.$$

For a homogeneous Markov chain, $\mathbf{P}(n) := \mathbf{P}\ \forall n$, so that

$$\mathbb{P}\Big(X_{n+k} = j \,\Big|\, X_k = i\Big) = (\mathbf{P}^n)^i_j. \tag{5.19}$$

Exercise 5.17 *Let $\{X_n\}$ be a sequence of random variables with values in a finite or denumerable set S. Prove the following.*

(i) If the random variables X_n are independent, then $\{X_n\}$ is a Markov chain.

(ii) If $\{X_n\}$ is a Markov chain and for each integer n, the random variables X_n and X_{n+1} are independent, then all the X_n's are independent.

Exercise 5.18 *Assume you need a certain piece of machinery for your work. Every time you switch on this machinery it may either be broken or start working. At the beginning of each working day, you switch it on. Assume the following:*

(i) At the first switch on, the probability that the machinery works is a, and the probability that it is broken is $b = 1 - a$.

(ii) At each of the subsequent times it is switched on:

- *If the machinery is broken, then it will be fixed for the next day with probability q.*

- *If the machinery works, then it is going to break down without being repaired on time for the next day with probability p.*

Describe the process.

Solution. Let X_n denote the state of the machinery at the nth switch on. The state-space has only two points (working or broken), say $S = \{0, 1\}$: $X_n = 0$ if the machinery is broken, $X_n = 1$ otherwise. The process is evidently a homogeneous Markov chain: what happens at each switch on depends only on the previous switch on. The process is then described by the initial condition,

$$\mathbb{P}(X_0 = 0) = 1 - a, \qquad \mathbb{P}(X_0 = 1) = a$$

and by the transition matrix

$$\mathbf{P} = \begin{pmatrix} 1-q & q \\ p & 1-p \end{pmatrix}. \tag{5.20}$$

Matrix $\mathbf{P}$ is schematically represented in Figure 5.3.

The probabilities that the machinery is either broken or not on the nth day are the components of the vector $\pi(n) = (\mathbb{P}(X_n = 0), \mathbb{P}(X_n = 1))$ and

$$\begin{cases} \pi(n+1) = \pi(n)\mathbf{P} & \forall n \geq 0, \\ \pi(0) = (b, a). \end{cases}$$

i.e. $\pi(n) = \pi(0)\mathbf{P}^n$.

Let us compute $\mathbf{P}^n$. If $p = q = 0$, then obviously

$$\mathbf{P}^n = \begin{pmatrix} 1 & 0 \\ 0 & 1 \end{pmatrix} \qquad \forall n.$$

If either $p > 0$ or $q > 0$, then $\mathbf{P}$ has two different eigenvalues. One can compute, or prove by induction, that

$$\mathbf{P}^n = \frac{1}{p+q}\left(\begin{pmatrix} p & q \\ p & q \end{pmatrix} + (1-p-q)^n \begin{pmatrix} q & -q \\ -p & p \end{pmatrix}\right). \tag{5.21}$$

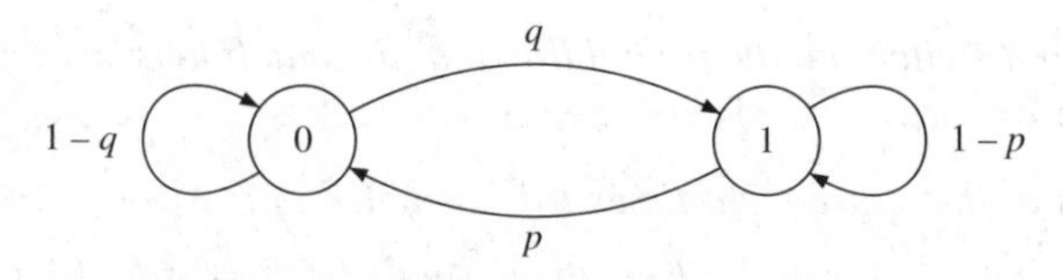

Figure 5.3 A graph representing the transition matrix in (5.20).

In particular, the probability that the machinery is working at the nth switch on, knowing that it was working at the first switch on is

$$\mathbb{P}\big(X_n = 1 \mid X_0 = 1\big) = \pi(n)_1 = \frac{1}{p+q}\Big((ap+bq) + (1-p-q)^n(ap-bq)\Big)$$
$$= \frac{1}{p+q}\Big(q + (1-p-q)^n(ap-bq)\Big).$$

Let us compute the *steady state* or the *asymptotic behaviour* of the chain, i.e. the limit of $\pi(0)\mathbf{P}^n$ as $n \to \infty$.

- If $(p, q) = (0, 0)$, then $\mathbf{P}^n = \mathrm{Id}\ \forall n$, so that the probabilities that the machinery is either broken or working are the same at every switch on.
- If $(p, q) \neq (0, 0)$ and $(p, q) \neq (1, 1)$, then $p + q < 2$, hence $|1 - p - q| < 1$, thus, by (5.21) $\mathbf{P}^n$ converges to the matrix

$$\mathbf{W} := \frac{1}{p+q}\begin{pmatrix} p & q \\ p & q \end{pmatrix}.$$

  The asymptotic probabilities are $\pi(0)\mathbf{W} = \left(\frac{p}{p+q}, \frac{q}{p+q}\right)$ which are independent of the initial probabilities a, $1 - a$.

- If $p = q = 1$, then $\mathbf{P} = \begin{pmatrix} 0 & 1 \\ 1 & 0 \end{pmatrix}$ and

$$\mathbf{P}^n = \begin{cases} \begin{pmatrix} 1 & 0 \\ 0 & 1 \end{pmatrix} & \text{if } n \text{ is even,} \\ \begin{pmatrix} 0 & 1 \\ 1 & 0 \end{pmatrix} & \text{if } n \text{ is odd.} \end{cases}$$

  In this case the limit of $\mathbf{P}^n$ as $n \to \infty$ does not exist.

Exercise 5.19 (Random walks) *Let us go back to Example 5.12 and let us assume that the set of all possible positions is finite, $S = \{0, \ldots, N\}$. We must specify how to proceed when $X_n = 0$ or $X_n = N$. Typical examples are an* absorbing walk *as in Figure 5.4, a* reflecting walk *as in Figure 5.5 or a* cyclic walk *as in Figure 5.6. For each of the figures, write the corresponding transition matrix.*

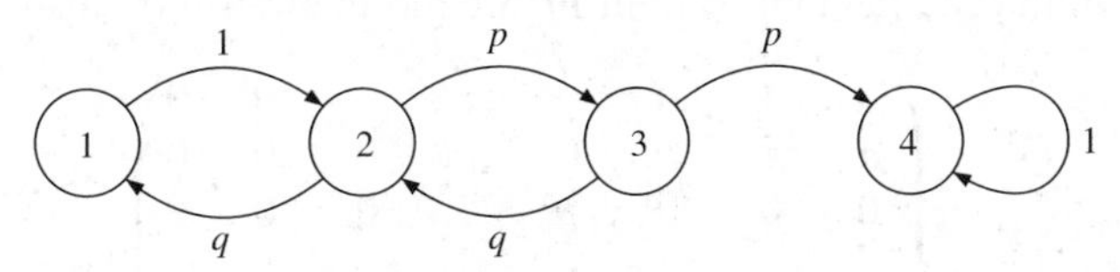

Figure 5.4 A graph representing the transition matrix of an absorbing Markov chain.

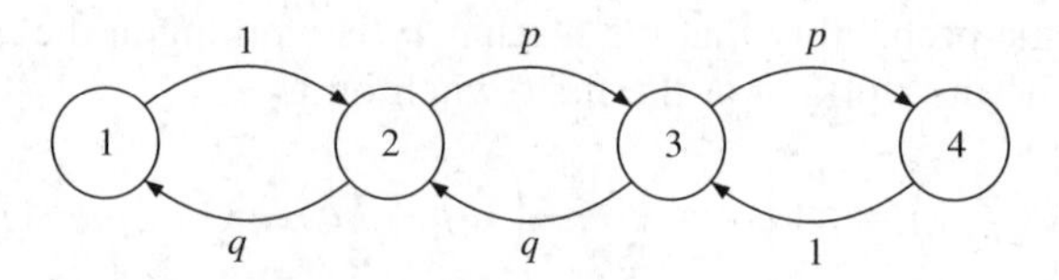

Figure 5.5 *A graph representing the transition matrix of a reflecting chain.*

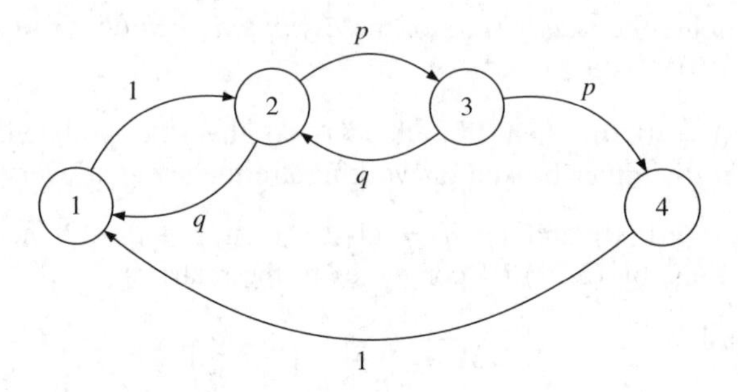

Figure 5.6 *A graph representing the transition matrix of a cyclic chain.*

Solution. For the random walk in Figure 5.4, which is absorbing at the right end, the transition matrix is

$$\mathbf{P} = \begin{pmatrix} 0 & 1 & 0 & 0 & \ldots & 0 & 0 & 0 \\ q & 0 & p & 0 & \ldots & 0 & 0 & 0 \\ 0 & q & 0 & p & \ldots & 0 & 0 & 0 \\ \ldots & \ldots & \ldots & \ldots & \ldots & \ldots & \ldots & \\ 0 & 0 & 0 & 0 & \ldots & q & 0 & p \\ 0 & 0 & 0 & 0 & \ldots & 0 & 0 & 1 \end{pmatrix}.$$

For the reflecting random walk in Figure 5.5 the transition matrix is

$$\mathbf{P} = \begin{pmatrix} 0 & 1 & 0 & 0 & \ldots & 0 & 0 & 0 \\ q & 0 & p & 0 & \ldots & 0 & 0 & 0 \\ 0 & q & 0 & p & \ldots & 0 & 0 & 0 \\ \ldots & \ldots & \ldots & \ldots & \ldots & \ldots & \ldots & \\ 0 & 0 & 0 & 0 & \ldots & q & 0 & p \\ 0 & 0 & 0 & 0 & \ldots & 0 & 1 & 0 \end{pmatrix}.$$

Finally, for the cyclic random walk in Figure 5.6 the transition matrix is

$$\mathbf{P} = \begin{pmatrix} 0 & 1 & 0 & 0 & \ldots & 0 & 0 & 0 \\ q & 0 & p & 0 & \ldots & 0 & 0 & 0 \\ 0 & q & 0 & p & \ldots & 0 & 0 & 0 \\ \ldots & \ldots & \ldots & \ldots & \ldots & \ldots & \ldots & \\ 0 & 0 & 0 & 0 & \ldots & q & 0 & p \\ 1 & 0 & 0 & 0 & \ldots & 0 & 0 & 0 \end{pmatrix}.$$

Exercise 5.20 (Ehrenfest diffusion model) *Assume that c molecules, $c \geq 1$, are distributed in two boxes A and B. At each step a molecule is randomly chosen (with probability $1/c$ for each molecule) and moved from its box to the other one. Describe the process.*

Solution. The state of the system is described by the number of molecules in the box A. If, at a certain step, A contains i molecules, then at the next step we pick a molecule from A with probability i/c or from B with probability $1 - i/c$ and we move it to the other box. Therefore, A will contain either $i - 1$ molecules with probability i/c or $i + 1$ molecules with probability $1 - i/c$. Since the state at each step depends only on the state at the previous step, the process is a Markov chain, see Figure 5.7. The transition matrix is the same at each step and is given by the $(c + 1) \times (c + 1)$ matrix

$$\mathbf{P} = \begin{pmatrix} 0 & 1 & 0 & 0 & \dots & 0 & 0 \\ \frac{1}{c} & 0 & 1-\frac{1}{c} & 0 & \dots & 0 & 0 \\ 0 & \frac{2}{c} & 0 & 1-\frac{2}{c} & \dots & 0 & 0 \\ \dots & \dots & \dots & \dots & \dots & \dots & \dots \\ 0 & 0 & 0 & 0 & \dots & 0 & \frac{1}{c} \\ 0 & 0 & 0 & 0 & \dots & 1 & 0 \end{pmatrix}.$$

Exercise 5.21 *A repair workshop receives Z_n objects to be repaired and repairs one object per unit of time. At each time n, let X_n be the number of broken objects which are in the workshop. Describe the process $\{X_n\}$.*

Solution. Clearly, $X_{n+1} = \max(X_n - 1, 0) + Z_{n+1}$ $\forall n \geq 0$. If the random variables $\{Z_n\}$ are independent, identically distributed and also independent of X_0, then $\{X_n\}$ is a homogeneous Markov chain, see Theorem 5.13. Let us compute the transition matrix. If $i = 0$ (no object to be repaired is in the workshop)

$$\mathbf{P}^0_j = \mathbb{P}\Big(X_1 = j \,\Big|\, X_0 = 0\Big) = \mathbb{P}\Big(Z_1 = j \,\Big|\, X_0 = 0\Big) = \mathbb{P}(Z_1 = j).$$

If $i = 1$, then

$$\mathbf{P}^1_j = \mathbb{P}\Big(X_1 = j \,\Big|\, X_0 = 1\Big) = \mathbb{P}\Big(Z_1 = j \,\Big|\, X_0 = 1\Big) = \mathbb{P}(Z_1 = j)$$

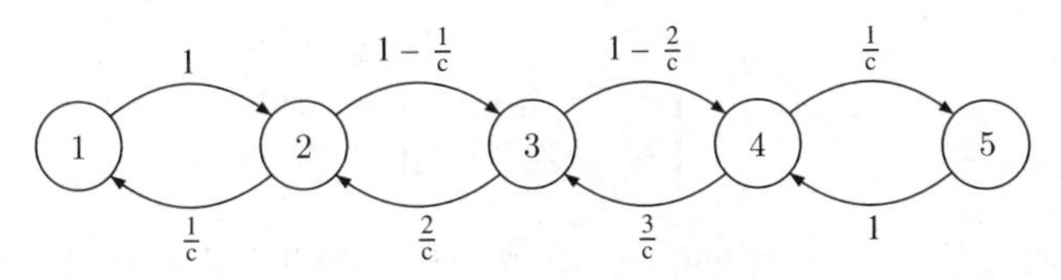

Figure 5.7 A graph representing the transition matrix of the Ehrenfest diffusion model with $c = 4$.

and, if $i > 1$, then

$$\mathbf{P}^1_j = \mathbb{P}\left(X_1 = j \,\middle|\, X_0 = i\right)$$
$$= \mathbb{P}\left(Z_1 = j - i + 1 \,\middle|\, X_0 = i\right) = \mathbb{P}(Z_1 = j - i + 1).$$

Therefore, setting $a_j := \mathbb{P}(Z_1 = j)$ the transition matrix of $\{X_n\}$ is

$$\mathbf{P} = \begin{pmatrix} a_0 & a_1 & a_2 & a_3 & \dots \\ a_0 & a_1 & a_2 & a_3 & \dots \\ 0 & a_0 & a_1 & a_2 & \dots \\ 0 & 0 & a_0 & a_1 & \dots \\ \dots & \dots & \dots & \dots & \dots \end{pmatrix}.$$

Exercise 5.22 (Queues) *In a shop customers arrive at the pay desk at times $0, 1, \dots$. Let ξ_n be the number of customers arriving at the desk at time n. At each time one customer (if there is at least one) pays and leaves the shop. Let $\{X_n\}$ be the number of customers queuing at the pay desk at time n. Describe the process.*

Solution. Clearly,

$$X_{n+1} = \begin{cases} X_n + \xi_{n+1} - 1 & \text{if } X_n \geq 1, \\ \xi_n & \text{if } X_n = 0. \end{cases}$$

or, equivalently, $X_n := \max(X_n - 1, 0) + \xi_{n+1}$. It is the same process as described in Exercise 5.21.

Exercise 5.23 (Inventory) *A logistic company manages the stock of a certain good with the following strategy: the company fixes two quantities m and M, $m < M$; every day the company delivers the orders and at the end of the day the people in charge check the stock. If, after delivering all the pending orders, the stock is more than m they do not do anything. Otherwise, they buy immediately enough supplies to fulfil all the pending orders and get the stock back to level M. Describe the process, assuming that the initial stock amount is less than M.*

Solution. For every $n \geq 0$, let Z_n be the aggregated demand of goods at day n and let X_n be the current stock at the end of day n. Then $Z_n \geq 0$, $X_0 \leq M$ and, clearly,

$$X_{n+1} = \begin{cases} X_n - Z_{n+1} & \text{if } X_n > m, \\ M - Z_{n+1} & \text{if } X_n \leq m, \end{cases}$$

for every $n \geq 0$. In particular, $X_n \leq M$ $\forall n$. Assuming that the aggregated demands Z_n are independent, identically distributed random variables and also independent of the initial stock X_0, then $\{X_n\}$ is a homogeneous Markov chain, see Theorem 5.13. Let us compute the transition matrix of the $\{X_n\}$. If $-\infty < i \le m$, then for every $-\infty < j \le M$

$$\mathbf{P}^i_j = \mathbb{P}\Big(X_1 = j \,\Big|\, X_0 = i\Big) = \mathbb{P}\Big(Z_1 = M - j \,\Big|\, X_0 = i) = \mathbb{P}(Z_1 = M - j)$$

while, if $m < i \le M$, then for every $-\infty < j \le M$

$$\mathbf{P}^i_j = \mathbb{P}\Big(X_1 = j \,\Big|\, X_0 = i\Big) = \mathbb{P}\Big(Z_1 = i - j \,\Big|\, X_0 = i) = \mathbb{P}(Z_1 = i - j).$$

5.3 Some characteristic parameters

In this section, we compute a few characteristic parameters of homogeneous Markov chains.

5.3.1 Steps for a first visit

Let $\{X_n\}$ be a homogeneous Markov chain on $(\Omega, \mathcal{E}, \mathbb{P})$ with a finite or denumerable state-space S and let $C \subset S$. We recall that one says that $x \in \Omega$ *visits* C *at step* n if $X_n(x) \in C$. We also say that $x \in \Omega$ *visits* C if $X_n(x) \in C$ for some $n \ge 1$.

Consider the minimum number of steps to visit C,

$$t_C(x) := \min\Big\{n \ge 1 \,\Big|\, X_n(x) \in C\Big\},$$

called the *first passage time* or the *waiting time* to get to C. We set $t_C(x) = +\infty$ if $X_n(x) \notin C\ \forall n$. Trivially, the waiting time $t_C(x)$ is a random variable with values in $\{1, 2, \dots\} \cup \{+\infty\}$; its distribution is obtained by computing its mass distribution, that is the probabilities $\mathbb{P}(t_C(x) = k)$ that the waiting time to get to C occurs at step k for every k.

For any $i \in S$ we introduce the *probabilities of visiting* C at step k (*starting*) *from* i, or *first passage time probabilities*

$$\begin{aligned} f^{(k)}_{iC} &:= \mathbb{P}\Big(t_C(x) = k \,\Big|\, X_0 = i\Big) \\ &= \mathbb{P}\Big(X_k \in C, X_{k-1} \notin C, \dots, X_1 \notin C \,\Big|\, X_0 = i\Big). \end{aligned} \tag{5.22}$$

Thus, the total probability law gives

$$\mathbb{P}(t_C = k) = \sum_{i \in S} f^{(k)}_{iC}\, \mathbb{P}(X_0 = i).$$

Since the sets $E_k := \{x \in \Omega \,|\, t_C(x) = k\}$ are pairwise disjoint, the *probability of visiting* C (at least once) starting from i is

$$f_{iC} := \mathbb{P}\Big(t_C < +\infty \,\Big|\, X_0 = i\Big) = \sum_{k=1}^{\infty} \mathbb{P}\Big(t_C = k \,\Big|\, X_0 = i\Big) = \sum_{k=1}^{\infty} f^{(k)}_{iC}. \tag{5.23}$$

If C is a singleton, $C = \{j\}$, we write $t_j(x)$, $f_{ij}^{(k)}$, f_{ij} instead of $t_{\{j\}}(x)$, $f_{i\{j\}}^{(k)}$ and $f_{i\{j\}}$, respectively. If $i = j$, f_{jj} and $f_{jj}^{(k)}$ are more appropriately called the *return probability* to state j and the *return probability* to state j at step k, respectively. The return probabilities $f_{jj}^{(k)}$ play a crucial role in what follows.

Proposition 5.24 *The first passage time probabilities $f_{iC}^{(k)}$, $k \geq 1$, satisfy the recurrence property*

$$f_{iC}^{(k)} = \begin{cases} \displaystyle\sum_{\ell \in C} p_{i\ell} & \text{if } k = 1, \\ \displaystyle\sum_{\ell \notin C} p_{i\ell} f_{\ell C}^{(k-1)} & \text{if } k \geq 2. \end{cases} \tag{5.24}$$

Proof. Let us prove the claim by induction on k. For $k = 1$, observe that $f_{iC}^{(1)}$ is the probabilty to visit C at the first step starting from i,

$$f_{iC}^{(1)} = \sum_{\ell \in C} \mathbb{P}\Big(X_1 = \ell \,\Big|\, X_0 = i\Big) = \sum_{\ell \in C} p_{i\ell}.$$

Let $k \geq 2$, and set

$$E_C^k := \Big\{x \in \Omega \,\Big|\, X_1(x) \notin C, \ldots, X_{k-1}(x) \notin C, X_k(x) \in C\Big\}$$

so that $E_C^k = \big\{x \,|\, t_C(x) = k\big\}$. If a path x from i visits C at step k, then the first step is outside C, $X_1(x) \notin C$, the next $k - 2$ steps are again outside C and finally the path hits C at the kth step. Set

$$E := \Big\{x \in \Omega \,\Big|\, X_2(x) \notin C, \ldots, X_{k-1}(x) \notin C, X_k(x) \in C\Big\},$$

$$F_\ell := \Big\{x \in \Omega \,\Big|\, X_1(x) = \ell\Big\}, \ell \notin C.$$

Of course, $E_C^k = E \cap (\cup_{\ell \notin C} F_\ell)$ and, by (5.11), we get

$$\begin{aligned} f_{iC}^{(k)} &= \mathbb{P}\Big(E_C^k \,\Big|\, X_0 = i\Big) = \mathbb{P}\Big(E \cap (\cup_{\ell \notin C} F_\ell) \,\Big|\, X_0 = i\Big) \\ &= \sum_{\ell \notin C} \mathbb{P}(E \,|\, F_\ell)\mathbb{P}(F_\ell \,|\, X_0 = i). \end{aligned} \tag{5.25}$$

Since $\mathbb{P}\Big(F_\ell \,\Big|\, X_0 = i\Big) = \mathbb{P}\Big(X_1 = \ell \,\Big|\, X_0 = i\Big) = p_{i\ell}$ and

$$\begin{aligned} \mathbb{P}\Big(E \,\Big|\, F_\ell\Big) &= \mathbb{P}\Big(X_2 \notin C, \ldots, X_{k-1} \notin C, X_k \in C \,\Big|\, X_1 = \ell\Big) \\ &= \mathbb{P}\Big(X_1 \notin C, \ldots, X_{k-2} \notin C, X_{k-1} \in C \,\Big|\, X_0 = \ell\Big) \\ &= f_{\ell C}^{(k-1)} \end{aligned}$$

by the inductive assumption, (5.24) follows from (5.25).

5.3.2 Probability of (at least) r visits

Let $\{X_n\}$ be a homogeneous Markov chain on $(\Omega, \mathcal{E}, \mathbb{P})$ with a finite or denumerable state-space S. For $n \geq 1$ and $C \subset S$, consider the event $E_n := \{x \in \Omega \mid X_n(x) \in C\}$ and let $\mathbb{1}_{E_n}$ be its characteristic function. Since $\mathbb{1}_{E_n}(x) = 1$ if x visits C at step n and is zero otherwise, the random variables

$$V_C^{(n)}(x) := \sum_{k=1}^{n} \mathbb{1}_{E_k}(x) \qquad \text{and} \qquad V_C(x) := \sum_{k=1}^{\infty} \mathbb{1}_{E_k}(x) \tag{5.26}$$

give the *number of visits* of x in C in the first n steps and the total number of visits of x in C, respectively. Clearly, we may have $V_C(x) = +\infty$. Moreover,

$$\begin{aligned} &\left\{x \in \Omega \,\middle|\, V_C^{(n)}(x) \geq 1\right\} = \left\{x \in \Omega \,\middle|\, t_C \leq n\right\}, \\ &\left\{x \in \Omega \,\middle|\, V_C(x) \geq 1\right\} = \left\{x \in \Omega \,\middle|\, t_C < +\infty\right\}. \end{aligned} \tag{5.27}$$

Similarly, for each non-negative integer r the *rth passage time in* C is defined recursively by setting $T_C^{(0)}(x) = 0$, $T_C^{(1)}(x) = t_C(x)$ and, for $r \geq 2$,

$$T_C^{(r)}(x) = \inf\left\{n \geq T_C^{(r-1)}(x) + 1 \,\middle|\, X_n(x) \in C\right\}, \qquad x \in \Omega.$$

If no $n \geq T_C^{(r-1)} + 1$ such that $X_n \in C$ exists, we set $T_C^{(r)}(x) = +\infty$. The random variables $T^{(r)}(x)$ are stopping times since, for each positive integer k, the event $\{x \mid T_C^{(r)} = k\}$ is detected by the random variables $X_0, \ldots, X_k$. The relations between the random variables $V_C^{(n)}$ and $T_C^{(r)}$ and between the random variables V_C and T_C are quite clear:

$$\begin{aligned} &\left\{x \in \Omega \,\middle|\, V_C^{(n)}(x) \geq r\right\} = \left\{x \in \Omega \,\middle|\, T_C^{(r)} \leq n\right\}, \\ &\left\{x \in \Omega \,\middle|\, V_C(x) \geq r\right\} = \left\{x \in \Omega \,\middle|\, T_C^{(r)} < +\infty\right\}. \end{aligned}$$

If C is a singleton, $C = \{j\}$, we use the shorter notation $T_j^{(r)}$, T_j, $V_j^{(n)}$, V_j instead of $T_{\{j\}}^{(r)}$, $T_{\{j\}}$, $V_{\{j\}}^{(n)}$, $V_{\{j\}}$, respectively. Finally, we point out that

$$V_C^{(n)}(x) = \sum_{j \in C} V_j^{(n)}(x) \qquad \text{and} \qquad V_C(x) = \sum_{j \in C} V_j(x).$$

Let $\{X_n\}$ be a Markov chain on $(\Omega, \mathcal{E}, \mathbb{P})$ with a finite or denumerable state-space S. For any $i, j \in S$ and $r \geq 1$, we look for the probabilities of visiting j at least r times starting from i:

$$\mathbb{P}\left(V_j \geq r \,\middle|\, X_0 = i\right).$$

Theorem 5.25 *For any positive integer r and for any pair of states $i, j \in S$ we have*

$$\mathbb{P}\Big(V_j \geq r \,\Big|\, X_0 = i\Big) = f_{ij}\, f_{jj}^{r-1}.$$

In particular, the probability of having at least r visits in j starting from j is

$$\mathbb{P}\Big(V_j \geq r \,\Big|\, X_0 = j\Big) = f_{jj}^{r}. \tag{5.28}$$

Proof. If $r = 1$, then (5.23) and (5.27) yield

$$\mathbb{P}\Big(V_j \geq 1 \,\Big|\, X_0 = i\Big) = \mathbb{P}\Big(t_j < +\infty \,\Big|\, X_0 = i\Big) = f_{ij}.$$

Assume now $r \geq 2$. Let $E_n := \big\{x \in \Omega \,|\, X_n(x) = j\big\}$ and, for $h \geq 1$, let $R_h(x) := \sum_{k=h+1}^{\infty} \mathbb{1}_{E_k}(x)$ so that

$$\Big\{x \in \Omega \,\Big|\, V_j(x) \geq r\Big\} = \bigcup_{h=1}^{\infty} \Big\{x \in \Omega \,\Big|\, R_h(x) \geq r-1, t_j(x) = h\Big\} \qquad \forall r \geq 2.$$

Set $E := \big\{x \in \Omega \,|\, R_h(x) \geq 2\big\}$, $F = \big\{x \in \Omega \,|\, t_j(x) = h\big\}$ and $G = \{x \in \Omega \,|\, X_0 = i\}$. From the disintegration formula, the renewal property (5.15) and the homogeneity of the chain, we obtain

$$\begin{aligned}
&\mathbb{P}\Big(R_h \geq r-1, t_j = h \,\Big|\, X_0 = i\Big)\\
&\quad = \mathbb{P}\Big(E \cap F \,\Big|\, G\Big) = \mathbb{P}\Big(E \,\Big|\, F \cap G\Big)\mathbb{P}\Big(F \cap G\Big)\\
&\quad = \mathbb{P}\Big(R_h \geq r-1 \,\Big|\, t_j = h, X_0 = i\Big)\mathbb{P}\Big(t_j = h \,\Big|\, X_0 = i\Big)\\
&\quad = \mathbb{P}\Big(R_h \geq r-1 \,\Big|\, X_h = j\Big)\mathbb{P}\Big(t_j = h \,\Big|\, X_0 = i\Big)\\
&\quad = \mathbb{P}\Big(V_j \geq r-1 \,\Big|\, X_0 = j\Big) f_{ij}^{(h)}.
\end{aligned}$$

Thus,

$$\mathbb{P}\Big(V_j \geq r \,\Big|\, X_0 = i\Big) = \sum_{h=1}^{\infty} \mathbb{P}\Big(R_h \geq r-1, t_j = h \,\Big|\, X_0 = i\Big)$$

$$= \Big(\sum_{h=1}^{\infty} f_{ij}^{h}\Big)\mathbb{P}\Big(V_j \geq r-1 \,\Big|\, X_0 = j\Big) = f_{ij}\, \mathbb{P}\Big(V_j \geq r-1 \,\Big|\, X_0 = j\Big).$$

Then the claim follows by an induction argument.

5.3.3 Recurrent and transient states

Let $\{X_n\}$ be a homogeneous Markov chain on $(\Omega, \mathcal{E}, \mathbb{P})$ with a finite or denumerable state-space S. For any $i, j \in S$, we now compute the *expected number of visits* in the state j starting from the state i, i.e. the averaged number of visits with respect to the probability measure $\mathbb{P}_i(A) := \mathbb{P}(A \mid X_0 = i)$, i.e.

$$\mathbb{E}_i[V_j] := \frac{1}{\mathbb{P}(X_0 = i)} \int_{\{X_0=i\}} V_j(x)\, \mathbb{P}(dx).$$

Theorem 5.26 *The following hold:*

$$\mathbb{E}_i[V_j] = \begin{cases} +\infty & \text{if } \mathbb{P}\Big(V_j = +\infty \,\Big|\, X_0 = i\Big) > 0, \\ \sum_{k=0}^{\infty} k\mathbb{P}\Big(V_j = k \,\Big|\, X_0 = i\Big) & \text{otherwise.} \end{cases} \tag{5.29}$$

$$\mathbb{E}_i[V_j] = \sum_{n=0}^{\infty} p_{ij}^{(n)}. \tag{5.30}$$

$$\mathbb{E}_i[V_j] = \begin{cases} +\infty & \text{if } f_{jj} = 1, \\ \frac{f_{ij}}{1-f_{jj}} & \text{if } f_{jj} < 1. \end{cases} \tag{5.31}$$

In particular,

$$\sum_{n=0}^{\infty} p_{ij}^{(n)} = \begin{cases} +\infty & \text{if } f_{jj} = 1, \\ \dfrac{f_{ij}}{1 - f_{jj}} & \text{if } f_{jj} < 1. \end{cases} \tag{5.32}$$

Proof. The Beppo Levi theorem, see Exercise B.33, gives

$$\int_{\{X_0=i\}} V_j(x)\mathbb{P}(dx) = \begin{cases} +\infty & \text{if } \mathbb{P}\Big(V_j = +\infty, X_0 = i\Big) > 0, \\ \sum_{k=0}^{\infty} k\mathbb{P}\Big(V_j = k, X_0 = i\Big) & \text{otherwise,} \end{cases}$$

hence (5.29) is proven. Taking advantage of (5.26) and (5.19), the Beppo Levi theorem also gives

$$\begin{aligned} \mathbb{E}_i[V_j] := \mathbb{E}_i\left[\sum_{n=0}^{\infty} \mathbb{1}_{E_n}\right] &= \sum_{n=0}^{\infty} \mathbb{E}_i[\mathbb{1}_{E_n}] \\ &= \sum_{n=0}^{\infty} \mathbb{P}\Big(X_n = j \,\Big|\, X_0 = i\Big) = \sum_{n=0}^{\infty} p_{ij}^{(n)}. \end{aligned}$$

Finally, the Cavalieri formula yields

$$\mathbb{E}_i[V_j] = \frac{1}{\mathbb{P}(X_0 = i)} \sum_{k=0}^{\infty} \mathbb{P}\Big(V_j > k, X_0 = i\Big)$$
$$= \sum_{k=0}^{\infty} \mathbb{P}\Big(V_j > k \Big| X_0 = i\Big) = \sum_{k=1}^{\infty} \mathbb{P}\Big(V_j \geq k \Big| X_0 = i\Big)$$
$$= \sum_{k=1}^{\infty} f_{ij} f_{jj}^{k-1} = \begin{cases} +\infty & \text{if } f_{jj} = 1, \\ \frac{f_{ij}}{1-f_{jj}} & \text{if } f_{jj} < 1. \end{cases}$$

Definition 5.27 *Let $\{X_n\}$ be a homogeneous Markov chain with a finite or denumerable state-space S. Let $\mathbf{P} = (p_{ij})$ be its transition matrix:*

(i) $j \in S$ is called a recurrent state *if $\sum_{n=1}^{\infty} p_{jj}^{(n)} = +\infty$,*

(ii) $j \in S$ is called a transient state *if $\sum_{n=1}^{\infty} p_{jj}^{(n)} < +\infty$.*

Thus, a state can be either transient or recurrent. These terms are justified by the following propositions.

Proposition 5.28 *The following are equivalent:*

(i) j is a recurrent state.

(ii) Paths from j visit j almost surely, $f_{jj} = 1$.

(iii) Paths from j visit j an infinite number of times almost surely.

If j is recurrent, paths from i visit j an infinite number of times with probability f_{ij}, that is with the same probability of visiting j at least once.

Proof. (i) and (ii) are equivalent by (5.30) and (5.31). Moreover, (iii) implies (ii) and, if (ii) holds, i.e. $f_{jj} = 1$, then

$$\mathbb{P}\Big(V_j = +\infty \Big| X_0 = i\Big) = \lim_{r\to\infty} \mathbb{P}\Big(V_j \geq r \Big| X_0 = j\Big) = \lim_{r\to\infty} f_{ij} f_{jj}^{r-1} = f_{ij}.$$

In particular, choosing $i = j$, we get $\mathbb{P}\Big(V_j = +\infty \Big| X_0 = j\Big) = f_{jj} = 1$, i.e. (iii).

Proposition 5.29 *The following are equivalent:*

(i) j is a transient state.

(ii) The probability that paths from j visit j is less than 1, $f_{jj} < 1$.

(iii) The probability that paths from j visit j an infinite number of times is zero.

(iv) Paths from j visit j at most a finite number of times almost surely.

If j is transient, paths from i visit j an infinite number of times with zero probability.

Proof. (i) and (ii) are equivalent by (5.30) and (5.31). Moreover,

$$\mathbb{P}\Big(V_j = +\infty \,\Big|\, X_0 = i\Big) = \lim_{r\to\infty} \mathbb{P}\Big(V_j \geq r \,\Big|\, X_0 = i\Big) = \lim_{r\to\infty} f_{ij} f_{jj}^{r-1}.$$

Thus, $\mathbb{P}(V_j = +\infty \mid X_0 = j) = 0$ if and only if $f_{jj} < 1$, so that (ii) and (iii) are equivalent, and $\mathbb{P}(V_j = +\infty \mid X_0 = i) = 0$ if j is transient. Finally, the equivalence between (iii) and (iv) is obvious.

Proposition 5.30 *Let $\{X_n\}$ be a homogeneous Markov chain with a finite or denumerable state-space S. Let $\mathbf{P} = (p_{ij})$ be its transition matrix. Then the following hold:*

(i) If $i \to j$ and j is recurrent, then $\sum_{k=1}^{\infty} p_{ij}^{(k)} = +\infty$. Therefore, if $f_{jj} = 1$ and $i \to j$, then $f_{ij} = 1$.

(ii) If i and j communicate, $i \leftrightarrow j$, then either both i and j are transient or both i and j are recurrent. That is, if $i \leftrightarrow j$, then $f_{ii} = 1$ if and only if $f_{jj} = 1$.

Proof. (i) If j is recurrrent and $i \to j$, then there exists k such that $p_{ij}^{(k)} > 0$. Thus, for each $n \geq 0$

$$p_{ij}^{(k+n)} = \sum_{\alpha\in S} p_{i\alpha}^{(k)} p_{\alpha j}^{(n)} \geq p_{ij}^{(k)} p_{jj}^{(n)}.$$

In particular, $\sum_{n=1}^{\infty} p_{ij}^{(n)} = +\infty$ if $\sum_{n=1}^{\infty} p_{jj}^{(n)} = +\infty$.

(ii) If $i \leftrightarrow j$, then there exist h, k such that $p_{ij}^{(h)} > 0$ and $p_{ji}^{(k)} > 0$. Thus, for each $n \geq 0$

$$p_{ii}^{(h+k+n)} = \sum_{\alpha,\beta\in S} p_{i\alpha}^{(h)} p_{\alpha\beta}^{(n)} p_{\beta i}^{(k)} \geq p_{ij}^{(h)} p_{jj}^{(n)} p_{ji}^{(k)}.$$

In particular, $\sum_{n=1}^{\infty} p_{ii}^{(n)} = +\infty$ if $\sum_{n=1}^{\infty} p_{jj}^{(n)} = +\infty$. Since the roles of i and j can be interchanged, either both series converge or both series diverge.

5.3.4 Mean first passage time

Let $\{X_n\}$ be a homogeneous Markov chain with a finite or denumerable state-space S and let $C \subset S$. The *mean first passage time*, or *expected waiting time*, to get to C starting from $i \in S$ is the expected number of steps to get to C starting from the state i, i.e. the averaged waiting time $t_C(x)$ with respect to the probability measure $\mathbb{P}_i(A) := \mathbb{P}(A \mid X_0 = i)$,

$$\mathbb{E}_i[t_C] := \frac{1}{\mathbb{P}(X_0 = i)} \int_{\{X_0=i\}} t_C(x)\, \mathbb{P}(dx).$$

If C is a singleton, $C = \{j\}$, then $\mathbb{E}_j[t_C]$ is briefly denoted by $\overline{T}_{jj}$,

$$\overline{T}_{jj} := \mathbb{E}_j[t_j]$$

and is called the *expected return time* to j.

The Beppo Levi theorem, see B.33, yields

$$\mathbb{E}_i[t_C] = \begin{cases} +\infty & \text{if } \mathbb{P}\Big(t_C = \infty \,\Big|\, X_0 = i\Big) > 0, \\ \sum_{k=1}^{\infty} k\mathbb{P}\Big(t_C = k \,\Big|\, X_0 = i\Big) & \text{otherwise.} \end{cases} \tag{5.33}$$

Since the probability of visiting C starting from i is

$$\mathbb{P}\Big(t_C < +\infty \,\Big|\, X_0 = i\Big) = \sum_{k=1}^{\infty} \mathbb{P}\Big(t_C = k \,\Big|\, X_0 = i\Big) = f_{iC},$$

see (5.22), we have

$$\mathbb{E}_i[t_C] := \begin{cases} +\infty & \text{if } f_{iC} < 1, \\ \sum_{k=1}^{\infty} k f_{iC}^{(k)} & \text{if } f_{iC} = 1. \end{cases} \tag{5.34}$$

In other words, the expected time to get to C is infinite if and only if either $f_{iC} < 1$ or $f_{iC} = 1$ and $\sum_{k=1}^{\infty} k f_{iC}^{(k)} = +\infty$. In particular,

$$\overline{T}_{jj} = \begin{cases} +\infty & \text{if } f_{jj} < 1, \\ \sum_{k=1}^{\infty} k f_{jj}^{(k)} & \text{if } f_{jj} = 1. \end{cases} \tag{5.35}$$

Summarizing, *the average return time $\overline{T}_{jj}$ in j*:

- *is infinite either if j is transient, $f_{jj} < 1$, or if j is recurrent, $f_{jj} = 1$, and $\sum_{k=1}^{\infty} k f_{jj}^{(k)} = \infty$;*
- *is finite if and only if j is recurrent, $f_{jj} = 1$, and $\sum_{k=1}^{\infty} k f_{jj}^{(k)} < \infty$.*

If the expected return time of state j is finite, then the state j is called *positive recurrent*. We shall see later, see Theorem 5.71 and 5.82, that the states of a Markov chain with finite state-space whose transition matrix is irreducible are positive recurrent.

Proposition 5.31 *Let $C \subset S$ and $i \in S$. Then*

$$\mathbb{E}_i[t_C] = 1 + \sum_{\ell \notin C} p_{i\ell} \mathbb{E}_\ell[t_C]. \tag{5.36}$$

Proof. It suffices to assume $\mathbb{E}_i[t_C] < +\infty$ and, consequently, $f_{iC} = 1$. Recurrence equation (5.24) gives

$$\mathbb{E}_i[t_C] = f_{iC}^{(1)} + \sum_{k=2}^{\infty} k f_{iC}^{(k)} = f_{iC}^{(1)} + \sum_{k=1}^{\infty} (k+1) \Big(\sum_{\ell \notin C} p_{i\ell} f_{\ell C}^{(k)} \Big)$$

$$= f_{iC}^{(1)} + \sum_{k=1}^{\infty} \sum_{\ell \notin C} p_{i\ell} f_{\ell C}^{(k)} + \sum_{\ell \notin C} p_{i\ell} \Big(\sum_{k=1}^{\infty} k f_{\ell C}^{(k)} \Big)$$

$$= f_{iC}^{(1)} + \sum_{k=2}^{\infty} f_{iC}^{(k)} + \sum_{\ell \notin C} p_{i\ell} \mathbb{E}_\ell[t_C]$$

$$= f_{iC} + \sum_{\ell \notin C} p_{i\ell} \mathbb{E}_\ell[t_C].$$

Since $f_{iC} = 1$, the claim is proven.

5.3.5 Hitting time and hitting probabilities

Let $\{X_n\}$ be a homogeneous Markov chain on a probability space $(\Omega, \mathcal{E}, \mathbb{P})$ with a finite or denumerate state-space S, and let $\mathbf{P} = (p_{ij})$ be its transition matrix. For each $C \subset S$ define

$$\tau_C(x) := \min \{ n \geq 0 \mid X_n(x) \in C \}$$

where, of course, we set $\tau_C(x) = +\infty$ if $X_n(x) \notin C$ for any $n \geq 0$. The random variable $\tau_C(x)$ is called the *hitting time* or *first time to absorption* in C. Obviously, $\tau_C(x) = 0$ if and only if $x \in C$, and $\tau_C(x) = t_C(x)$ if and only if $X_0(x) \notin C$. The mass density of τ_C with respect to $\mathbb{P}_i$ is the sequence $\{h_{iC}^{(k)}\}$ where

$$h_{iC}^{(k)} := \mathbb{P}\Big(\tau_C(x) = k \,\Big|\, X_0 = i\Big)$$

is the probability that state i hits the set C at time k. Obviously,

$$h_{iC}^{(0)} = \begin{cases} 1 & \text{if } i \in C, \\ 0 & \text{if } i \notin C, \end{cases} \tag{5.37}$$

and, for each $k \geq 1$,

$$h_{iC}^{(k)} = \begin{cases} 0 & \text{if } i \in C, \\ f_{iC}^{(k)} & \text{if } i \notin C. \end{cases} \tag{5.38}$$

Starting from the mass density $\{h_{iC}^{(k)}\}$ of τ_C we can compute the probability that state i hits C,

$$h_{iC} := \mathbb{P}\Big(\tau_C < \infty \,\Big|\, X_0 = i\Big) = \sum_{k=0}^{\infty} h_{iC}^{(k)}$$

and the *expected hitting time* in C

$$\mathbb{E}_i[\tau_C] := \frac{1}{\mathbb{P}(X_0 = i)} \int_{\{X_0 = i\}} \tau_C(x)\, \mathbb{P}(dx).$$

Thus, (5.37) and (5.38) give

$$h_{iC} = \begin{cases} 1 & \text{if } i \in C, \\ f_{iC} & \text{if } i \notin C, \end{cases} \tag{5.39}$$

see (5.70), and

$$\mathbb{E}_i[\tau_C] = \begin{cases} \infty & \text{if } h_{iC} < 1, \\ \sum_{k=1}^{\infty} k h_{iC}^{(k)} & \text{if } h_{iC} = 1. \end{cases} \tag{5.40}$$

In particular, see (5.34),

$$\mathbb{E}_i[\tau_C] = \begin{cases} 0 & \text{if } i \in C, \\ \mathbb{E}_i[t_C] & \text{if } i \notin C. \end{cases}$$

In the next two theorems we characterize the hitting probabilities and the expected hitting times.

Theorem 5.32 *The vector $(h_{iC})_{i \in S}$ of hitting probabilities is the minimal non-negative solution $x = (x_i)_{i \in S}$ of the system*

$$x_i = \begin{cases} 1 & \textit{if } i \in C, \\ \displaystyle\sum_{\ell \in S} p_{i\ell} x_\ell & \textit{if } i \notin C. \end{cases} \tag{5.41}$$

That is, (h_{iC}) is a solution of (5.41) and, if $(x_i)_{i \in S}$ is any further non-negative solution of (5.41), then $x_i \geq h_{iC}$ $\forall i \in S$.

Proof. (i) We first show that $(h_{iC})_{i \in S}$ is a solution of system (5.41). If $i \in C$, then $\tau_C(x) = 0$ $\forall x \in \Omega$, so that $h_{iC}^{(0)} = 1$ and $h_{iC} = 1$. If $i \notin C$, then $h_{iC}^{(k)} = f_{iC}^{(k)}$, hence by (5.24), we get

$$\begin{aligned} h_{iC} &= \sum_{k=1}^{\infty} h_{iC}^{(k)} = h_{iC}^{(1)} + \sum_{k=2}^{\infty} \sum_{\ell \notin C} p_{i\ell} h_{\ell C}^{(k-1)} = \sum_{\ell \in C} p_{i\ell} + \sum_{\ell \notin C} p_{i\ell} h_{\ell C} \\ &= \sum_{\ell \in S} p_{i\ell} h_{\ell C}. \end{aligned}$$

This proves that $(h_{iC})_{i \in S}$ is a solution to system (5.41).

(ii) Assume $(x_i)_{i \in S}$ is a solution to (5.41). Trivially, $x_i = 1 = h_{iC}$ if $i \in C$. We now prove by an induction argument that

$$x_i \geq \sum_{k=0}^{n} h_{iC}^{(k)} \qquad \forall n \in \mathbb{N}, \forall i \notin C. \tag{5.42}$$

Since $x_\ell = 1\ \forall \ell \in C$ and $(x_i)_{i\in S}$ is non-negative, we get

$$x_i = \sum_{\ell\in C} p_{i\ell} + \sum_{\ell\notin C} p_{i\ell} x_\ell \ge h_{iC}^{(1)},$$

i.e. the claim holds true for $n = 1$.

Assume $n > 1$. By (5.41), the induction argument and since $x_i \ge 0\ \forall i$, we get

$$\begin{aligned} x_i &= \sum_{\ell\in C} p_{i\ell} + \sum_{\ell\notin C} p_{i\ell} x_\ell \ge h_{iC}^{(1)} + \sum_{\ell\notin C} p_{i\ell} x_\ell \ge h_{iC}^{(1)} + \sum_{\ell\notin C} p_{i\ell}\Big(\sum_{k=1}^{n} h_{iC}^{(k)}\Big) \\ &= h_{iC}^{(1)} + \sum_{k=1}^{n}\Big(\sum_{\ell\notin C} p_{i\ell} h_{\ell C}^{(k)}\Big) = h_{iC}^{(1)} + \sum_{k=2}^{n+1} h_{iC}^{(k)} = \sum_{k=1}^{n+1} h_{iC}^{(k)}, \end{aligned}$$

i.e. (5.42) is proven. If we now let n go to ∞ in (5.42), we get $x_i \ge \sum_{k=0}^{\infty} h_{iC}^{(k)} = h_{iC}$.

Theorem 5.33 *The vector $(\mathbb{E}_i[\tau_C])_{i\in S}$ of the expected hitting times in C is a solution of the system*

$$y_i = \begin{cases} 0 & \text{if } i \in C, \\ 1 + \sum_{\ell\notin C} p_{i\ell} y_\ell & \text{if } i \notin C, \end{cases} \tag{5.43}$$

where we accept $y_i = +\infty$. If $h_{iC} = 1$ and $(y_i)_{i\in S}$ is any further solution of (5.43), then $y_i \ge \mathbb{E}_i[\tau_C]$.

Proof. We first show that $\mathbb{E}_i[\tau_C]$ solves (5.43). If $i \in C$, then $\mathbb{E}_i[\tau_C] = 0$. If $i \notin C$, then $\mathbb{E}_i[\tau_C] = \mathbb{E}_i[t_C]$, hence $\mathbb{E}_i[\tau_C]$ solves (5.43) by Proposition 5.31.

Assume $(y_i)_{i\in S}$ is a solution of (5.43). If $i \in C$, then $y_i = 0 = \mathbb{E}_i[\tau_C]$. If $i \notin C$ and $h_{iC} = 1$, we show by induction that

$$y_i \ge \sum_{k=1}^{n} k h_{iC}^{(k)} \tag{5.44}$$

so that, as n goes to infinity, we get $y_i \ge \sum_{k=1}^{\infty} k h_{iC}^{(k)} = \mathbb{E}_i[\tau_C]$, see (5.34).

If $n = 1$, then (5.44) holds trivially: $y_i \ge 1 = h_{iC} \ge h_{iC}^{(1)}$. Let $n \neq 2$. From the induction argument and (5.24) we infer

$$\begin{aligned} y_i &= 1 + \sum_{\ell\notin C} p_{i\ell} y_\ell \ge 1 + \sum_{\ell\notin C} p_{i\ell}\Big(\sum_{k=1}^{n} k h_{\ell C}^{(k)}\Big) = 1 + \sum_{k=1}^{n} k\Big(\sum_{\ell\notin C} p_{i\ell} h_{\ell C}^{(k)}\Big) \\ &= 1 + \sum_{k=1}^{n} k h_{iC}^{(k+1)} = 1 + \sum_{k=2}^{n} (k-1) h_{iC}^{(k)} = 1 - \sum_{k=2}^{n} h_{iC}^{(k)} + \sum_{k=2}^{n+1} k h_{iC}^{(k)} \\ &\ge h_{iC}^{(1)} + \sum_{k=2}^{n+1} k h_{iC}^{(k)} = \sum_{k=1}^{n+1} k h_{iC}^{(k)}. \end{aligned}$$

5.3.6 Exercises

For the reader's convenience the main definitions on discrete time Markov chains are summarized in Figure 5.8.

Exercise 5.34 (Holding time) *Let $\{X_n\}$ be a homogeneous Markov chain with a finite or denumerable state-space S and transition matrix $\mathbf{P} = (p_{ij})$. Compute the probability distribution of the holding time in a state. We regard the paths starting from j as staying at j at least once. Compute also the expected holding time in a state.* $\left[\mathbb{E}_j[SJ_j] = \frac{1}{1-p_{jj}}\right]$

Solution. The holding time at $j \in S$ is the random variable defined as

$$SJ_j(x) = \min\left\{n \,\middle|\, X_n(x) \neq j\right\}$$

($SJ_j(x) = +\infty$ if $X_n(x) = j\ \forall n$). For $k \geq 1$, set

$$E_k := \left\{x \in \Omega \,\middle|\, X_1(x) = j, \dots, X_{k-1}(x) = j, X_k(x) \neq j\right\}.$$

We want to compute the holding time probabilities $\mathbb{P}\left(SJ_j = k \,\middle|\, X_0 = j\right) = \mathbb{P}\left(E_k \,\middle|\, X_0 = j\right)$. If $k = 1$, then

$$\mathbb{P}\left(SJ_j = 1 \,\middle|\, X_0 = j\right) = \mathbb{P}\left(X_1 \neq j \,\middle|\, X_0 = j\right) = 1 - p_{jj}.$$

If $k \geq 2$, (5.8) implies

$$\begin{aligned}\mathbb{P}\left(SJ_j = k \,\middle|\, X_0 = j\right) &= \mathbb{P}\left(E_k \,\middle|\, X_0 = j\right)\\ &= \mathbb{P}\left(X_k \neq j \,\middle|\, X_{k-1} = j\right)\prod_{h=1}^{k-1}\mathbb{P}\left(X_h = j \,\middle|\, X_{h-1} = j\right)\\ &= (1 - p_{jj})p_{jj}^{k-1}.\end{aligned}$$

Thus, the holding time in state j follows the geometric distribution of parameter $1 - p_{jj}$, i.e. it is a memoryless random variable.

Integrating $SJ_j(x)$ one gets the expected holding time in state j starting from j:

$$\begin{aligned}\mathbb{E}_j[SJ_j] &= \sum_{k=1}^{\infty} k\mathbb{P}\left(SJ_j = k \,\middle|\, X_0 = j\right) = (1 - p_{jj})\sum_{k=1}^{\infty} kp_{jj}^{k-1}\\ &= (1 - p_{jj})\frac{1}{(1 - p_{jj})^2} = \frac{1}{1 - p_{jj}}.\end{aligned}$$

Exercise 5.35 (Gambler's ruin paradox) *A player possesses a capital $i \in \mathbb{N}$, $i \geq 1$ and plays a sequence of games. In each game, he either wins or loses*

Characteristic parameters of discrete Markov chains

Let $\{X_n\}$ be a homogeneous Markov chain with a finite or denumerable state-space S and let $i, j \in S$ and $C \subset S$.

- *Conditional probability given the event* $\{X_0 = i\}$

$$\mathbb{P}_i(A) := \mathbb{P}(A \mid X_0 = i).$$

- *Conditional expectation given the event* $\{X_0 = i\}$

$$\mathbb{E}_i[f] := \frac{1}{\mathbb{P}(X_0 = i)} \int_{\{X_0=i\}} f(x)\, \mathbb{P}(dx).$$

- *Waiting time to get to* C

$$t_C(x) := \min\{n \geq 1 \mid X_n \in C\}, \qquad t_j(x) \text{ if } C = \{j\}.$$

- *Probability to get to* C *for the first time at step* k *starting from* i

$$f_{iC}^{(k)} := \mathbb{P}\Big(t_C = k \,\Big|\, X_0 = i\Big), \qquad f_{ij}^{(k)} \text{ if } C = \{j\}.$$

- *Probability to get to* C *starting from* i

$$f_{iC} := \mathbb{P}\Big(t_C < +\infty \,\Big|\, X_0 = i\Big), \qquad f_{ij} \text{ if } C = \{j\}.$$

- *Number of visits at* j *in the first* n *steps*

$$V_C^{(n)}(x), \qquad V_j^{(n)}(x) \text{ if } C = \{j\}.$$

- Expected number of visits at j starting from i $\mathbb{E}_i[V_j]$.
- *Expected waiting time to get to* C *from* i

$$\mathbb{E}_i[t_C], \qquad \mathbb{E}_i[t_j] \text{ if } C = \{j\}.$$

- *Expected return time to* j $\overline{T}_{jj} := \mathbb{E}_j[t_j]$.

Figure 5.8 Some characteristic parameters for a homogeneous Markov chain with finite or denumerable state-space.

1 with probabilities p and $q := 1 - p$, respectively. Compute the probability of losing the capital and the expected number of matches to be played in order to lose all the capital.

Solution. Let X_n be the capital available before the nth play. The game is a one-dimensional random walk where we move forwards (the capital increases

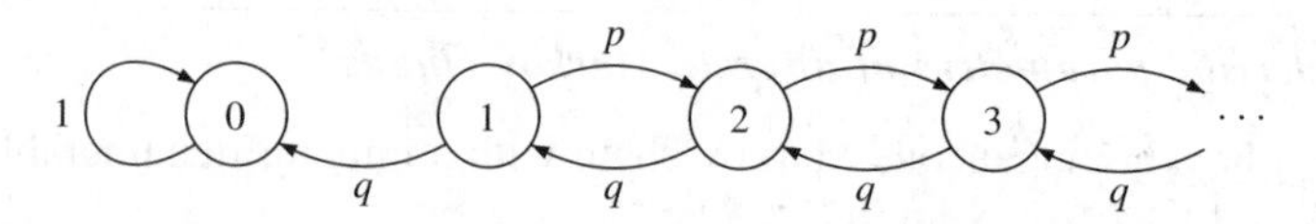

Figure 5.9 A graph representing the transition matrix **P** *in Exercise 5.35.*

of 1 unit) with probability p and backwards with probability $q := 1 - p$. The sequence $\{X_n\}$ is a homogeneous Markov chain with transition matrix

$$\mathbf{P} = \begin{pmatrix} 1 & 0 & 0 & 0 & \dots & 0 & \dots \\ q & 0 & p & 0 & \dots & 0 & \dots \\ 0 & q & 0 & p & \dots & 0 & \dots \\ \vdots & \vdots & \vdots & \vdots & \vdots & \vdots & \ddots \end{pmatrix}$$

We want to compute the probability f_{i0} that, starting with a capital i the player hits 0, $h_{i0} = \mathbb{P}(\tau_0 < \infty \mid X_0 = 1)$. If $i = 0$ then trivially $h_{i0} = 1$. If $i \neq 0$, then $h_{i0} = f_{i0}$. Since 0 is recurrent and $i \to j$, see Figure 5.9, then $f_{i0} = 1$. So, whatever the initial capital and the seeming fortune of the game are, the player will almost surely end up with no money left.

Let us now compute the expected number of matches the player has to play before getting ruined. We denote by t_i such expected time, when the player starts with capital i, i.e. $t_i := h_{i0}$. Thus, see Theorem 5.33, the vector (t_i) is the minimal non-negative solution of the following system

$$\begin{cases} t_0 = 0, \\ t_i = 1 + q t_{i-1} + p t_{i+1} \quad i > 0. \end{cases} \tag{5.45}$$

- If $p \neq q$, then the solutions of the system (5.45) are the one parameter family of sequences (t_i), $t_i = 1 - a + a\left(\frac{q}{p}\right)^i + \frac{i}{q-p}$.
- If $q < p$, then t_i diverges to $-\infty$ as $i \to +\infty$, i.e. system (5.45) admits no non-negative solution. Hence $t_i = +\infty \; \forall i$.
- If $p < q$, then the minimal non-negative solution is $t_i = \frac{i}{q-p} \; \forall i$.
- If $p = q$, then the solutions of system (5.45) are the sequences (t_i), $t_i = a + bi - i^2$, $a, b \in \mathbb{R}$. There is no non-negative finite solution, so that $t_i = +\infty \forall i$.

Summarizing, the expected number of matches the gambler has to play, starting with capital i is

$$t_i = \begin{cases} \dfrac{i}{q-p} & \text{if } q > p, \\ +\infty & \text{if } q \leq p. \end{cases}$$

5.4 Finite stochastic matrices

In this section we go back to finite stochastic matrices defined in Section 5.1 presenting their *canonical representation*, a very useful tool for describing the powers of the matrix itself.

5.4.1 Canonical representation

5.4.1.1 Minimal absorbing classes

Let $\mathbf{P}$ be a $N \times N$ stochastic matrix. For notational convenience, for every $n \in \mathbb{N}$, we write $p_{ij}^{(n)}$ instead of $(\mathbf{P}^n)^i_j$ and $p_{ij} := p_{ij}^{(1)} = \mathbf{P}^i{}_j$. Of course $p_{ij}^{(0)} = \delta_{ij}$.

Definition 5.36 *A nonempty subset C of $\{1, \dots, N\}$ is called a* closed class *or an* absorbing class *for* $\mathbf{P}$ *if for any $i \in C$ and for any $j \notin C$, j is not accessible from i, $i \not\rightarrow j$; equivalently, C is a closed class if and only if $i \in C$ and $i \to j$ imply $j \in C$.*

In other words, C is a closed class if there is no path starting from any point of C and arriving to points outside C. Notice that the points of C are not required to be pairwise connected.

The simplest way to enumerate the closed classes of $\{1, \dots, N\}$ is to consider for any $i \in \{1, \dots, N\}$ the set of nodes that are accessible from i

$$\Big\{ j \in \{1, \dots, N\} \Big| i \to j \Big\},$$

and which can be computed, e.g. via the matrix $\mathbf{B}$ in (5.2), see (5.3).

Trivially, the whole set of indexes is a closed class for $\mathbf{P}$. Moreover, it is easy to show that if C and D are closed classes, then so is $C \cup D$, and so is also $C \cap D$ provided it is nonempty. Thus, closed classes are partially ordered with respect to inclusion, and minimal closed classes $C_1, \dots, C_r$, $r \geq 1$ exist since $\mathbf{P}$ is finite; each C_i is a closed class such that if D is a closed class and $D \subset C_i$, then $D = C_i$. In particular, minimal closed classes are pairwise disjoint. Minimal closed classes can be found in the set of all closed classes. Using the partial order with respect to inclusion, write the *tree diagram* of the closed classes starting from the maximal closed class $\{1, \dots, N\}$; then the leaves of the tree are the closed minimal classes.

Minimal closed classes can be also detected by the following proposition.

Proposition 5.37 *A closed class is minimal if and only if all its elements are pairwise communicating.*

Proof. Let C be a closed class whose elements are pairwise comunicating. Assume, by contradiction, that C is not minimal: then there exists a minimal closed class $D \subset C$, $D \neq C$, and each element of D comunicates with at least one element in $C \setminus D \subset D^c$, a contradiction.

Conversely, assume C is a minimal closed class. Let $j \in C$ and define

$$D_j := \left\{ i \,\middle|\, i \not\to j \right\}. \tag{5.46}$$

We want to prove that $D_j = \emptyset$. Clearly, $j \notin D_j$, hence D_j is a proper subset of C. We show that if $D_j \neq \emptyset$, then D_j is a closed class, a contradiction since $D_j \subset C$, $D_j \neq C$ and C is minimal.

Assume $D_j \neq \emptyset$. In order to prove that D_j is a closed class, we only need to show that $i \not\to k$ for $i \in D_j$ and $k \notin D_j$. We distinguish between two cases:

- if $k \notin C$, then $i \not\to k$ since C is a closed class;
- if $k \in C \setminus D_j$, then $k \to j$ by the definition of D_j. If $i \to k$, then by transitivity $i \to j$, a contradiction since $i \in D_j$.

Minimal closed classes for $\mathbf{P}$ can be also detected via the matrix $\mathbf{B}$ in (5.2). In fact, $i \leftrightarrow j$ if and only if $\mathbf{B}^i_j > 0$ and $\mathbf{B}^j_i > 0$, cf. (5.3), so that the following holds.

Proposition 5.38 *The index $i \in \{1, \dots, N\}$ belongs to a minimal closed class if and only if for any j such that $\mathbf{B}^i_j > 0$ then also $\mathbf{B}^j_i > 0$.*

Proof. In fact, by Proposition 5.37, a closed class $C = \left\{ j \mid \mathbf{B}^i_j > 0 \right\}$ containing i is minimal if and only if

$$\left\{ j \,\middle|\, \mathbf{B}^i_j > 0 \right\} = \left\{ j \,\middle|\, \mathbf{B}^i_j > 0, \mathbf{B}^j_i > 0 \right\}.$$

We then summarize the above in the following theorem.

Theorem 5.39 Canonical representation *Let $\mathbf{P} \in M_{N,N}$ be stochastic, $S = \{1, \dots, N\}$. Let $C_1, \dots, C_r$ be the minimal closed classes for $\mathbf{P}$ and for $i = 1, \dots .r$ denote by k_i the cardinality of C_i. Finally, let $T := \{1, \dots, N\} \setminus (C_1 \cup \dots \cup C_r)$. Up to a reordering of rows and columns, a stochastic matrix $\mathbf{P}$ has the* canonical form

$$\mathbf{P} = \begin{pmatrix} \boxed{\mathbf{B}_1} & 0 & 0 & \cdots & 0 & 0 \\ 0 & \boxed{\mathbf{B}_2} & 0 & \cdots & 0 & 0 \\ \vdots & \vdots & \vdots & \ddots & \vdots & \vdots \\ 0 & 0 & 0 & \cdots & \boxed{\mathbf{B}_r} & 0 \\ \multicolumn{5}{c}{\boxed{\quad\quad\quad \mathbf{R} \quad\quad\quad}} & \boxed{\mathbf{V}} \end{pmatrix} \tag{5.47}$$

where $\mathbf{B}_1, \mathbf{B}_2, \dots;, \mathbf{B}_r$ are irreducible stochastic matrices of dimension $k_1 \times k_1$, $\dots$, $k_r \times k_r$, respectively, see Proposition 5.37.

5.4.2 States classification

Following the Markov chains terminology, we call *states* the indexes in $\{1, \dots, N\}$. Given a $N \times N$ stochastic matrix $\mathbf{P} = (p_{ij})$, we recall that a state $j \in \{1, \dots, N\}$ is *recurrent* if $\sum_{k=1}^{\infty} p_{jj}^{(k)} = +\infty$ and *transient* if $\sum_{k=1}^{\infty} p_{jj}^{(k)} < +\infty$, cf. Definition 5.27.

Proposition 5.40 *Let* $\mathbf{P}$ *be a* $N \times N$ *stochastic matrix,* $S = \{1, \dots, N\}$*, and let* $C_1, \dots, C_r$ *be the minimal closed classes in* S*. Set*

$$C := C_1 \cup \dots \cup C_r \qquad \textit{and} \qquad T := \{1, \dots, N\} \setminus C$$

and let $i, j \in \{1, \dots, N\}$*. We have the following:*

(i) *If* $j \in C$*, then* j *is recurrent. Therefore, if* $i \to j$*, then* $\sum_{k=1}^{\infty} p_{ij}^{(k)} = +\infty$*.*

(ii) *If* $j \in T$ *then* j *is transient. More precisely, for all* $i, j \in T$*,* $p_{ij}^{(k)}$ *converges to* 0 *exponentially fast as* $k \to +\infty$*, hence* $\sum_{k=1}^{\infty} p_{ij}^{(k)} < +\infty$*.*

Proof. (i) Let $j \in C_s$ for some s. Let us prove that j is recurrent. Since $p_{jh}^{(n)} = 0$ for any $h \notin C_s$ and any $n \geq 0$, $\sum_{h \in C_s} p_{jh}^{(n)} = 1$, hence $\sum_{h \in C_s} \sum_{n=0}^{\infty} p_{jh}^{(n)} = \sum_{n=0}^{\infty} \sum_{h \in C_s} p_{jh}^{(n)} = +\infty$. In particular, there exists $j_0 \in C_s$ such that $\sum_{n=1}^{\infty} p_{jj_0}^{(n)} = +\infty$. Since $j_0 \leftrightarrow j$, there exists k such that $p_{j_0 j}^{(k)} > 0$, so that from

$$p_{jj}^{(n+k)} = \sum_{\alpha} p_{j\alpha}^{(n)} p_{\alpha j}^{(k)} \geq p_{jj_0}^{(n)} p_{j_0 j}^{(k)} \qquad \forall n$$

we infer $\sum_{n=1}^{\infty} p_{jj}^{(n)} = +\infty$, i.e. j is recurrent. Now, if $i \to j \in C$, then $\sum_{k=1}^{\infty} p_{ij}^{(k)} = +\infty$ by (i) of Proposition 5.30.

(ii) Let $i \in T$. The set

$$D := \left\{ j \in \{1, \dots, N\} \,\middle|\, i \to j \right\}$$

is a closed class, so that $D \cap C \neq \emptyset$. Thus, setting

$$p_{iC}^{(n)} := \sum_{j \in C} p_{ij}^{(n)},$$

there exists $\overline{k} = \overline{k}(i)$ such that $p_{iC}^{(\overline{k})} > 0$. We claim that the sequence $\left\{ p_{iC}^{(k)} \right\}$ is monotone nondecreasing with respect to k. In fact, for any $\alpha \in C$ since $p_{\alpha j} = 0$ $\forall j \notin C$, we have $\sum_{j \in C} p_{\alpha j} = 1$. Therefore,

$$p_{iC}^{(k+1)} = \sum_{j \in C} \sum_{\alpha=1}^{N} p_{i\alpha}^{(k)} p_{\alpha j} = \sum_{\alpha=1}^{N} p_{i\alpha}^{(k)} \Big(\sum_{j \in C} p_{\alpha j} \Big) \geq \sum_{\alpha \in C} p_{i\alpha}^{(k)} \Big(\sum_{j \in C} p_{\alpha j} \Big) = p_{iC}^{(k)}.$$

Thus, for any $i \in T$ $p_{iC}^{(k)} \geq p_{iC}^{(\overline{k})} > 0$ if $k \geq \overline{k}$. Since T is finite, there exist $p_0 > 0$ and k_0 such that $p_{iC}^{(k)} \geq p_0$ for any $k \geq k_0$ and every $i \in T$. In particular,

$$p_{iT}^{(k_0)} = \sum_{j \in T} p_{ij}^{(k_0)} = 1 - p_{iC}^{(k_0)} \leq 1 - p_0.$$

The equality $p_{\alpha j}^{(k)} = 0$ if $\alpha \in C$, $j \notin C$ also implies

$$p_{iT}^{(2k_0)} = \sum_{j \in T} p_{ij}^{(2k_0)} = \sum_{j \in T} \sum_{\alpha=1}^{N} p_{i\alpha}^{k_0} p_{\alpha j}^{k_0} = \sum_{\alpha \in T} p_{i\alpha}^{k_0} \sum_{j \in T} p_{\alpha j}^{k_0} \leq (1 - p_0)^2$$

and, moreover, by an induction argument,

$$p_{iT}^{(qk_0)} \leq (1 - p_0)^q \qquad \text{for any positive integer } q.$$

Divide k by k_0, $k = qk_0 + r$, $0 \leq r < k_0$. Since the map $k \to p_{iT}^{(k)} = 1 - p_{iC}^{(k)}$ is monotone decreasing, we finally get

$$p_{iT}^{(k)} \leq p_{iT}^{(qk_0)} \leq (1 - p_0)^q \leq (1 - p_0)^{k/k_0 - 1},$$

and

$$\sum_{k=1}^{\infty} p_{iT}^{(k)} \leq k_0 + k_0 \sum_{q=1}^{\infty} p_{iT}^{(qk_0)} \leq k_0 + k_0 \sum_{q=1}^{\infty} (1 - p_0)^q = \frac{k_0}{p_0}.$$

We summarize Proposition 5.40 as follows. Let $C_1, \dots, C_r$ be the minimal closed classes in $\{1, \dots, N\}$ for $\mathbf{P}$ and, for $i = 1, \dots .r$, let k_i be the cardinality of C_i. As we have seen in Proposition 5.37, up to a relabelling of the states, we may assume that $\mathbf{P}$ has the canonical form (5.47). Consequently, for each integer $n \geq 1$,

$$\mathbf{P}^n = \begin{pmatrix} \boxed{\mathbf{B}_1^n} & 0 & 0 & \cdots & 0 & 0 \\ 0 & \boxed{\mathbf{B}_2^n} & 0 & \cdots & 0 & 0 \\ \vdots & \vdots & \vdots & \ddots & \vdots & \vdots \\ 0 & 0 & 0 & \cdots & \boxed{\mathbf{B}_r^n} & 0 \\ \multicolumn{5}{c}{\boxed{\quad\quad\quad \mathbf{R}_n \quad\quad\quad}} & \boxed{\mathbf{V}^n} \end{pmatrix}$$

so that, by (i) and (ii) of Proposition 5.40, setting for each $i \in T$ and $j \notin T$

$$(\mathbf{R}_\infty)^i_j = \begin{cases} +\infty & \text{if } i \to j, \\ 0 & \text{otherwise} \end{cases}$$

we have

$$\sum_{n=0}^{\infty} \mathbf{P}^n = \begin{pmatrix} \boxed{+\infty} & 0 & 0 & \cdots & 0 & 0 \\ 0 & \boxed{+\infty} & 0 & \cdots & 0 & 0 \\ \vdots & \vdots & \vdots & \ddots & \vdots & \vdots \\ 0 & 0 & 0 & \cdots & \boxed{+\infty} & 0 \\ \multicolumn{5}{c}{\boxed{\mathbf{R}_\infty}} & \boxed{\sum_{n=0}^{\infty} \mathbf{V}^n} \end{pmatrix}. \tag{5.48}$$

Finally, we point out that the matrix $\sum_{n=0}^{\infty} \mathbf{V}^n$ can be easily computed starting from (5.47). Consider the identity

$$(\text{Id} - \mathbf{V})(\text{Id} + \mathbf{V} + \cdots + \mathbf{V}^n) = \text{Id} - \mathbf{V}^{n+1}. \tag{5.49}$$

Since $\mathbf{V}^n$ converges to the zero matrix exponentially fast as n goes to infinity, see (ii) of Proposition 5.40, the series $\sum_{n=0}^{\infty} \mathbf{V}^n$ converges. Passing to the limit as $n \to \infty$, (5.40) yields $(\text{Id} - \mathbf{V}) \sum_{n=0}^{\infty} \mathbf{V}^n = \text{Id}$. In other words, $\text{Id} - \mathbf{V}$ is invertible and

$$\sum_{n=0}^{\infty} \mathbf{V}^n = (\text{Id} - \mathbf{V})^{-1}. \tag{5.50}$$

5.4.3 Exercises

Exercise 5.41 *Prove the following:*

(i) The matrix $\mathbf{P} = \begin{pmatrix} 1 & 0 \\ 0 & 1 \end{pmatrix}$ *has two minimal closed classes and no transient state.*

(ii) The matrix $\mathbf{P} := \begin{pmatrix} 0 & 1 \\ 1/2 & 1/2 \end{pmatrix}$ *is irreducible.*

(iii) The matrix $\mathbf{P} := \begin{pmatrix} 1 & 0 \\ 1/2 & 1/2 \end{pmatrix}$ *has one minimal closed class,* $\{1\}$*, and* 2 *is a transient state.*

(iv) The matrix $\mathbf{P} := \begin{pmatrix} 0 & 1 \\ 1 & 0 \end{pmatrix}$ *is irreducible.*

Exercise 5.42 *Find the canonical form of the stochastic matrix*

$$\mathbf{P} = \begin{pmatrix} 0 & 0 & 0 & 0 & 0 & 0 & 0.5 & 0 & 0.5 & 0 \\ 0 & 0.3 & 0 & 0.7 & 0 & 0 & 0 & 0 & 0 & 0 \\ 0 & 0 & 0 & 0 & 0.5 & 0 & 0 & 0.5 & 0 & 0 \\ 0 & 0 & 0 & 1 & 0 & 0 & 0 & 0 & 0 & 0 \\ 0 & 0.1 & 0.1 & 0.1 & 0.1 & 0 & 0 & 0 & 0 & 0.6 \\ 0.1 & 0 & 0 & 0 & 0.9 & 0 & 0 & 0 & 0 & 0 \\ 0.2 & 0 & 0 & 0 & 0 & 0 & 0 & 0 & 0.8 & 0 \\ 0 & 0 & 1 & 0 & 0 & 0 & 0 & 0 & 0 & 0 \\ 0 & 0 & 0 & 0 & 0 & 0 & 0.6 & 0 & 0.4 & 0 \\ 0 & 0 & 0 & 0 & 0 & 0 & 0 & 0 & 0 & 1 \end{pmatrix}.$$

Exercise 5.43 *Let $\mathbf{P} = (p_{ij})$ be a $N \times N$ irreducible stochastic matrix and let $C \subset \{1, \dots, N\}$, $C \neq \emptyset$. Show that there exist $h \in C$ and $k \notin C$ such that $p_{hk} > 0$.*

Solution. Without loss of generality, assume $C = \{1, \dots, s\}$. If $p_{hk} = 0$ $\forall h \in C$ and $\forall k \notin C$, then $\mathbf{P}$ has the form

$$\mathbf{P} = \begin{pmatrix} \boxed{\mathbf{A}} & 0 \\ \multicolumn{2}{c}{\boxed{\quad\mathbf{B}\quad}} \end{pmatrix},$$

a contradiction, since $\mathbf{P}$ is irreducible.

5.5 Regular stochastic matrices

In this section we deal with the convergence of the sequence of the powers of the transition matrix for a particular class of homogenenous Markov chains with finite state-space $S = \{1, \dots, N\}$.

A fundamental difference exists between finite dimensional stochastic vectors and infinite dimensional ones. Let $\{x^{(n)}\} \subset [0, 1]^\infty$ be the sequence of stochastic vectors defined by $x^{(n)} = \{x_j^{(n)}\}_j$, $x_j^{(n)} := \delta_{j,n}$. For each fixed j, $x_j^{(n)} \to 0$ as $n \to \infty$, i.e. $\{x^{(n)}\}$ converges to $(0, 0, 0, 0, \dots)$ componentwise. This shows that, in general, a converging sequence of stochastic vectors does not converge to a stochastic vector. On the other hand, the set of stochastic vectors in $\mathbb{R}^{N*}$

$$\mathcal{T} := \left\{ x = (x_1, \dots, x_n) \in \mathbb{R}^N \,\middle|\, x_i \geq 0, \sum_{i=1}^{N} x_i = 1 \right\}$$

is a closed set: if $\{x_n\} \subset \mathcal{T}$ and $x_n \to x$, then $x \in \mathcal{T}$.

5.5.1 Iterated maps

We now recall a few facts from the theory of iterated maps.

Let X be a metric space and let $f : X \to X$ be a continuous map from X into itself. For every $k \in \mathbb{N}$, we denote by $f^k : X \to X$ the kth iterate of f, i.e. $f^k := f \circ f \circ \cdots \circ f$ k times. For the sake of convenience we also set $f^0(x) = x$. For every $x \in X$, the sequence $\{f^k(x)\}_k$ is called the *path* of x. We say that $p \in X$ is a fixed point for f if $f(p) = p$. Finally, if for every $x \in X$ $f^k(x) \to p$ we say that p is the *sink* of f.

Simple examples of the above can be drawn considering affine maps $f : [0, 1] \to [0, 1]$. The following claims are easily proved, see Exercise 5.63.

Proposition 5.44 *The following hold:*

(i) *If p is a fixed point for f, then, for every $k \geq 0$, $f^k(p) = p$, i.e. p is a fixed point for f^k.*

(ii) *Let $n > 1$ and let p be a fixed point for f^n. Then $p, f(p), \ldots, f^{n-1}(p)$ are fixed points for f^n, too. In particular, if f^n has a unique fixed point, then p is also the unique fixed point of f.*

(iii) *If for some $x \in X$, $f^k(x) \to p$ as $k \to \infty$, then p is a fixed point for f.*

(iv) *If p is the sink of f, then p is the unique fixed point of f.*

(v) *There exist maps f that have fixed points but no sink.*

(vi) *Let $n \geq 1$. A point $p \in X$ is the sink of f if and only if p is the sink of f^n.*

The existence of fixed points can be granted under quite general assumptions. We have the following.

Theorem 5.45 (Brouwer) *Let X be a compact and convex set in $\mathbb{R}^N$ and let $f : X \to X$ be a continuous map. Then f has a fixed point.*

Proof. Let $x \in X \subset \mathbb{R}^N$. Consider the path $\{f^n(x)\}$ of x. For every n

$$w_n := \frac{1}{n+1} \sum_{k=0}^{n} f^n(x)$$

is a convex combination of $f^0(x), f^1(x), \ldots, f^n(x)$, hence $w_n \in X$ since X is convex. Since X is also compact, there exist $w \in X$ and a subsequence of $\{w_n\}$, that for convenience we still denote by $\{w_n\}$, such that $w_n \to w$. Let us prove that $f(w) = w$. For every n we have

$$\begin{aligned} \left| f(w_n) - w_n \right| &= \frac{1}{n+1} \left| \sum_{j=0}^{n} \left(f^{j+1}(x) - f^j(x) \right) \right| \\ &= \frac{1}{n+1} \left| f^{n+1}(x) - x \right| \leq \frac{\operatorname{diam}(X)}{n+1}. \end{aligned}$$

Passing to the limit as $n \to \infty$, continuity of f yields $|f(w) - w| = 0$, i.e. $f(w) = w$.

Existence of a sink for f requires different assumptions on f. Let us start with a definition.

A *contraction map*, or simply a *contraction*, on X is a map $f : X \to X$ which shrinks distances uniformly, i.e. a map for which there exists a constant L, $0 < L < 1$, such that $|f(x) - f(y)| \leq L|x - y|$ $\forall x, y \in X$. Of course, a contraction map is continuous on X.

Theorem 5.46 (Banach fixed point theorem) *Let X be a complete metric space with distance d and let $f : X \to X$ be a contraction with contraction factor $L < 1$. Then f has an unique sink $p \in X$. More precisely, for every $x \in X$ the path $\{f^n(x)\}$ converges to p at least exponentially fast,*

$$d(f^n(x), p) \leq \frac{L^n}{1-L} d(f(x), x) \qquad \forall n. \tag{5.51}$$

Proof. Uniqueness. If $p, q \in X$ are two fixed points, then

$$d(p, q) = d(f(p), f(q)) \leq L\, d(p, q),$$

hence $d(p, q) = 0$, since $L < 1$.

Existence. For any $x \in X$, let $x_k := f^k(x)$, where, we recall, $f^k := f \circ \cdots \circ f$ k times. We then have

$$d(x_{k+1}, x_k) \leq L d(x_k, x_{k-1}) \leq \cdots \leq L^k d(x_1, x_0) = L^k d(f(x), x)$$

hence, for $q \geq p \geq 1$,

$$d(x_q, x_p) \leq \sum_{k=p}^{q-1} d(x_{k+1}, x_k) \leq \sum_{k=p}^{q-1} L^k\, d(f(x), x) \leq \frac{L^p}{1-L} d(f(x), x) \tag{5.52}$$

since $L < 1$. The right-hand side of (5.52) converges to zero as $p \to \infty$, so that $\{x_n\}$ is a Cauchy sequence and thus converges to some $p = p(x) \in X$ which is a fixed point for f, see (iii) of Proposition 5.44. Since f has a unique fixed point, $p(x) = p$ $\forall x$ so that for every $x \in X$, $f^k(x) \to p$, i.e. p is the sink of f.

Combining the Banach fixed point theorem with (vi) of Proposition 5.44, we have the following.

Corollary 5.47 *Let X be a complete metric space and let $f : X \to X$ be continuous. If, for some $n \geq 1$, f^n is a contraction with contraction factor $L < 1$, then f has a unique sink $p \in X$ and for every $x \in X$*

$$d(f^k(x)), p) \leq \frac{L^{\left\lfloor \frac{k}{n} \right\rfloor}}{1-L} \max_{0 \leq j < n} d(f^n(x_j), x_j) \tag{5.53}$$

where for $j = 0, \dots, n-1$, we set $x_j := f^j(x)$. Moreover, if X is bounded, diam$(X) < \infty$, then

$$d(f^k(x)), p) \leq \text{diam}(X) L^{\lfloor \frac{k}{n} \rfloor}. \tag{5.54}$$

Proof. By the Banach fixed point theorem, f^n has a sink p and for every $x \in X$

$$d(f^n(x), p) \leq \frac{L^n}{1-L} d(f^n(x), x).$$

Consequently, p is the sink of f by (vi) of Proposition 5.44. In order to prove estimate (5.53), write every k as $k = qn + j$, $0 \leq j < n$, $q = \lfloor \frac{k}{n} \rfloor$. Then

$$d(f^k(x), p) = d(f^{qn}(f^j(x)), p) \leq \frac{L^q}{1-L} d(f^n(x_j), x_j).$$

Taking the maximum of the right-hand sides for $j = 0, \dots, n-1$, we get (5.53). Finally, with f^n being a contraction and p being a fixed point for f^n, we have

$$\begin{aligned} d(f^k(x)), p) = d(f^{qn}(f^j(x)), p) &\leq L\, d(f^{(q-1)n}(f^j(x)), p) \\ &\leq \cdots \leq L^q\, d(f^j(x), p) \leq L^q \,\text{diam}(X). \end{aligned}$$

5.5.2 Existence of fixed points

Let $\mathcal{T} \subset \mathbb{R}^N$ be the set of stochastic vectors and let $\mathbf{P}$ be a stochastic matrix, so that the linear map $P(x) := x\mathbf{P}$ maps $\mathcal{T}$ into itself. Since $\mathcal{T}$ is convex and compact, the Brouwer fixed point theorem applies to the map P and we have the following.

Theorem 5.48 (Perron–Frobenius) *There exists at least a stochastic vector w such that $w := w\mathbf{P}$.*

Example 5.49 Let $\mathbf{P} = \begin{pmatrix} 1 & 0 \\ 0 & 1 \end{pmatrix}$. Each vector of $\mathbb{R}^{2*}$ is a fixed point for the associated linear map $P(x) = x\mathbf{P}$ and P has no sink.

Let $\mathbf{P} = \begin{pmatrix} 0 & 1 \\ 1 & 0 \end{pmatrix}$, then the only fixed point for the map $P(x) := x\mathbf{P}$ in $\mathcal{T}$ is $w := (1/2, 1/2)$. Moreover,

$$\mathbf{P}^n = \begin{cases} \text{Id} & \text{if } n \text{ is even,} \\ \mathbf{P} & \text{if } n \text{ is odd,} \end{cases}$$

so that, if $x \in \mathcal{T}$, then $x\mathbf{P}^n$ converges to w if and only if $x = w$ and the map $P(x) := x\mathbf{P}$ has no sink.

Example 5.49 shows that, in general, a map can have more than one fixed point and, in general, the sequence $\{x\mathbf{P}^n\}$ does not converge.

Example 5.50 Let $\mathbf{P} = \begin{pmatrix} 1-\alpha & \alpha \\ \beta & 1-\beta \end{pmatrix}$ with $0 < \alpha, \beta < 1$. The only fixed point in $\mathcal{T}$ for $P(x) = x\mathbf{P}$ is $w := \frac{1}{\alpha+\beta}(\beta, \alpha)$. Moreover, computing $\mathbf{P}^n$ one can show that w is a sink for P,

$$x\mathbf{P}^n \to \frac{1}{\alpha+\beta}(\beta, \alpha) \qquad \forall x \in \mathcal{T}.$$

5.5.3 Regular stochastic matrices

Definition 5.51 *A finite stochastic matrix* $\mathbf{P}$ *is said to be* regular *if there exists* k_0 *such that all the entries of the matrix* $\mathbf{P}^{k_0}$ *are nonzero.*

Trivially, a regular stochastic matrix is irreducible.

5.5.3.1 Convergence

We now prove that if $\mathbf{P}$ is regular then $P(x) := x\mathbf{P}$ has a unique sink $w \in \mathcal{T}$, that is, $x\mathbf{P}^n \to w$ $\forall x \in \mathcal{T}$. In order to do that, we need to introduce some definitions.

For each $x \in \mathbb{R}^{N*}$, set

$$||x||_1 := \sum_{i=1}^{N} |x_i|. \tag{5.55}$$

Thus, $x \mapsto ||x||_1$ is a *norm* in $\mathbb{R}^{N*}$ and for each stochastic vector x, $||x||_1 = 1$.

With an induction argument on the dimension N, it can be also easily proven that

$$||x||_2 \le ||x||_1 \le \sqrt{N}||x||_2 \qquad \forall x \in \mathbb{R}^{N*}, \tag{5.56}$$

where $||x||_2 := \sqrt{\sum_{i=1}^{N} |x_i|^2}$ is the Euclidean norm of x.

Lemma 5.52 *The space* $\mathbb{R}^{N*}$*, equipped with the norm* $||\ ||_1$*, is complete. Thus* $\mathcal{T}$*, a closed subset of* $\mathbb{R}^{N*}$*, is a complete metric space.*

Proof. By inequality (5.56) a sequence in $\mathbb{R}^{N*}$ converges with respect to the $||\ ||_1$ norm if and only if it converges with respect to the $||\ ||_2$ norm. Similarly, a sequence in $\mathbb{R}^{N*}$ is a Cauchy sequence with respect to the $||\ ||_1$ norm if and only if it is a Cauchy sequence with respect to the $||\ ||_2$ norm. Thus, since $\mathbb{R}^{N*}$ is complete with respect to the Euclidean norm, it is also complete with respect to the $||\ ||_1$ norm.

Proposition 5.53 *Let* $\mathbf{P}$ *be a stochastic* $N \times N$ *matrix. Denote by* $R^1, \ldots, R^N$ *the rows of* $\mathbf{P}$ *and let* $K \subset \mathbb{R}^{N*}$ *be their convex envelope. Set*

$$C := \frac{1}{2} \max_{i,j} ||R^i - R^j||_1.$$

Then:

(i) $C \leq 1$.

(ii) If all the entries of $\mathbf{P}$ *are nonzero, then* $C < 1$.

(iii) $diam_1(K) := \sup_{x,y \in K} ||x - y||_1 = 2C$.

(iv) Let $P(x) := x\mathbf{P}$. *Then* $||P(x) - P(y)||_1 \leq C||x - y||_1 \ \forall x, y \in \mathcal{T}$.

Proof. (i) Each row of $\mathbf{P}$ belongs to $\mathcal{T}$, hence $||R^i - R^j||_1 \leq ||R^i||_1 + ||R^j||_1 = 2 \ \forall i, j \in \{1, \dots, N\}$.

(ii) Let $x = (x^1, \dots, x^N)$ and $y = (y^1, \dots, y^N) \in \mathcal{T}$ be such that $||x - y||_1 = 2 = ||x||_1 + ||y||_1$. Let $A = \{i \in \{1, \dots, N\} \,|\, x_i > y_i\}$ and $B := \{1, \dots, N\} \setminus A$. We claim that both A and B are nonempty sets. Assume by contradiction that A is empty, so that $x_i \leq y_i \ \forall i$. Since x and y are stochastic, then $x = y$, hence $||x - y||_1 = 0 \neq 2$, a contradiction. If B is empty, then $x_i > y_i \ \forall i$, so that $1 = ||x||_1 = \sum_{i=1}^N x_i > \sum_{i=}^N y_i = ||y||_1 = 1$, a contradiction.

We show that some component of both x and y is null. In fact,

$$\sum_{i=1}^N x_i + \sum_{i=1}^N y_i = 2 = ||x - y||_1 = \sum_{i \in A} (x_i - y_i) + \sum_{i \in B} (y_i - x_i)$$

i.e.

$$\sum_{i \in A} y_i + \sum_{i \in B} x_i = 0.$$

In particular, $y_i = 0 \ \forall i \in A$ and $x_i = 0 \ \forall i \in B$.

If each entry of $\mathbf{P}$ is nonzero, then, from the above, $||R^i - R^j||_1 < 2$ for any $i, j = 1, \dots, N$, i.e. $C < 1$.

(iii) Let $\delta := \text{diam}_1(K)$. By definition, $2C \leq \delta$. Let us show that $\delta \leq 2C$. Let $B(x, r) := \{y \,|\, ||y - x||_1 \leq r\}$ be the closed ball in $\mathbb{R}^{N*}$ with respect to the $|| \ ||_1$ norm centred at x and with radius r. By definition of C, $R^i \in B(R^j, 2C)$ for any $i, j \in \{1, \dots, N\}$. Since $B(R^j, 2C)$ is convex, then $K \subset B(R^j, 2C)$ for any j, or, equivalently, $R^j \in B(x, 2C) \ \forall x \in K$. Since $B(x, 2C)$ is convex, $K \subset B(x, 2C)$ $\forall x \in K$, i.e. $||x - y||_1 \leq 2C \ \forall x, y \in K$.

(iv) We show that, if $a = (a_1, \dots, a_N)$ is such that $\sum_{i=1}^N a_i = 0$, then

$$||P(a)||_1 \leq C \, ||a||_1. \tag{5.57}$$

Without loss of generality, we may assume $||a||_1 = 2$. For any $x \in \mathbb{R}^{N*}$ let $\sum_i' x_i = \sum_{x_i > 0} x_i$ and $\sum_i'' x_i = \sum_{x_i < 0} x_i$, so that $\sum_i' a^i = -\sum_i'' a^i = 1$. Thus, both the convex combinations $x' = \Sigma_i' a^i R^i$ and $x'' = -\Sigma_i'' a^i R^i$ belong to K, so that by (iii)

$$||P(a)||_1 = ||\sum_i a_i R^i||_1 = ||x' - x''||_1 \leq 2C = C \, ||a||_1.$$

Claim (iv) follows by choosing $a = x - y$ in (5.57).

Theorem 5.54 *Let* $\mathbf{P}$ *be a finite stochastic matrix. The following are equivalent:*

(i) $\mathbf{P}$ *is regular,*

(ii) $\mathbf{P}$ *is irreducible and the map* $P : \mathcal{T} \to \mathcal{T},\ P(x) = x\mathbf{P}$ *has a sink* $w \in \mathcal{T}$, *i.e.* $x\mathbf{P}^n \to w\ \forall x \in \mathcal{T}$.

(iii) $\exists \overline{n}$ *such that* $\forall n \geq \overline{n}$ *all the entries of* $\mathbf{P}^n$ *are nonzero.*

Moreover, in this case:

(a) w *is the only fixed point in* $\mathcal{T}$ *of the map* $x \mapsto x\mathbf{P}$.

(b) All the components of w *are nonzero.*

(c) Let $R^1, \dots, R^N$ *be the rows of* $\mathbf{P}^{k_0}$. *Then*

$$C := \frac{1}{2} \max_{i,j=1,\dots,N} ||R^i - R^j||_1 < 1$$

and, for any $x \in \mathcal{T}$,

$$||x\mathbf{P}^n - w||_1 \leq 2\,C^{\lfloor n/k_0 \rfloor}. \tag{5.58}$$

Proof. (i) $\Rightarrow$ (ii). If all the entries of $\mathbf{P}^{k_0}$ are positive, then all the states $1, \dots, N$ are pairwise communicating, i.e. $\mathbf{P}$ is irreducible. By Proposition 5.53 $C < 1$ so that $P^{k_0}(x) := x\mathbf{P}^{k_0}$ is a contraction on $\mathcal{T}$ with contraction factor less than or equal to C. Thus, applying the Banach fixed point theorem, P^{k_0} admits a unique sink w. Since $\text{diam}(\mathcal{T}) \leq 2$, Corollary 5.47 yields

$$||P^n(x) - w||_1 \leq 2C^{\left\lfloor \frac{n}{k_0} \right\rfloor}.$$

Finally, the components of $w = (w_1, \dots, w_n)$ are positive, see Exercise 5.65.

(ii) $\Rightarrow$ (iii). $P^n(x) = x\mathbf{P}^n$ converges to w for any $x \in \mathcal{T}$ if and only if $e^i\mathbf{P}^n \to w$ for any $i = 1, \dots, N$, i.e. if and only if each row of $\mathbf{P}^n$ converges w. Since $\mathbf{P}$ is irreducible, all the components of w are positive, see Exercise 5.65, so that all the entries of $\mathbf{P}^n$ are positive for n large enough.

(iii) $\Rightarrow$ (i). Obvious.

Remark 5.55 If $\mathbf{P}$ denotes the transition matrix of a Markov chain, the convergence of $x\mathbf{P}^n$ to $w\ \forall x \in \mathcal{T}$ reads as follows: for any $j \in \{1, \dots, N\}$ we have

$$\mathbb{P}(X_n = j) = \sum_{i=1}^{N} \mathbb{P}(X_0 = i)\, (\mathbf{P}^n)^i_j \longrightarrow w_j;$$

in other words, *whatever the distribution of the initial random variable* X_0, *the mass distributions of the* X_n*'s converge to the stochastic vector* $w = (w_j)$.

We may rewrite the claim of Theorem 5.54 in terms of matrices. With the notation above, let $w \in \mathcal{T}$ be the fixed point of the map $P(x) := x\mathbf{P}$ and let $\mathbf{W}$ be the $N \times N$ stochastic matrix whose rows coincide with w,

$$\mathbf{W} := w(1, 1, \ldots, 1)^T.$$

Then:

(i) $x\mathbf{P}^n \to w \ \forall x \in \mathcal{T}$ if and only if $\mathbf{P}^n \to \mathbf{W}$.

(ii) The exponential rate of convergence in (5.58) reads

$$||\mathbf{P}^n - \mathbf{W}|| \leq 2\, C^{\lfloor n/k_0 \rfloor}.$$

where

$$||\mathbf{A}|| = \max_{i=1,\ldots,N} ||A^i||_1.$$

(iii) $w\mathbf{P} = w$ if and only if $\mathbf{WP} = \mathbf{W}$.

Finally, since all the rows of $\mathbf{W}$ are the same stochastic vector w and $\mathbf{W} = \mathbf{WP}$, the following equalities can be easily proven

$$\mathbf{W} = \mathbf{W}^2 = \mathbf{WP} = \mathbf{PW}. \tag{5.59}$$

5.5.3.2 Eigenvalues

Proposition 5.56 *Let* $\mathbf{P}$ *be a* $N \times N$ *stochastic matrix. Then* 1 *is an eigenvalue for* $\mathbf{P}$ *with corresponding eigenvector* $v := (1, 1, \ldots, 1)^T$. *Moreover, for any eigenvalue* $\lambda \in \mathbb{C}$ *of* $\mathbf{P}$ *we have* $|\lambda| \leq 1$.

Proof. Since $\mathbf{P}$ is stochastic, trivially, $\mathbf{P}v = v$.

Let $\lambda \in \mathbb{C}$ be an eigenvalue of $\mathbf{P}$ and $z = (z^1, \ldots, z^N) \neq 0$ be an eigenvector of λ, $\lambda z = \mathbf{P}z$. Let i_0 be such that $|z^{i_0}| = \max_i |z^i|$ so that $|z^{i_0}| \neq 0$ and

$$|\lambda|\,|z^{i_0}| = |(\mathbf{P}z)^{i_0}| = \left|\sum_{j=1}^N \mathbf{P}_j^{i_0} z^j\right| \leq \sum_{i=1}^N \mathbf{P}_j^{i_0} |z^j| \leq |z^{i_0}| \sum_{j=1}^N \mathbf{P}_j^{i_0} = |z^{i_0}|.$$

Therefore $|\lambda| \leq 1$.

The next theorem characterizes the eigenvalues of regular stochastic matrices. We omit the proof, which can be found, e.g. in [16] Proposition 9.2.

Theorem 5.57 *Let* $\mathbf{P}$ *be a regular stochastic matrix. Then* 1 *is the only eigenvalue of* $\mathbf{P}$ *of modulus* 1. *Its algebraic and geometric multiplicities are* 1. *Consequently, all the eigenvalues of* $\mathbf{P}$ *but* 1 *have modulus strictly less than* 1.

Since $\mathbf{P}$ is regular, one can easily compute the powers of $\mathbf{P}$ and their limit as $n \to \infty$ by Theorem 5.57. In fact, Theorem 5.57 implies that $\mathbf{P}$ has a Jordan decomposition $\mathbf{P} = \mathbf{SJS}^{-1}$ where the first column of $\mathbf{S}$ is an eigenvector of $\mathbf{P}$ relative to the eigenvalue 1, i.e. $(1, 1, \dots, 1)^T$,

$$\mathbf{J} = \begin{pmatrix} 1 & 0 & 0 & \dots & 0 \\ 0 & \boxed{\mathbf{J}_1} & 0 & \dots & 0 \\ 0 & 0 & \boxed{\mathbf{J}_2} & \dots & 0 \\ \vdots & \vdots & \vdots & \ddots & 0 \\ 0 & 0 & 0 & \dots & \boxed{\mathbf{J}_p} \end{pmatrix}$$

and $\mathbf{J}_1, \mathbf{J}_2, \dots, \mathbf{J}_p$ are Jordan blocks relative to the eigenvalues of $\mathbf{P}$ but 1. Therefore, from $\mathbf{P} = \mathbf{SJS}^{-1}$ we infer $\mathbf{P}^n = \mathbf{SJ}^n\mathbf{S}^{-1}$ where

$$\mathbf{J}^n = \begin{pmatrix} 1 & 0 & 0 & \dots & 0 \\ 0 & \boxed{\mathbf{J}_1^n} & 0 & \dots & 0 \\ 0 & 0 & \boxed{\mathbf{J}_2^n} & \dots & 0 \\ \vdots & \vdots & \vdots & \ddots & 0 \\ 0 & 0 & 0 & \dots & \boxed{\mathbf{J}_r^n} \end{pmatrix}.$$

Since all the eigenvalues but 1 have modulus strictly smaller than 1, $\mathbf{J}^n$ converges to

$$\mathbf{K} := \begin{pmatrix} 1 & 0 & \dots & 0 \\ 0 & 0 & \dots & 0 \\ \vdots & \vdots & \ddots & \vdots \\ 0 & 0 & \dots & 0 \end{pmatrix}$$

and, for each $\epsilon > 0$ there exists a constant C_ϵ such that

$$||\mathbf{J}^n - \mathbf{K}|| \le C_\epsilon(\lambda_* + \epsilon)^n \qquad \text{as } n \to \infty$$

where

$$\lambda_* := \max\left\{|\lambda| \,\middle|\, \lambda \text{eigenvalue of } \mathbf{P}, \, |\lambda| < 1\right\}.$$

We then infer

$$(\mathbf{P}^n)^i_j \to (\mathbf{SKS}^{-1})^i_j = \mathbf{S}^i_1(\mathbf{S}^{-1})^1_j = (1, 1, \dots, 1)^T w, \tag{5.60}$$

where $w = (w_1, \dots, w_N)$ is the row vector defined by $w_j := (\mathbf{S}^{-1})^1_j \ \forall j$. Denoting by $\mathbf{W}$ the matrix with each row equal to w, we conclude from (5.60) that

$$\mathbf{P}^n \to \mathbf{W}$$

with an exponential rate of convergence driven by λ_*: for any $\epsilon > 0$ there exists C_ϵ such that

$$||\mathbf{P}^n - \mathbf{W}|| \le C_\epsilon(\lambda_* + \epsilon)^n.$$

Finally, observe that $\mathbf{W}$ is a stochastic matrix, being the limit of stochastic matrices. Moreover, cf. Exercise 5.65, all its entries are positive.

5.5.3.3 Fixed point and expected return times

Let $\{X_n\}$ be a homogeneous Markov chain with a finite or denumerable state-space S and transition matrix $\mathbf{P} = (p_{ij})$. Recall that for $i, j \in S$, $f_{ij}^{(k)}$ denotes the probability that a path from i visits for the first time j at step k.

Proposition 5.58 *For any $n \geq 1$ we have*

$$p_{ij}^{(n)} = \sum_{k=1}^{n} f_{ij}^{(k)} p_{jj}^{(n-k)}. \tag{5.61}$$

Proof. Let $k \in \{1, \dots, n\}$ and let

$$F_{k,n} := \left\{x \in \Omega \,\middle|\, X_1(x) \neq j, \dots, X_{k-1}(x) \neq j, X_k(x) = j, X_n(x) = j\right\}$$

be the set of paths that visit state j for the first time at step k and then return to j again at time n. Let

$$E := \left\{x \in \Omega \,\middle|\, X_n = j\right\},$$

$$F := \left\{x \in \Omega \,\middle|\, X_k(x) = j, X_{k-1}(x) \neq j, \dots, X_1(x) \neq j\right\},$$

$$G := \left\{x \in \Omega \,\middle|\, X_0 = i\right\}.$$

By (5.8) and the homogeneity of the chain, we infer

$$\begin{aligned}
\mathbb{P}\left(F_{k,n} \,\middle|\, X_0 = i\right) &= \mathbb{P}(E \mid F)\mathbb{P}(F \mid G) \\
&= \mathbb{P}\left(X_n = j \,\middle|\, X_k = j\right) \\
&\qquad \mathbb{P}\left(X_k = j, X_{k-1} \neq j, \dots, X_1 \neq j \,\middle|\, X_0 = i\right) \\
&= p_{jj}^{(n-k)} f_{ij}^{(k)}.
\end{aligned}$$

Since the sets $\{F_{k,n}\}$, $k = 1, \dots, n$, are a partition of the set $\{x \in \Omega \mid X_n = j\}$, we get

$$p_{ij}^{(n)} = \mathbb{P}\left(X_n = j \,\middle|\, X_0 = i\right) = \sum_{k=1}^{n} \mathbb{P}\left(F_{k,n} \,\middle|\, X_0 = i\right) = \sum_{k=1}^{n} f_{ij}^{(k)} p_{jj}^{(n-k)}.$$

Lemma 5.59 *Let $\{p_n\}$ and $\{f_n\}$ be two sequences of non-negative numbers such that $p_0 = 1$, $\sum_{i=1}^{\infty} f_i = 1$ and*

$$p_n = \sum_{j=1}^{n} f_j p_{n-j} \qquad \forall n \geq 1. \tag{5.62}$$

If $p_n \to a \in \overline{\mathbb{R}}_+$ (finite or infinite), then

$$a = \frac{1}{T} \qquad T := \sum_{k=1}^{\infty} k f_k.$$

Proof. Summing equalities (5.62) from 1 to N, we get

$$\sum_{n=1}^{N} p_n = \sum_{n=1}^{N} \sum_{j=1}^{n} f_j p_{n-j} = \sum_{n=1}^{N} \Big(\sum_{j=0}^{n-1} f_{n-j} p_j \Big), \tag{5.63}$$

or, equivalently,

$$p_N + \sum_{j=1}^{N-1} \Big(1 - \sum_{i=1}^{N-j} f_i \Big) p_j = \sum_{j=1}^{N} f_j. \tag{5.64}$$

or, taking into account that $\sum_{j=1}^{\infty} f_j = 1$,

$$\sum_{j=1}^{N} p_j \Big(\sum_{i=N-j+1}^{\infty} f_i \Big) = \sum_{j=1}^{N} f_j. \tag{5.65}$$

For each $j \geq 1$, let $T_j := \sum_{i=j}^{\infty} f_i$. Then $T_1 = \sum_{i=1}^{\infty} f_i = 1$, and

$$\sum_{j=1}^{\infty} T_j = \sum_{j=1}^{\infty} \sum_{i=j}^{\infty} f_i = \sum_{i=1}^{\infty} \sum_{j=1}^{i} f_i = \sum_{i=1}^{\infty} i f_i = T.$$

Thus, (5.64) or (5.65) reads $\sum_{j=1}^{N} p_j T_{N-j-1} = \sum_{j=1}^{N} f_j$, i.e.

$$\sum_{j=1}^{N} T_j p_{N+1-j} = \sum_{j=1}^{N} p_j T_{N-j+1} = \sum_{j=1}^{N} f_j. \tag{5.66}$$

Let $\{\varphi_N(j)\}$ be the double indexed sequence defined as

$$\varphi_N(j) := \begin{cases} T_j p_{N+1-j} & \text{if } 1 \leq j \leq N, \\ 0 & \text{if } j > N. \end{cases}$$

Then equality (5.66) reads

$$\sum_{j=1}^{\infty} \varphi_N(j) = \sum_{j=1}^{N} f_j. \tag{5.67}$$

Since, by assumption, for any $j \in S$ $\varphi_N(j) \to aT_j$ as $N \to \infty$, the following hold:

- If $T = \sum_{j=1}^{\infty} T_j = +\infty$, then the Fatou lemma, Lemma B.51, together with (5.67), gives

$$\begin{aligned} a\,T = a\sum_{j=1}^{\infty} T_j &= \sum_{j=1}^{\infty} \lim_{N\to\infty} \varphi_N(j) \\ &\leq \liminf_{N\to\infty} \sum_{j=1}^{\infty} \varphi_N(j) = \sum_{i=1}^{\infty} f_i = 1 \end{aligned}$$

  so that $a = 0$.

- If $T < +\infty$, then $\sup_N \varphi_N(j) \leq T_j \ \forall j$, and $\sum_{j=1}^{\infty} T_j = T < +\infty$. Thus, the dominated convergence theorem for series, see Example B.56, yields

$$a\,T = a\sum_{j=1}^{\infty} T_j = \sum_{j=1}^{\infty} \lim_{N\to\infty} \varphi_N(j) = \lim_{N\to\infty} \sum_{j=1}^{\infty} \varphi_N(j) = \sum_{i=1}^{\infty} f_i = 1.$$

The proof is thus complete.

Lemma 5.60 *Let $\{X_n\}$ be a homogeneous Markov chain with finite or denumerable state-space S and transition matrix $\mathbf{P} = (p_{ij})$ and let $j \in S$. If $p_{jj}^{(n)} \to \lambda$, then*

$$p_{ij}^{(n)} \to f_{ij}\lambda;$$

Similarly, if $\frac{1}{n}\sum_{k=1}^{n} p_{jj}^{(k)} \to \lambda$, then $\frac{1}{n}\sum_{k=1}^{n} p_{ij}^{(k)} \to f_{ij}\lambda$.

Proof. Equality (5.61) can be written as

$$p_{ij}^{(n)} = \sum_{k=1}^{n} f_{ij}^{(k)} p_{jj}^{(n-k)} = \sum_{k=1}^{\infty} \varphi_n(k)$$

where

$$\varphi_n(k) := \mathbb{1}_{\{1,\dots,n\}}(k) f_{ij}^{(k)} p_{jj}^{(n-k)}.$$

Since for any fixed k, $\varphi_n(k) \to f_{ij}^{(k)}\lambda$ as $n \to \infty$, applying the dominated convergence theorem for series, Example B.56, we get

$$\begin{aligned} \lim_{n\to\infty} p_{ij}^{(n)} = \lim_{n\to\infty} \sum_{k=1}^{\infty} \varphi_n(k) &= \sum_{k=1}^{\infty} \lim_{n\to\infty} \varphi_n(k) \\ &= \Big(\sum_{k=1}^{\infty} f_{ij}^{(k)}\Big)\lambda = f_{ij}\lambda. \end{aligned}$$

Similarly, one proves the last claim, writing this time

$$\frac{1}{n}\sum_{k=1}^{n} p_{ij}^{(k)} = \sum_{r=1}^{\infty} \psi_n(r)$$

where $\psi_n(r) := \mathbb{1}_{\{1,\dots,n\}}(r) f_{ij}^{(r)} \Big(\frac{1}{n}\sum_{k=1}^{n-r} p_{jj}^{(k)}\Big)$.

Theorem 5.61 *Let $\{X_n\}$ be a homogeneous Markov chain with finite or denumerable state-space S and transition matrix $\mathbf{P} = (p_{ij})$. For any $i, j \in S$ let $\overline{T}_{jj} \in \overline{\mathbb{R}}$ and f_{ij} be the expected return time in state j and the probability of visiting j starting from i, respectively. If $p_{jj}^{(n)} \to w_j$, then*

$$w_j = \frac{1}{\overline{T}_{jj}} \qquad \textit{and} \qquad p_{ij}^{(n)} \to f_{ij} w_j.$$

Notice that, if j is transient, then trivially $w_j = 0 = f_{ij} w_j$. Moreover, if j is recurrent and $i \to j$, then $f_{ij} = 1$ so that $f_{ij} w_j = w_j$.

Proof. In order to prove that $w_j := \frac{1}{\overline{T}_{jj}}$, we distinguish two cases:

- Assume j is transient, i.e. $f_{jj} < 1$. Thus, $\overline{T}_{jj} = +\infty$ by (5.34) and $\sum_{n=i}^{\infty} p_{jj}^{(n)} < +\infty$ by (5.32). Hence, $p_{jj}^{(n)} \to 0 = 1/\overline{T}_{jj} = w_j$.
- Assume j is recurrent, $f_{jj} = 1$. Then, by (5.34) $\overline{T}_{jj} = \sum_{j=1}^{\infty} k f_{jj}^{(k)}$ and, by Proposition 5.58,

$$p_{jj}^{(n)} = \sum_{k=1}^{n} f_{jj}^{(k)} p_{jj}^{(n-k)} \qquad \forall n \geq 1.$$

The claim now follows applying Lemma 5.59 to the sequences $\{p_n\}$ and $\{f_k\}$ where $p_n := p_{jj}^{(n)}$ and $f_k := f_{jj}^{(k)}$.

The last claim follows from Lemma 5.60.

5.5.4 Characteristic parameters

For the reader's convenience we summarize the formulas that can be used to compute the characteristic parameters introduced in Section 5.3 in terms of the transition matrix $\mathbf{P}$. Assuming $\mathbf{P}$ is reordered in a canonical form (5.47), then (5.48) gives:

- the expected number of visits, see (5.30);
- the probability that a path from i visits j at least once, see (5.32);
- the probability of many passages in j starting from i, see (5.28).

5.5.4.1 Waiting time matrix

In Section 5.3.4 we have proven that if j is transient then the expected return time $\overline{T}_{jj}$ is infinite, while, if j is recurrent, then $\overline{T}_{jj} = \sum_{k=1}^{\infty} k f_{jj}^{(k)}$, see formula (5.35).

We now deal with the waiting time matrix $\overline{T} = (\overline{T}_j^i)$, where $\overline{T}_j^i := \mathbb{E}_i[t_j]$ is the conditional expected first passage time to j starting from i. We only consider homogeneous Markov chains $\{X_n\}$ with a *finite* state-space S so that we can take advantage of the canonical decomposition of its transition matrix $\mathbf{P}$ and therefore confine ourselves to the case when $\mathbf{P}$ is irreducible.

We have already shown that, if $\mathbf{P}$ is regular, then $\mathbf{P}$ has a unique stochastic fixed point, $w = (w_j) \in \mathcal{T}$ with non-zero components, cf. Theorem 5.54. Moreover, the expected return time to the state j is $\overline{T}_{jj} = 1/w_j$, in particular, j is positive recurrent. This fact has already been proven if $\mathbf{P}$ is regular, see Theorems 5.54 and 5.61, but the same holds if $\mathbf{P}$ is irreducible, see Theorem 5.71 and Remark 5.82. In other words, the computation of the diagonal entries of $\mathbf{T}$, i.e. of the expected return times, boils down to the computation of the unique solution $w \in \mathcal{T}$ of the equation $w = w\mathbf{P}$.

In order to write an explicit formula for all the entries of $\mathbf{T}$, let us introduce a few notations. For any square matrix $\mathbf{A} = (a_{ij})$, let $\mathbf{A}_d := (a_{ij}\delta_{ij})$. Also let $\mathbf{E}$ be the square matrix whose entries are all equal to 1. Clearly, $(\mathbf{AB}_d)_d = \mathbf{A}_d\mathbf{B}_d$ for any $\mathbf{A}$ and $\mathbf{B}$, and $\mathbf{EA}_d$ is such that all its rows are equal to the row vector of the diagonal elements of $\mathbf{A}$. Finally, let $\mathbf{W}$ be the matrix whose rows agree with the stochastic vector $w \in \mathcal{T}$ such that $w = w\mathbf{P}$. Then

$$\mathbf{T}_d = ((1/\mathbf{W}^i_j)\delta_{ij}) = (\delta_{ij}/w_j).$$

Proposition 5.62 (Expected first passage time matrix) *The matrix*

$$\mathrm{Id} - (\mathbf{P} - \mathbf{W})$$

is invertible. Denoting $\mathbf{Z} := (\mathrm{Id} - (\mathbf{P} - \mathbf{W}))^{-1}$, *we have*

$$\mathbf{T} = (\mathrm{Id} - \mathbf{Z} + \mathbf{EZ}_d)\mathbf{T}_d, \tag{5.68}$$

i.e. $\mathbf{T}^i_j = (\delta^i_j - \mathbf{Z}^i_j + \mathbf{Z}^j_j)\frac{1}{w_j}$ $\forall i, j \in S$.

Proof. Since $\mathbf{W} = \mathbf{W}^2 = \mathbf{PW} = \mathbf{WP}$, cf. (5.59), we have $(\mathbf{P} - \mathbf{W})^2 = \mathbf{P}^2 - 2\mathbf{PW} + \mathbf{W}^2 = \mathbf{P}^2 - \mathbf{W}$. Moreover, it can be proved (either with an induction argument or using Newton binomial theorem) that

$$(\mathbf{P} - \mathbf{W})^k = \mathbf{P}^k - \mathbf{W} \qquad \forall k.$$

Since $\mathbf{P}^n \to \mathbf{W}$, the previous formula yields $(\mathbf{P} - \mathbf{W})^k \to 0$. Therefore, one can pass to the limit as $n \to \infty$ in the identity

$$(\mathrm{Id} - (\mathbf{P} - \mathbf{W}))\sum_{k=0}^{n}(\mathbf{P} - \mathbf{W})^k = \mathrm{Id} - (\mathbf{P} - \mathbf{W})^{n+1}$$

to get both the invertibility of $\mathrm{Id} - (\mathbf{P} - \mathbf{W})$ and a formula for its inverse $\mathbf{Z}$,

$$\mathbf{Z} = \sum_{k=0}^{\infty}(\mathbf{P} - \mathbf{W})^k = \mathrm{Id} + \sum_{k=1}^{\infty}(\mathbf{P}^k - \mathbf{W}).$$

We now prove (5.68) in three steps.

Step 1. From Proposition 5.31 we infer

$$\mathbb{E}_i[t_j] = 1 + \sum_{\ell \neq j} p_{i\ell}\mathbb{E}_\ell[t_j],$$

i.e.

$$\mathbf{T} = \mathbf{E} + \mathbf{P}(\mathbf{T} - \mathbf{T}_d).$$

Therefore, $\mathbf{T}$ is a solution of the system

$$\begin{cases} \mathbf{X}_d = \mathbf{T}_d, \\ \mathbf{X} = \mathbf{P}(\mathbf{X} - \mathbf{T}_d) + \mathbf{E}. \end{cases} \tag{5.69}$$

Step 2. Let $\mathbf{M}$ be the right-hand side of (5.68),

$$\mathbf{M} := (\mathrm{Id} - \mathbf{Z} + \mathbf{E}\mathbf{Z}_d)\mathbf{T}_d.$$

We claim that the matrix $\mathbf{M}$ satisfies system (5.69). In fact, $\mathbf{M}_d = \mathbf{T}_d$ since all the diagonal entries of $(\mathrm{Id} - \mathbf{Z} + \mathbf{E}\mathbf{Z}_d)$ are equal to 1. In order to prove that $\mathbf{M}$ satisfies also the second equation in (5.69), let us point out the followings:

(i) $\mathrm{Id} - \mathbf{Z} = \mathbf{W} - \mathbf{P}\mathbf{Z}$: to prove it, multiply on the right by $\mathbf{Z}$ both sides of the equality $\mathbf{Z}^{-1} = \mathrm{Id} - \mathbf{P} + \mathbf{W}$ and recall that $\mathbf{W}\mathbf{Z}^{-1} = \mathbf{W}$, i.e $\mathbf{W} = \mathbf{W}\mathbf{Z}$.

(ii) $\mathbf{W}\mathbf{T}_d = \mathbf{E}$, which follows from the definition of $\mathbf{W}$.

(iii) $\mathbf{P}\mathbf{E}\mathbf{Z}_d = \mathbf{E}\mathbf{Z}_d$ which easily follows from the identities $\sum_{j=1}^{N} p_{ij} = 1$ $\forall i \in S$.

Thus we have

$$\begin{aligned} \mathbf{P}(\mathbf{M} - \mathbf{T}_d) + \mathbf{E} &= \mathbf{P}(-\mathbf{Z} + \mathbf{E}\mathbf{Z}_d)\mathbf{T}_d + \mathbf{E} = (-\mathbf{P}\mathbf{Z} + \mathbf{E}\mathbf{Z}_d)\mathbf{T}_d + \mathbf{E} \\ &= \mathbf{M} - \mathbf{W}\mathbf{T}_d + \mathbf{E} = \mathbf{M}. \end{aligned}$$

Step 3. We now show that system (5.69) has a unique solution $\mathbf{R}$, so that $\mathbf{T} = \mathbf{M}$ as claimed. Let $\mathbf{M}_1$ and $\mathbf{M}_2$ be two different solutions to system (5.69) and let $\mathbf{R} := \mathbf{M}_1 - \mathbf{M}_2$. Then, the diagonal entries of $\mathbf{R}$ are zero and $\mathbf{R} = \mathbf{P}\mathbf{R}$. Let r be a column of $\mathbf{R}$ so that $r = \mathbf{P}r$. Consequently, $r = \mathbf{P}^k r$ for any integer k and, passing to the limit as $k \to \infty$, we get $\mathbf{W}r = r$. Since the rows of $\mathbf{W}$ are equal, then so are the components of r, and since the diagonal of $\mathbf{R}$ is null, then all the components of r are null. Hence r is the null vector, and since r is an arbitrary column of $\mathbf{R}$, we conclude that $\mathbf{R} = 0$, i.e. $\mathbf{M}_1 = \mathbf{M}_2$.

5.5.5 Exercises

Exercise 5.63 *Prove Proposition 5.44.*

Solution. (i) Trivial.

(ii) If $f^n(p) = p$, then for every $j \geq 1$ $f^n(f^j(p)) = f^j(f^n(p)) = f^j(p)$, i.e. $f^j(p)$ is a fixed point for f^n. If f^n has a unique fixed point p, then $p = f(p)$, i.e. p is a fixed point for f, actually the unique fixed point by (i).

(iii) If $f^k(x) \to p$, then $f^{k+1}(x) \to p$ and $f^{k+1}(x) = f(f^k(x)) \to f(p)$ by continuity, hence $f(p) = p$.

(iv) Let p_1 be another fixed point for f. Then $f^k(p_1) = p_1$ for every k, i.e. $f^k(p_1)$ is constant and converges to p, hence $p = p_1$.

(v) Consider $f(x) = -x$, $x \in [-1, 1]$. Then 0 is a fixed point but for every $x \neq 0$ $f^k(x) = (-1)^k x$.

(vi) If $f^k(x) \to p$, then trivially $(f^n)^k(x) = f^{kn}(x) \to p$. Conversely, if p is the sink of f^n, then p is the unique fixed point of f^n by (iv) and consequently the unique fixed point of f by (ii). Moreover, for every $j \geq 0$, $f^{hn+j}(x) = f^j(f^{hn}(x)) \to f^j(p) = p$ as $h \to \infty$. Writing every k as $k = hn + j$, $0 \leq j < n$, we conclude that $f^k(x) \to p$ as $k \to +\infty$.

Exercise 5.64 *Let* $\mathbf{P} = (p_{ij})$ *be a* $N \times N$ *irreducible stochastic matrix. If there exists* $h \in \{1, \ldots, N\}$ *such that* $p_{hh} > 0$, *then* $\mathbf{P}$ *is regular.*

Solution. $\mathbf{P}$ is irreducible, hence for every $\alpha, \beta \in \{1, \ldots, N\}$ there exists $n = n(\alpha, \beta)$ such that $p_{\alpha\beta}^{(n(\alpha,\beta))} > 0$. Let $k_0 := 2\max_{\alpha,\beta} n(\alpha, \beta) + 1$. Then, for any $i, j \in \{1, \ldots, N\}$, we get

$$p_{ij}^{(k_0)} \geq p_{ih}^{(n(i,h))} p_{hh}^{(q)} p_{hj}^{(n(h,j))} \geq p_{ih}^{(n(i,h))} (p_{hh})^q p_{hj}^{(n(h,j))} > 0$$

where $q := k_0 - n(i, h) - n(h, j) \geq 1$.

Exercise 5.65 *Let* $\mathbf{P} = (p_{ij})$ *be an irreducible* $N \times N$ *stochastic matrix and let* $j \in \{1, \ldots, N\}$. *Show that, if* $\frac{1}{n}\sum_{k=1}^{n} p_{jj}^{(k)}$ *converges to* λ *(in particular, if* $p_{jj}^{(n)} \to \lambda$*), then* $\lambda > 0$.

Solution. Assume, by contradiction, that $\frac{1}{n}\sum_{k=1}^{n} p_{jj}^{(k)} \to 0$ converges to zero. Since $\mathbf{P}$ is irreducible, for any $i \in \{1, \ldots, N\}$ there exists $k_0 = k_0(i)$ such that $p_{ij}^{(k_0)} > 0$. Thus $p_{ji}^{(k)} p_{ij}^{(k_0)} \leq p_{jj}^{(k+k_0)}$ and

$$\lim_{n\to\infty} \left(\frac{1}{n}\sum_{k=1}^{n} p_{ji}^{(k)}\right) p_{ij}^{(k_0)} \leq \lim_{n\to\infty} \left(\frac{1}{n}\sum_{k=1}^{n} p_{jj}^{(k+k_0)}\right) = 0.$$

Therefore, $\frac{1}{n}\sum_{k=1}^{n} p_{ji}^{(k)} \to 0$ $\forall i$, a contradiction since $\sum_{i=1}^{N} p_{ji}^{(k)} = 1$.

Exercise 5.66 *Let* $\{X_n\}$ *be a homogeneous Markov chain with a* finite *state-space* S *and a* regular *transition matrix* $\mathbf{P} = (p_{ij})$. *For* $j \in S$, *let* $\overline{T}_{jj} \in \overline{\mathbb{R}}$ *be the* *expected return time to j. Show that $p_{ij}^{(n)} \to \frac{1}{\overline{T}_{jj}}$ $\forall i, j \in S$. Conclude that each state is positive recurrent, $\overline{T}_{jj} < +\infty$ $\forall j \in S$.*

Solution. Since $\mathbf{P}$ is regular, $\mathbf{P}$ has a unique sink $w = (w_j)$, i.e. $p_{ij}^{(n)} \to w_j$, $\forall i, j \in S$, and, moreover, $w_j > 0$ $\forall j$, cf. Theorem 5.54. Finally, Theorem 5.61 yields $w_j = \frac{1}{\overline{T}_{jj}}$.

Exercise 5.67 *Prove the following: let $\mathbf{A}$ be a $N \times N$ complex matrix. For any $i = 1, \dots N$ set $x_i := \mathbf{A}_i^i$, $r_i := \sum_{j \neq i} |\mathbf{A}_j^i|$ and let B_i be the closed ball $B_i := \{z \in \mathbb{C} \,|\, |z - x_i| \leq r_i\}$. Then the eigenvalues of $\mathbf{A}$ are contained in $\bigcup_{i=1,\dots,N} B_i$.*

Solution. Let λ be an eigenvalue of $\mathbf{A}$ and let $z = (z^1, \dots, z^N)$ be an eigenvector of λ, $\mathbf{A}z = \lambda z$. Let $h = h(\lambda)$ be such that $|z^h| = \max_i |z^i|$. Then

$$|\lambda - x_h|\,|z^h| = |(\mathbf{A}z)^h - \mathbf{A}_h^h z^h| = \Big|\sum_{j \neq h} \mathbf{A}_j^h z^j\Big|$$

$$\leq \Big(\sum_{j \neq h} |\mathbf{A}_j^h|\Big) \max_i |z^i| \leq r_h |z^h|.$$

Therefore $|\lambda - x_h| \leq r_h$, i.e. $\lambda \in B_h$.

5.6 Ergodic property

Let $\{X_n\}$ be a homogeneous Markov chain whose state-space S is either finite or denumerable. Assume its transition matrix $\mathbf{P}$ is irreducible. Let $\overline{T}_{jj}$ be the expected return time to state j and let $V_j^{(n)}(x)$ be the number of visits in j made by the path x in the first n steps. Roughly speaking, the *ergodic theorem* says that in the long run the frequency of the visits in state j converges almost surely to $\frac{1}{\overline{T}_{jj}}$,

$$\frac{V_j^{(n)}(x)}{n} \longrightarrow \frac{1}{\overline{T}_{jj}} \qquad \mathbb{P}\text{-almost surely.}$$

5.6.1 Number of steps between consecutive visits

Let $\{X_n\}$ be a homogeneous Markov chain on a probability space $(\Omega, \mathcal{E}, \mathbb{P})$ with a finite or denumerable state-space S and let $j \in S$. Closely related to the sequence of passage times $\{T_j^{(r)}\}_r$ in j are the random variables $\{S_j^{(r)}(x)\}$ $r = 1, \dots,$ where $S_j^{(1)} = T_j^{(1)} = t_j$ and, for $r \geq 2$,

$$S_j^{(r)}(x) := \begin{cases} T_j^{(r)}(x) - T_j^{(r-1)}(x) & \text{if } T_j^{(r-1)}(x) < +\infty, \\ 0 & \text{otherwise.} \end{cases}$$

$S_j^{(r)}(x)$ counts the number of steps between the $(r-1)$th visit and the rth visit in j.

Theorem 5.68 *Let $j \in S$ be a recurrent state and let $i \to j$. Set $\Omega_i := \{x \mid X_0(x) = i\}$. Assume $\mathbb{P}(\Omega_i) > 0$ and let $\mathbb{P}_i(A) := \mathbb{P}(A \mid X_0 = i)$. Then:*

(i) $S_j^{(r)}(x) \ge 1$ *for almost any* $x \in \Omega_i$ *and for all* $r \ge 1$.

(ii) $\left\{S_j^{(r)}\right\}_r$ *is a sequence of independent random variables with respect to* $\mathbb{P}_i$.

(iii) The variables $S_j^{(r)}$, $r \ge 2$, *are identically distributed with respect to the probability measure* $\mathbb{P}_i$ *and, for every* $n \ge 1$ *and* $r \ge 2$,

$$\mathbb{P}_i(S_j^{(r)} = n) = f_{jj}^{(n)} \qquad \textit{and} \qquad \mathbb{E}_i[S_j^{(r)}] = \overline{T}_{jj}.$$

Proof. (i) Since j is recurrent and $i \to j$, paths from i visit j infinitely many times almost surely, see Proposition 5.30. Therefore, for any $r \ge 2$, $T_j^{(r)}(x) < +\infty$ for a.e. $x \in \Omega_i$. In particular, $1 \le S_j^{(r)}(x) < +\infty$ for almost every $x \in \Omega_i$.

(ii) Let $r \ge 2$, $T := T_j^{(r-1)}$, $h_1, \dots, h_{r-1} \ge 1$, $h = (h_1, \dots, h_{r-1}) \in \mathbb{N}^{r-1}$, $r \ge 1$ and set

$$H_h := \left\{S_j^{(r-1)} = h_{r-1}, S_j^{(r-2)} = h_{r-2}, \dots, S_j^{(1)} = h_1\right\}.$$

Observe that:

- The events $\left\{x \in \Omega_i \mid S_j^{(r)}(x) \ge 1\right\}$ are detected by the random variables $X_{T+1}, X_{T+2}, \dots$.
- The event H_h is generated by the random variables $X_0, \dots, X_k$ where $k := h_1 + \dots + h_{r-1} \le T$, and $H_h \subset \{X_k = j\}$.
- T is a stopping time and $T(x) < \infty$ for a.e. $x \in \Omega_i$, that is, $\mathbb{P}_i$-almost surely.

Then the strong Markov property, Theorem 5.11, gives

$$\begin{aligned}
\mathbb{P}_i\left(S_j^{(r)} = n \,\middle|\, H_h\right) &= \mathbb{P}_i\left(S_j^{(r)} = n \,\middle|\, \{X_k = j\} \cap H_h \cap \{T < \infty\}\right) \\
&= \mathbb{P}_i\left(S_j^{(r)} = n \,\middle|\, T < \infty, X_k = j\right) \\
&= \mathbb{P}\left(S_j^{(r)} = n \,\middle|\, T < +\infty, X_k = j\right) \\
&= \mathbb{P}\left(S_j^{(1)} = n \,\middle|\, X_0 = j\right) = f_{jj}^{(n)}.
\end{aligned} \tag{5.70}$$

Therefore we conclude that the events detected by $S_j^{(r)}$ are independent of h, that is, independent of the events detected by $S_j^{(r-1)}, \dots, S_j^{(1)}$, if $\mathbb{P}_i(H_h) > 0$. Thus, $S_j^{(r)}$ is independent of $S_j^{(r-1)}, \dots, S_j^{(1)}$, and, since r is arbitrary, $\left\{S_j^{(r)}\right\}$ is a sequence of independent random variables with respect to $\mathbb{P}_i$.

(iii) From (5.70), we get for $r \geq 2$ and $n \geq 1$

$$\mathbb{P}_i(S_j^{(r)} = n) = \sum_h \mathbb{P}_i\Big(S_j^{(r)} = n \,\Big|\, H_h\Big)\mathbb{P}_i(H_h) = f_{jj}^{(n)} \sum_{h \in \mathbb{R}^{r-1}} \mathbb{P}_i(H_h) = f_{jj}^{(n)},$$

hence

$$\mathbb{E}_i[S_j^{(r)}] = \sum_{k=0}^{\infty} k\mathbb{P}_i(S_j^{(r)} = k) = \sum_{k=1}^{\infty} kf_{jj}^{(k)} = \overline{T}_{jj}.$$

5.6.2 Ergodic theorem

Let $\{X_n\}$ be a homogeneous Markov chain on $(\Omega, \mathcal{E}, \mathbb{P})$ with a finite or denumerable state-space S and let $j \in S$. Set

$$w_j := \frac{1}{\overline{T}_{jj}}$$

where $\overline{T}_{jj}$ is the expected return time to j. For any $i \in S$, set $\Omega_i := \{x \in \Omega \mid X_0(x) = i\}$. If $\mathbb{P}(\Omega_i) > 0$, let $\mathbb{P}_i$ be the probability measure defined by $\mathbb{P}_i(A) := \mathbb{P}(A \mid \Omega_i)$.

Proposition 5.69 *The following hold:*

(i) Assume $i \to j$. Then the frequency of visits in the first n steps, $\frac{V_j^{(n)}(x)}{n}$, converges to w_j for a.e. $x \in \Omega_i$, that is

$$\mathbb{P}_i\Big(\frac{V_j^{(n)}(x)}{n} \to w_j\Big) = 1. \tag{5.71}$$

(ii) For every $i, j \in S$ we have

$$\frac{1}{n}\sum_{k=1}^{n} p_{ij}^{(k)} \to f_{ij} w_j. \tag{5.72}$$

Proof. (i) If j is transient, then $w_j = 0$ by (5.35) and paths from i visit j a finite number of times almost surely, see Proposition 5.37. Therefore

$$\frac{V_j^{(n)}(x)}{n} \leq \frac{V_j(x)}{n} \to 0 = w_j \qquad \mathbb{P}\text{-almost surely.}$$

If $j \in S$ is recurrent and $i \to j$, then the random variables $S_j^{(r)}$, $r \geq 2$, are independent and identically distributed with respect to $\mathbb{P}_i$ and $\mathbb{E}_i[S_j^{(r)}] = \overline{T}_{jj} \in \overline{\mathbb{R}}_+$, see Theorem 5.68. Moreover,

$$V_j^{(n)}(x) = \sup\left\{k \,\middle|\, \sum_{r=2}^{k} S_j^{(r)}(x) \leq n\right\}$$

almost surely. Thus, a consequence of the strong law of large numbers, Proposition 4.93, yields

$$\mathbb{P}_i\left(\frac{V_j^{(n)}(x)}{n} \to w_j\right) = 1.$$

(ii) Since $\frac{V_j^{(n)}(x)}{n} \le 1$ from (5.71) with $i = j$ we get

$$\int_{\Omega_j} \left|\frac{V_j^{(n)}}{n} - w_j\right| \mathbb{P}(dx) \to 0,$$

by the dominated convergence theorem, see Theorem B.54, hence by (5.30),

$$\frac{1}{n}\sum_{k=1}^{n} p_{jj}^{(k)} = \mathbb{E}_j\left[\frac{V_j^{(n)}}{n}\right] \to w_j.$$

The conclusion (5.72) then follows from Lemma 5.60.

Theorem 5.70 (Ergodic theorem) *Let $\{X_n\}$ be a homogeneous Markov chain with a finite or denumerable state-space and with an irreducible transition matrix* **P**. *Then the following hold:*

(i) For every state $j \in S$ we have

$$\frac{V_j^{(n)}(x)}{n} \to w_j \qquad \mathbb{P}\text{-almost surely.} \tag{5.73}$$

(ii) Let $\phi : S \to \mathbb{R}$ be a function. Then

$$\frac{1}{n}\sum_{k=1}^{n} \phi(X_k) \to \sum_{j \in S} \phi(j) w_j \qquad \mathbb{P}\text{-almost surely.} \tag{5.74}$$

Proof. (i) By (i) of Proposition 5.69 $\frac{V_j^{(n)}(x)}{n} \to w_j$ almost surely in $\Omega_i := \{x \in \Omega \mid X_0(x) = i\}$ for every $i \in S$, hence $\frac{V_j^{(n)}(x)}{n} \to w_j$ for $\mathbb{P}$-a.e. $x \in \Omega$.

(ii) Set $E_{j,k} := \{x \mid X_k(x) = j\}$. Trivially, $V_j^{(n)}(x) = \sum_{k=1}^{n} \mathbb{1}_{E_{k,j}}(x)$ and

$$\phi(X_k(x)) = \sum_{j \in S} \phi(j) \mathbb{1}_{E_{j,k}}(x)$$

Therefore,

$$\begin{aligned} \frac{1}{n}\sum_{k=1}^{n} \phi(X_k(x)) &= \frac{1}{n}\sum_{k=1}^{n}\Big(\sum_{j \in S} \phi(j)\mathbb{1}_{E_{j,k}}(x)\Big) \\ &= \sum_{j \in S} \phi(j) \frac{1}{n}\sum_{k=1}^{n} \mathbb{1}_{E_{j,k}}(x) = \sum_{j \in S} \phi(j)\frac{V_j^{(n)}(x)}{n}, \end{aligned}$$

hence, by the dominated convergence theorem, Example B.56, and (5.73) one concludes

$$\lim_{n\to\infty}\frac{1}{n}\sum_{k=1}^{n}\phi(X_k(x)) = \lim_{n\to\infty}\sum_{j\in S}\phi(j)\frac{V_j^{(n)}(x)}{n}$$

$$= \sum_{j\in S}\phi(j)\Big(\lim_{n\to\infty}\frac{V_j^{(n)}(x)}{n}\Big) = \sum_{j\in S}\phi(j)w_j \qquad \mathbb{P}\text{-almost surely.}$$

Notice that in the previous computation the sum and the limit order can be interchanged also if S is denumerable; in fact, since ϕ is bounded and $V_j^{(n)}/n \le 1$, one can apply Lebesgue dominated convergence theorem, see Example B.56.

5.6.3 Powers of irreducible stochastic matrices

Let $\{X_n\}$ be a homogeneous Markov chain with a finite or denumerable state-space and an *irreducible* transition matrix $\mathbf{P}$. Then from (5.72)

$$\frac{1}{n}\sum_{k=1}^{n}p_{ij}^{(k)} \to w_j \qquad \forall i, j \in S \tag{5.75}$$

since $f_{ij} = 1$ $\forall i, j \in S$.

Equation (5.75) allows the conclusions of Theorem 5.61 and Exercise 5.66 to be extended to irreducible matrices. We have the following.

Theorem 5.71 *Let $\{X_n\}$ be a homogeneous Markov chain with a finite or denumerable state-space S and an irreducible transition matrix $\mathbf{P}$. For every $j \in S$, let $w_j := \frac{1}{T_{jj}}$. Then the following hold:*

(i) The vector $w = (w_j)_{j\in S}$ is such that $w = w\mathbf{P}$. Moreover, if $z := (z_i)_{i\in S}$ is such that $\sum_{i\in S}|z_i| < +\infty$ and $z = z\mathbf{P}$, then $z = \lambda w$ for some $\lambda > 0$.

(ii) Assume the state-space S is finite. Then $w = (w_j)$ is a stochastic vector, $w \in \mathcal{T}$, hence the unique stochastic vector such that $z = z\mathbf{P}$. Moreover, all the components of w are nonzero, equivalently, every state is positive recurrent.

Proof. (i) From the Beppo Levi theorem and (5.75), we have, for any $j \in S$,

$$(w\mathbf{P})_j = \sum_{\ell\in S}w_\ell p_{\ell j} = \sum_{\ell\in S}\Big(\lim_{n\to\infty}\frac{1}{n}\sum_{k=0}^{n}p_{i\ell}^{(k)}\Big)p_{\ell j}$$

$$= \lim_{n\to\infty}\frac{1}{n}\sum_{k=0}^{n}\sum_{\ell\in S}p_{i\ell}^{(k)}p_{\ell j} = \lim_{n\to\infty}\frac{1}{n}\sum_{k=1}^{n+1}p_{ij}^{(k)} = w_j.$$

Let $z = (z_i)$ with $\sum_{i\in S} |z_i| < \infty$. and $z = z\mathbf{P}$. Then $z = z\mathbf{P} = (z\mathbf{P})\mathbf{P} = z\mathbf{P}^2$, and, by an induction argument, $z = z\mathbf{P}^k$ for any k, i.e. $z_j := \sum_{i\in S} z_i p_{ij}^{(k)}$ for any j and for any k. Therefore,

$$z_j = \frac{1}{n}\sum_{k=1}^{n}\sum_{i\in S} z_i p_{ij}^{(k)} = \sum_{i\in S} z_i \Big(\frac{1}{n}\sum_{k=1}^{n} p_{ij}^{(k)}\Big).$$

Since $\frac{1}{n}\sum_{k=1}^{n} p_{ij}^{(k)} \le 1$ and $\sum_{i\in S}|z_i| < +\infty$, one can apply the dominated convergence theorem, see Example B.56. Thus, taking into account also (5.75), as $n \to \infty$ one gets

$$z_j = \lim_{n\to\infty}\sum_{i\in S} z_i\Big(\frac{1}{n}\sum_{k=1}^{n} p_{ij}^{(k)}\Big) = \sum_{i\in S} z_i\Big(\lim_{n\to\infty}\frac{1}{n}\sum_{k=1}^{n} p_{ij}^{(k)}\Big) = \Big(\sum_{i\in S} z_i\Big) w_j$$

i.e. $z = \lambda w$, $\lambda := \sum_{i\in S} z_i \in \mathbb{R}$.

(ii) If S is finite, $S = \{1, \dots, N\}$, then

$$\begin{aligned}\sum_{j=1}^{N} w_j &= \sum_{j\in S}\Big(\lim_{n\to\infty}\frac{1}{n}\sum_{k=1}^{n} p_{ij}^{(k)}\Big) = \lim_{n\to\infty}\frac{1}{n}\sum_{j\in S}\Big(\sum_{k=1}^{n} p_{ij}^{(k)}\Big)\\ &= \lim_{n\to\infty}\frac{1}{n}\sum_{k=1}^{n}\Big(\sum_{j\in S} p_{ij}^{(k)}\Big) = \lim_{n\to\infty}\frac{1}{n}\sum_{k=1}^{n} 1 = 1.\end{aligned}$$

i.e. w is a stochastic vector. Finally, the components of w are nonzero by Exercise 5.65.

Remark 5.72 We point out that in the second part of the ergodic theorem, Theorem 5.71, the assumption that S if finite cannot be dropped: consider the unlimited one-dimensional random walk whose transition matrix is $\mathbf{P} = (p_{ij})$ where $p_{ij} = 1/2$ if $|j - i| = 1$ and zero otherwise. In this case the matrix is clearly irreducible, and for every $i, j \in \mathbb{Z}$, $(\mathbf{P}^n)^i_j \to 0$ as $n \to \infty$ since as n goes to infinity, more and more states are visited. What fails in extending the proof we have given when S is finite, is that if S is not finite, then we cannot exchange the limits:

$$\lim_{n\to\infty}\frac{1}{n}\sum_{k=1}^{n}\sum_{j\in S} p_{ij}^{(k)} \ne \sum_{j\in S}\lim_{n\to\infty}\frac{1}{n}\sum_{k=1}^{n} p_{ij}^{(k)}.$$

Finally, we observe that (5.75) and, consequently, Theorem 5.71 can also be proven without taking advantage of the ergodic property of the chain, see Theorem 5.81.

5.6.4 Markov chain Monte Carlo

In many applications one needs to evaluate the *weighted sum* of a certain quantity $f : S \to \mathbb{R}$ defined on a finite set S

$$E := \frac{1}{z} \sum_{j \in S} f(j)\pi_j, \qquad z := \sum_{j \in S} \pi_j \tag{5.76}$$

with respect to the weight $\pi = (\pi_j)$. With the language of probability, denoting by μ the probability measure on S with point mass density given by $\frac{1}{z}(\pi_j)_{j \in S}$, we want to evaluate

$$E := \int_S f(t)\mu(dt). \tag{5.77}$$

If the set S is small, then it is easy to compute E straightforwardly. If S is a large set (say $|E| = (50)!$), then the computation of E or even just of the total mass z cannot be performed directly, so that some strategy has to be devised in order to compute or at least approximate E.

5.6.4.1 Monte Carlo

One such strategy is the Monte Carlo method, see Section 4.4.5. In practice, one has to specify a probability space $(\Omega, \mathcal{E}, \mathbb{P})$ and a random variable $X : \Omega \to S$ such that $\mathbb{P}_X = \mu$; then one considers the Bernoulli scheme of Ω-valued sequences $(\Omega^\infty, \mathcal{E}^\infty, \mathbb{P}^\infty)$ and the random variables $X_n : \Omega^\infty \to \mathbb{R}$ defined for $n = 1, 2, \dots,$ by $X_n(x) := X(x_n)$ if $x = \{x_n\} \in \Omega^\infty$. Such random variables are independent and equidistributed as X, see Proposition 4.87. The strong law of large numbers says that, for almost any sequence of points $x = \{x_k\}$, $x_k \in \Omega$,

$$\frac{1}{n} \sum_{j=1}^{n} f(X(x_j)) = \frac{1}{n} \sum_{j=1}^{n} f(X_j(x)) \to E \tag{5.78}$$

and the weak estimate (4.9) holds.

In order to obtain an algorithm from the strong law of large numbers, we must face two issues:

(i) How to choose (as efficiently as possible) the probability space $(\Omega, \mathcal{E}, \mathbb{P})$ and the random variable $X : \Omega \to S$ such that $\mathbb{P}_X = \mu$.

(ii) Since the limit in (5.78) holds for almost any sequence $\{x_k\} \in \Omega^\infty$ (and not for every sequence), we must be sure to pick up a 'good' sequence. The 'risk' to pick up a 'bad' sequence is zero, but, nevertheless, it cannot be completely neglected.
If some sort of regularity (or pattern) appears in the sequence, then it may influence the value of the limit or even its existence. For example, assume we are flipping a fair coin an infinite number of times. If we get a success (head) at every flipping or every three flippings, then, in either

case, $\frac{1}{n}\sum_{k=1}^{n} X_k(x) = \frac{1}{x}\sum_{k=1}^{n} X(x_k)$ converges to 1 or 1/3 and not to the expected value 1/2. We should be able to pick up a sequence that is as 'irregular' or as 'random' as possible.

There are several ways to get sequences or, better, large finite tuples of random numbers uniformly distributed in the interval $[0, 1]$. We have to be cautious with the terminology since these numbers are in fact integer multiples of a fixed small number. However, if f is not fastly oscillating and the approximation needs not be too accurate, this would not be a problem. For instance, *physical random number generators* can be based on an essentially random atomic or subatomic physical phenomenon whose unpredictability can be traced to the laws of quantum mechanics. Potential sources include radioactive decay, thermal noise and shot noise. With some conditioning they are suitable sources of random samples that are uniformly distributed in $[0, 1]$. Also *pseudo-random number generators* can be useful. They are deterministic algorithms that can automatically create long runs of numbers with good random properties. They actually output periodic sequences, although with a hopefully very long period, so a complete 'randomless' is out of the question. Also, correlations between the outcomes, especially between far outcomes, or between groups of outcomes, usually involve surprises and must be carefully checked. Nevertheless, software libraries that contain pseudo-random generators with good and specified statistic properties are available. We do not enter into these aspects, which are or course relevant in simulation, see e.g. [17], but which are out of the scope of probability. For the purpose of this discussion, we simply assume the availability of generators of 'random sequences' uniformly distributed in the interval $[0, 1]$ and such that grouping them in N-tuples yields a 'random sequence of N-tuples' that are also uniformly distributed in $Q = [0, 1]^N$. Thus, typical choices for $(\Omega, \mathcal{E}, \mathbb{P})$ and X in computing E in (5.77) when $\mu = \mathcal{L}^N$ are $(Q, \mathcal{B}(\mathbb{R}^N), \mathcal{L}^N)$ and $X(x) := x$, respectively.

Integrating with respect to a nonuniform measure μ is harder. We give a hint in the scalar case. Assume we want to compute

$$E := \int_{-\infty}^{+\infty} f(t)\, d\mu(t)$$

where f is a bounded and Borel-regular function and $(\mathcal{B}(\mathbb{R}), \mu)$ is a measure on $\mathbb{R}$. In this case one takes $([0, 1], \mathcal{B}(\mathbb{R}), \mathcal{L}^1)$, considers the law $F(t) := \mu([-\infty, t])$ of μ and defines $X : [0, 1] \to \mathbb{R}$ by (3.10), i.e.

$$X(s) := \min\left\{t \,\middle|\, s \leq F(t)\right\}. \tag{5.79}$$

Since $F(t)$ is right-continuous, $X(s)$ is left-continuous, hence Borel-measurable. Moreover, $\{s \in [0, 1] \,|\, X(s) \leq t\} = \{s \in [0, 1] \,|\, 0 \leq s \leq F(t)\}$ for every $t \in \mathbb{R}$ so that

$$\mathcal{L}^1(X \leq t) = \mathcal{L}^1([0, F(t)]) = F(t) \qquad \forall t \in \mathbb{R},$$

i.e. X follows μ. Therefore, for almost any sequence $\{x_k\} \subset [0, 1]$ we get

$$\frac{1}{n}\sum_{k=1}^{n} f(X(x_k)) \to \int_{-\infty}^{\infty} f(t)\,\mu(dt) \qquad \text{as } n \to \infty.$$

The procedure described above cannot be implemented to evaluate (5.76) for large sets S, since the computation of the law of μ and of the random variable X in (5.79) are of the same order of complexity as the direct computation of E. When S can be seen as a product, $S = S_1 \times \cdots \times S_N$ and the measure μ with mass density $\frac{1}{z}(\pi_j)$ is a product,

$$\mu = \mu_1 \times \mu_2 \times \cdots \times \mu_N$$

where each μ_i is a measure on S_i, we can factorize the computation of X. For instance, if $N = 2$, we define two random variables $X : [0, 1] \to S_1$ and $Y : [0, 1] \to S_2$ following the distributions μ_1 and μ_2, respectively. Then

$$\int f(t, s)\mu_1(dt) \times \mu_2(dt) = \int \left(\int f(t, s)\mu_2(ds) \right) \mu_1(dt)$$
$$= \int \left(\int f(t, Y(y))\mathbb{P}(dy) \right) \mu_1(dt) = \iint g(X(x), Y(y))\, \mathbb{P} \times \mathbb{P}(dxdy),$$

so that the strong law of large numbers yields

$$\frac{1}{n}\sum_{k=1}^{n} f(X(x_k), Y(y_k)) \to E$$

for almost any sequence $\{(x_k, y_k)\} \in [0, 1]^2$.

5.6.4.2 Markov chain Monte Carlo

When μ is not a product measure, one takes advantage of the ergodic property of Markov chains and of the fact that arbitrary Markov chains can be run from uniformly distributed random variables.

The paradigm we pursue in computing E in (5.76) is the following: given the probability mass distribution $w = (w_j) \in S$, $w_j := \frac{1}{z}\pi_j$, we define a homogeneous Markov chain on a probability space $(\Omega, \mathcal{E}, \mathbb{P})$ with state-space S and such that w is a fixed point for its transition matrix $\mathbf{P}$, $w = w\mathbf{P}$, i.e. $\pi = \pi\mathbf{P}$.

We have already proven, Theorem 5.14, that given a stochastic matrix $\mathbf{P}$ one can always define a homogeneous Markov chain whose transition matrix is $\mathbf{P}$. Thus, it suffices to define a stochastic matrix $\mathbf{P}$ such that π is a fixed point for $\mathbf{P}$. This is realized by the *Hastings–Metropolis algorithm*.

Let $\mathbf{Q} = (q_{ij})$ be any stochastic matrix. For $i, j \in S$, $i \neq j$, let

$$p_{ij} := \frac{1}{\pi_i} \min\{\pi_i q_{ij}, \pi_j q_{ji}\}, \tag{5.80}$$

so that $0 \le p_{ij} \le 1$ and $\sum_{j \neq i} p_{ij} \le 1$. Define $\mathbf{P} = (p_{ij})$ as follows:

- if $i \neq j$, then p_{ij} is defined by (5.80);
- if $i = j$, set $p_{ii} = 1 - \sum_{j \neq i} p_{ij}$;

so that $\mathbf{P}$ is a stochastic matrix. The matrix $\mathbf{R} = (r_{ij})$, $r_{ij} := \min\{\pi_i q_{ij}, \pi_j q_{ji}\}$ is symmetric, hence

$$\pi_i p_{ij} = \pi_j p_{ji} \qquad \forall i, j \in S$$

and

$$\sum_{i=1}^{N} \pi_i p_{ij} = \sum_{i=1}^{N} \pi_j p_{ji} = \pi_j \qquad \forall j \in S$$

i.e. $\pi = \pi \mathbf{P}$.

With the Metropolis algorithm one chooses $\mathbf{Q} = (q_{ij})$ as an irreducible and symmetric matrix, so that if $i \neq j$, then

$$p_{ij} = \min\left\{1, \frac{\pi_j}{\pi_i}\right\} q_{ij},$$

hence $\mathbf{P} = (p_{ij})$ where

$$p_{ij} = \begin{cases} q_{ij} & \text{if } \pi_j \ge \pi_i, \\ \dfrac{\pi_j}{\pi_i} q_{ij} & \text{if } \pi_j < \pi_i, \\ 1 - \sum_{j \neq i} p_{ij} & \text{if } i = j. \end{cases} \tag{5.81}$$

Possible choices for $\mathbf{Q}$ are

$$q_{ij} = \begin{cases} 1/(N-1) & \text{if } i \neq j, \\ 0 & \text{if } i = j, \end{cases}$$

where $N = |S|$, or, for instance, $\mathbf{Q}$ can be chosen as the matrix associated with a symmetric circular random walk. Notice that, in general, in order to define $\mathbf{P}$ one does not need to know the total mass z or the cardinality N of S.

Now, let $\{X_n\}$ be a homogeneous Markov chain whose transition matrix is $\mathbf{P}$, see Theorem 5.14. If $\mathbf{P}$ *is irreducible*, then $\frac{1}{z}\pi$ is the only stochastic vector which is also a fixed point of $P(x) = x\mathbf{P}$. Moreover, for any $j \in S$ the number $(\frac{1}{z}\pi_j)^{-1}$ is the return time in j for $\{X_n\}$, see Theorem 5.71. The ergodic theorem, (ii) of Theorem 5.70, gives the convergence

$$\frac{1}{n} \sum_{k=1}^{n} f(X_k(x)) \to \frac{1}{z} \sum_{j \in S} f(j) \pi_j$$

for almost any $x \in \Omega$.

In order to simulate the chain, one only needs to have a large number of independent variables that follow the uniform distribution in $[0, 1]$, see Theorem 5.14, so that, this procedure, known as a *Markov chain Monte Carlo* procedure, may eventually lead to numerical algorithms for the approximate computation of E, see e.g. [18].

The simulation of a Markov chain with transition matrix $\mathbf{P}$ is just the transition from a position $i \in S$ at step n to a new position j at step $(n+1)$ following the probability density $(p_{ij})_j$. In this case

$$p_{ij} = \min\left\{1, \frac{\pi_j}{\pi_i}\right\} q_{ij}.$$

Assume we are in state i at step n. We pick $j \in S$ according to the mass distribution $(q_{ij})_j$. Then:

- if $\pi_i \le \pi_j$ we move to state j, $X_{n+1} = j$;
- if $\pi_i > \pi_j$ we play a Bernoulli trial of parameter $p = \pi_i/\pi_j$. Then we set $X_{n+1} = j$ if we obtain a success and $X_{n+1} = i$ otherwise.

Two facts are crucial for numerical applications. First of all, there is the irreducibility assumption on $\mathbf{P}$, which can be achieved in many applications. The second crucial point is the number of steps one needs to perform in order to obtain an estimate of E which is as accurate as wished. We shall not deal with these problems which involve both statistical and analytic estimates and are still an open field of research. We refer the reader to [18] and to the stimulating paper [19] where further applications, such as optimization problems, of the Markov chain Monte Carlo method, are investigated. Here, we only give the following result, see [20].

Theorem 5.73 (Metropolis) *If π is not constant and the matrix $\mathbf{P}$ defined in* (5.81) *in the Metropolis algorithm is irreducible, then $\mathbf{P}$ is regular.*

Proof. We prove that there exists $h \in S$ such that $p_{hh} > 0$. Let $C = \{i \mid \pi_i = \max_j \pi_j\}$. Obviously, $C \neq \emptyset$ and, by assumption, $C \neq S$. Thus, see Exercise 5.64, there exists $h \in C$ and $k \notin C$ such that $q_{hk} > 0$ and $\pi_h > \pi_k$. If $i \neq j$, then $p_{ij} \le q_{ij}$, and by definition of $\mathbf{P}$

$$\begin{aligned} p_{hh} &= 1 - \sum_{j \neq h} p_{hj} = 1 - \sum_{j \neq h,k} p_{hj} - p_{hk} \\ &\ge 1 - \sum_{j \neq h,k} q_{hj} - \frac{\pi_k}{\pi_h} q_{hk} = 1 - \sum_{j \neq h} q_{hj} + q_{hk}\left(1 - \frac{\pi_k}{\pi_h}\right) \\ &= q_{hh} + q_{hk}\left(1 - \frac{\pi_k}{\pi_h}\right) > 0. \end{aligned}$$

The claim thus follows from Exercise 5.64.

5.7 Renewal theorem

Let $\{X_n\}$ be a homogeneous Markov chain with a finite or denumerable state-space S and transition matrix $\mathbf{P}$. In this section, we discuss the convergence of the sequence $\{\mathbf{P}^n\}$ as $n \to \infty$. The results extend partially the results of Section 5.5: there S was supposed finite and $\mathbf{P}$ regular.

5.7.1 Periodicity

Given two positive integers d and n, we say that d *divides* n and write $d \perp n$ if $n = kd$ for some positive integer k. If N is a finite or denumerable set of positive integers, then the largest integer $d \geq 1$ that divides any element of N is called the *greatest common divisor* of N and is denoted by $\gcd(N)$. It can be easily proven that if N is infinite and $d = \gcd(N)$, then there exists a finite subset $M \subset N$ such that $d = \gcd(M)$.

Definition 5.74 *Let $\mathbf{P} = (p_{ij})$ be a stochastic matrix with a finite or denumerable set S of indices. The number*

$$d_j := \gcd\Big(\Big\{n \,\Big|\, p_{jj}^{(n)} > 0\Big\}\Big)$$

is called the period *of the state j. If $d_j = 1$, then the state j is said to be* aperiodic, *while if $d_j = d \geq 2$, then j is said to be d*-periodic.

By Definition 5.74, if d_j is d-periodic, then $p_{jj}^{(k)} = 0$ whenever k is not a multiple of d. Notice that it is not excluded that $p_{jj}^{(k)} = 0$ for some k such that $d \perp k$.

Exercise 5.75 *Prove the following:*

(i) If $\mathbf{P} = \mathrm{Id}$, then $\mathbf{P}^n = \mathrm{Id}\ \forall n$, so that all the states are aperiodic.

(ii) If $\mathbf{P}$ is irreducible and all its rows coincide, then $\mathbf{P}^n = \mathbf{P}$ for every n and all the states are aperiodic.

(iii) If $\mathbf{P} = \begin{pmatrix} 0 & 1 \\ 1 & 0 \end{pmatrix}$, then $\mathbf{P}^n = \begin{cases} \mathrm{Id} & \text{if } n \text{ is even,} \\ \mathbf{P} & \text{if } n \text{ is odd.} \end{cases}$ Both the states 1 and 2 have period 2. Moreover, the limit of $\mathbf{P}^n$ as $n \to \infty$ does not exist.

Proposition 5.76 *If two states i are j are communicating, $i \leftrightarrow j$, then they have the same period. Thus, the period of an irreducible stochastic matrix is well defined as the period of any of its states.*

Proof. Let $N_i = \big\{k \mid p_{ii}^{(k)} > 0\big\}$, $N_j = \big\{k \mid p_{jj}^{(k)} > 0\big\}$, $d_i = \gcd(N_i)$ and $d_j = \gcd(N_j)$. Let h, k be such that $p_{ij}^{(h)} > 0$ and $p_{ji}^{(k)} > 0$. Then $p_{ii}^{(h+k)} = \sum_{\ell=1}^{N} p_{i\ell}^{(h)} p_{\ell i}^{(k)} \geq p_{ij}^{(h)} p_{ji}^{(k)} > 0$, i.e. $h + k \in N_i$, so that $d_i \perp (h + k)$.

For each $n \in N_j$ we also have $p_{ii}^{(h+k+n)} \geq p_{ij}^{(h)} p_{jj}^{(n)} p_{ji}^{(k)} > 0$. Therefore, $h+k+n \in N_i$, hence $d_i \perp (h+k+n)$. Thus $d_i \perp n \ \forall n \in N_j$, so that $d_i \leq d_j$. Interchanging i and j, the previous argument gives also $d_j \leq d_i$, hence $d_i = d_j$.

Proposition 5.77 *Let $\{X_n\}$ be a homogeneous Markov chain with a finite or denumerable state-space S and transition matrix $\mathbf{P} = (p_{ij})$. Let $j \in S$ be a state, let $f_{jj}^{(k)}$ be the probability that a path from j visits j for the first time at step k and denote by $d_j \geq 1$ the period of j. Then*

$$\gcd\left\{k \geq 1 \,\middle|\, f_{jj}^{(k)} > 0\right\} = d_j.$$

Proof. Let us denote $p_k := p_{jj}^{(k)}$, $f_k := f_{jj}^{(k)}$ and let $P := \{k \mid p_k > 0\}$ and $F := \{k \mid f_k > 0\}$. It suffices to show that $d \perp k \ \forall k \in P$ if and only if $d \perp k \ \forall k \in F$.

Assume d divides any $h \in P$. Let $k \in F$. Since, see Proposition 5.58,

$$p_k = f_k + f_{k-1} p_1 + \cdots + f_1 p_{k-1} \tag{5.82}$$

we get $p_k \geq f_k > 0$ so that $k \in P$ and $d \perp k$.

Assume d divides any $h \in F$. We show, with an induction argument on k, that d divides any $h \in P$ such that $h \leq k$.

If $k = 1$, the claim is obviously true since $p_1 = f_1 > 0$. Assume the claim holds true for $k-1$. Let k be such that $p_k > 0$. Two cases may occur. Either $f_k > 0$, so that $d \perp k$, or $f_k = 0$. In the latter case (5.82) implies

$$f_{k-1} p_1 + \cdots + f_1 p_{k-1} = p_k > 0$$

so that at least one addendum must be positive: there exists $k' < k$ such that $p_{k'} > 0$ and $f_{k-k'} > 0$. Since $k' < k$, $d \perp k'$ and $d \perp (k-k')$, so that $d \perp k$.

5.7.2 Renewal theorem

Theorem 5.78 (Erdös–Feller–Pollard) *Let $\{f_n\}$, $n \geq 1$, and $\{p_n\}$, $n \geq 0$, be two non-negative sequences such that $p_0 = 1$ and $\sum_{n=1}^{\infty} f_n = 1$. Assume*

$$p_n = \sum_{j=1}^{n} f_j p_{n-j} \qquad \forall n \geq 1. \tag{5.83}$$

Let $T := \sum_{j=1}^{\infty} j f_j$ ($T = +\infty$ is allowed) and $\mathcal{S} := \{n \mid f_n > 0\}$. If $\gcd(\mathcal{S}) = 1$, then

$$\lim_{n \to +\infty} p_n = \frac{1}{T} \qquad (0 \textit{ if } T = +\infty).$$

Lemma 5.79 *Let $a_1, \dots, a_r$ be relatively prime integers. Then there exists n_0 such that any integer $n \geq n_0$ is a linear combination of $a_1, \dots, a_r$ with non-negative coefficients.*

Proof. Since $\gcd(a_1, \dots, a_r) = 1$, the Bezout identity holds: namely, *there exist* $\beta_1, \dots, \beta_r \in \mathbb{Z}$ *such that* $\sum_{i=1}^r \beta_i a_i = 1$. Let $a := \sum_{i=1}^r a_i$. Divide n by a, $n = qa + s$ with $0 \le s < a$, so that

$$n = q\sum_{i=1}^r a_i + s\sum_{i=1}^r \beta_i a_i = \sum_{i=1}^r (q + \beta_i s)a_i.$$

The right-hand side of the previous equality is a linear combination of $a_1, \dots, a_r$ with positive coefficients if n is such that $q \ge a\ \max_i(|\beta_i|)$.

Lemma 5.80 *Let* $\{f_j\}$, $j \ge 1$, *be a non-negative sequence such that* $\sum_{j=1}^\infty f_j = 1$, *and set* $\mathcal{S} := \{j \mid f_j > 0\}$. *Let* $a \in \mathbb{R}$ *and let* $\{q_n\}$, $n \in \mathbb{Z}$, *be a sequence such that*

$$\begin{cases} q_n = \sum_{j=1}^\infty f_j q_{n-j} & \forall n \in \mathbb{Z}, \\ q_0 = a, \\ q_n \le a & \forall n \le 0. \end{cases} \tag{5.84}$$

If $\gcd(\mathcal{S}) = 1$, *then* $q_n = a\ \forall n$. *The same holds if the assumption* $q_n \le a\ \forall n \le 0$ *in* (5.84) *is replaced by* $q_n \ge a\ \forall n \le 0$.

Proof. Up to a change of sign of $\{q_n\}$ we may assume $q_n \le a\ \forall n \le 0$. System (5.84) gives

$$a = q_0 = \sum_{j=1}^\infty f_j q_{-j} \le a\sum_{j=1}^\infty f_j = a.$$

Thus $\sum_{j=1}^\infty f_j(a - q_{-j}) = 0$ so that $q_{-n} = a$ whenever $f_n > 0$, i.e. for any $n \in \mathcal{S}$. For any $n \in \mathcal{S}$, (5.84) gives

$$a = q_{-n} = \sum_{j=1}^\infty f_j q_{-n-j} \le a\sum_{j=1}^\infty f_j = a$$

so that $\sum_{j=1}^\infty f_j(a - q_{-n-j}) = 0$, i.e. $q_{-n-m} = a$ if both $n, m \in \mathcal{S}$. With an induction argument, one can now show that $q_{-\alpha n - \beta m} = 0$ for any positive integers α, β, and then that $q_{-n} = 0$ for any integer n which is a linear combination with positive coefficients of integers in $\mathcal{S}$.

Since $\gcd(\mathcal{S}) = 1$, there exist $a_1, \dots, a_r \in \mathcal{S}$ such that $\gcd(a_1, \dots, a_r) = 1$ so that by Lemma 5.79 each integer $n \ge n_0$ is a linear combination of integers in $\mathcal{S}$ with positive coefficients. Therefore, we conclude that

$$q_{-n} = a \qquad \forall n \ge n_0.$$

It is now easy to check by an induction argument that $q_{-n_0+1}, q_{-n_0+2}, \dots = a$. In fact, if $q_n = a\ \forall n \le h$, then (5.84) reads

$$q_{h+1} = \sum_{j=1}^\infty f_j q_{h+1-j} = a\sum_{j=1}^\infty f_j = a.$$

Proof of Theorem 5.78. Step 1. Let $a \in \mathbb{R}$ and let $\{h_n\}$ be a sequence such that $p_{h_n} \to a$. We show that there exist a sequence $\{q_j\}_{j\in\mathbb{Z}}$ and a subsequence $\{k_n\}$ of $\{h_n\}$ such that

$$p_{k_n-j} \to q_j \qquad \text{for any } j \tag{5.85}$$

and

$$\begin{cases} q_n = \sum_{j=1}^{\infty} f_j q_{n-j} & \forall n \in \mathbb{Z}, \\ q_0 = a. \end{cases} \tag{5.86}$$

Let $a_{n,j} := p_{n-j}$. For any integer j, the subsequence $\{a_{n,j}\}_n$ is bounded, hence, by compactness, there exist a number q_j and a subsequence $k_n^{(j)}$ such that $a_{k_n^{(j)},j} \to q_j$ as $n \to \infty$. Iterating the procedure for $j = 0, \pm 1, \pm 2, \dots$, we may assume that $\left\{k_n^{(-j)}\right\} = \left\{k_n^{(j)}\right\}$ and that $\left\{k_n^{(j+1)}\right\}$ is a subsequence of $\left\{k_n^{(j)}\right\}$. Let $k_n := k_n^{(n)}$. Then, for every fixed j and n_j large enough, the sequence $\{k_n\}_{n\geq n_j}$ is a subsequence of $\left\{k_n^{(j)}\right\}$ so that $a_{k_n,j} \to q_j$. Thus the limit (5.85) holds.

Let us now prove (5.86). Equation (5.83) reads

$$p_{k_n-h} = \sum_{j=1}^{k_n-h} f_j p_{k_n-h-j}.$$

Since

- $f_j p_{k_n-h-j} \to f_j q_{h-j}$ as $n \to \infty$ for any j,
- $\sup_n (f_j p_{k_n-h-j}) \leq f_j$ for any j and $\sum_{j=1}^{\infty} f_j = 1$,

applying the dominated convergence theorem, see Example B.56, we get

$$q_h = \sum_{j=1}^{\infty} f_j q_{h-j} \qquad \forall h \in \mathbb{Z},$$

i.e. (5.86) holds since $q_0 = a$.

Step 2. Let $a = \limsup_n p_n$ and let $\{h_n\}$ be a sequence such that $p_{h_n} \to a$. Let $\{k_n\}$ and $\{q_j\}$ as in Step 1. Then $p_{k_n-j} \to q_j$ so that $q_j \leq a$ $\forall j$. Since $\{q_j\}$ satisfies system (5.86), applying Lemma 5.80 we get $q_j = a$ $\forall j \in \mathbb{Z}$, thus obtaining

$$\lim_{n\to\infty} p_{k_n-j} = a \qquad \forall j \in \mathbb{Z}.$$

Step 3. Summing up (5.83) for $1, 2, \dots, N$, we get

$$\sum_{n=1}^{N} p_n = \sum_{n=1}^{N}\sum_{j=1}^{n} f_j p_{n-j} = \sum_{n=1}^{N}\Big(\sum_{j=0}^{n-1} f_{n-j} p_j\Big). \tag{5.87}$$

or, equivalently,

$$p_N + \sum_{j=1}^{N-1} \Big(1 - \sum_{i=1}^{N-j} f_i\Big) p_j = \sum_{j=1}^{N} f_j. \tag{5.88}$$

For any $j \geq 1$, let $T_j := \sum_{i=j}^{\infty} f_i$. Then $T_1 = \sum_{i=1}^{\infty} f_i = 1$ and $\sum_{j=1}^{\infty} T_j = \sum_{k=1}^{\infty} k f_k = T$. Equation (5.88) can equivalently be written as

$$p_N T_1 + \sum_{j=1}^{N-1} \Big(1 - \sum_{i=1}^{N-j} f_i\Big) p_j = \sum_{j=1}^{N} f_j,$$

or as

$$\sum_{j=1}^{N} T_j p_{N+1-j} = \sum_{j=1}^{N} f_j, \tag{5.89}$$

see (5.64), (5.65) and (5.66). Consider now the double index sequence

$$\varphi_N(j) := \begin{cases} T_j p_{N+1-j} & \text{if } 1 \leq j \leq N, \\ 0 & \text{if } j > N. \end{cases}$$

Equation (5.89) or (5.87) thus reads

$$\sum_{j=1}^{\infty} \varphi_N(j) = \sum_{j=1}^{N} f_j. \tag{5.90}$$

Since $\varphi_{k_n-1}(j) \to T_j q_j = a T_j$ by Step 2, we get the following:

- If $T = +\infty$, then Step 2, (5.90) and the Fatou lemma yield

$$T\,a = \Big(\sum_{j=1}^{\infty} T_j\Big) a = \sum_{j=1}^{\infty} \lim_{n\to\infty} \varphi_{k_n-1}(j) \leq \liminf_{n\to\infty} \sum_{j=1}^{\infty} \varphi_{k_n-1}(j)$$
$$= \liminf_{n\to\infty} \sum_{j=1}^{k_n-1} f_j = \sum_{j=1}^{\infty} f_j = 1,$$

  so that $a = 0$, i.e. $\limsup_n p_n = 0$. Thus the claim is proven whenever $T = +\infty$.

- If $T < \infty$, then Step 2, (5.90) and the dominated convergence theorem, see Example B.56, yield

$$T\,a = \Big(\sum_{j=1}^{\infty} T_j\Big) a = \sum_{j=0}^{\infty} \lim_{n\to\infty} \varphi_{k_n-1}(j) = \lim_{n\to\infty} \sum_{j=1}^{\infty} \varphi_{k_n-1}(j)$$
$$= \liminf_{n\to\infty} \sum_{j=1}^{k_n-1} f_j = \sum_{j=1}^{\infty} f_j = 1,$$

so that $\limsup_n p_n = a = 1/T$.

Since we may repeat Step 2 for any sequence h_n such that $p_{h_n} \to b := \liminf_n p_n$, from the above we also get $\liminf_n p_n = 1/T$, thus concluding that $p_n \to 1/T$.

The renewal theorem has important consequences.

Theorem 5.81 *Let $\{X_n\}$ be a homogeneous Markov chain with a finite or denumerable space-state S and denote by $\mathbf{P} = (p_{ij})$ its transition matrix. For $j \in S$, let $\overline{T}_{jj}$ be the expected return time to j, $w_j := \frac{1}{\overline{T}_{jj}}$ and let d be the period of j. Then*

$$p_{jj}^{(nd)} \to d w_j \qquad \textit{and} \qquad \frac{1}{n}\sum_{k=1}^{n} p_{ij}^{(k)} \to f_{ij} w_j \quad \forall i \in S. \tag{5.91}$$

Proof. Step 1. Let us prove the first limit of (5.91) when j is aperiodic, $d = 1$.

If j is a transient state, then $w_j = 1/\overline{T}_{jj} = 0$, see (5.34) and $\sum_{n=1}^{\infty} p_{jj}^{(n)} < \infty$. Therefore, $p_{jj}^{(n)} \to 0 = w_j$. If j is recurrent, let $f_k := f_{jj}^{(k)}$ and $p_k := p_{jj}^{(k)}$ so that $f_{jj} = \sum_{k=1}^{\infty} f_k = 1$. Thus, Proposition 5.77 gives $\gcd\Big\{k \,\Big|\, f_k > 0\Big\} = 1$. Moreover, $p_0 = 1$, $p_n = \sum_{k=1}^{n} f_k p_{n-k}$ $\forall n \geq 1$ by (5.61) and $\overline{T}_{jj} = \sum_{k=1}^{\infty} k f_k$, cf. (5.35). The renewal theorem, Theorem 5.78, gives $p_{jj}^{(n)} \to w_j$.

Step 2. Let us prove the first limit of (5.91) assuming j is d-periodic, $d \geq 2$.

The sequence of random variables $\{X_{kd}\}_k$ is a homogeneous Markov chain with state-space S and transition matrix $\widetilde{\mathbf{P}} := \mathbf{P}^d$. For this chain, j is aperiodic since

$$\widetilde{d} = \gcd\Big\{k \,\Big|\, p_{jj}^{(kd)} > 0\Big\} = \frac{1}{d}\gcd\Big\{kd \,\Big|\, p_{jj}^{(kd)} > 0\Big\} = \frac{d}{d} = 1.$$

Let $\widetilde{T}_{jj}$ be the return time to j for the new chain and let $\widetilde{w}_j := 1/\widetilde{T}_{jj}$. From Step 1

$$p_{jj}^{(nd)} \to \widetilde{w}_j. \tag{5.92}$$

Denote by $\widetilde{f_{ij}^{(k)}}$ the probability that a path of the new chain from i visits j for the first time at step k. Trivially, $\widetilde{f_{jj}^{(k)}}$ is also the probability that the initial chain visits j starting from j for the first time at step kd, $\widetilde{f_{jj}^{(k)}} = f_{jj}^{(kd)}$. Moreover, $f_{jj}^{(n)} = 0$ if n is not a multiple of d. Therefore,

$$\widetilde{T}_{jj} = \sum_{k=1}^{\infty} k \widetilde{f_{jj}^{(k)}} = \frac{1}{d}\sum_{k=1}^{\infty} (kd) f_{jj}^{(kd)} = \frac{1}{d}\sum_{k=1}^{\infty} k f_{jj}^{(k)} = \frac{1}{d}\overline{T}_{jj}.$$

Therefore, (5.92) yields $p_{jj}^{(nd)} \to d w_j$.

Step 3. Let us prove that $\frac{1}{n}\sum_{k=1}^{n} p_{jj}^{(k)} \to w_j$. Since $p_{jj}^{(k)} = 0$ if k is not a multiple of d, by the Cesaro theorem,

$$\frac{1}{n}\sum_{k=1}^{n} p_{jj}^{(k)} = \frac{1}{n}\sum_{k=1}^{\lfloor n/d \rfloor} p_{jj}^{(kd)} \to \frac{1}{d} d w_j = w_j.$$

Step 4. The second limit in (5.91) then follows from Step 3 and (ii) of Lemma 5.60.

Remark 5.82 The second limit in (5.91) is sufficient to prove Theorem 5.71. Thus we have another proof of Theorem 5.71 that does not mention the ergodic property of Markov chains.

5.7.3 Exercises

Exercise 5.83 *Let $\{X_n\}$ be a homogeneous Markov chain with a finite state-space $S = \{1, \dots, N\}$ and irreducible transition matrix* **P**. *Let $M_{jj}(n)$ be the expected return time to j measured on a window of n steps, and let N_{jj} be the expected number of visits in j that paths starting from j do in the first n steps. Compute*

$$\lim_{n\to\infty} \frac{M_{jj}(n) N_{jj}(n)}{n}.$$

Solution. By definition, $N_{jj}(n) = \frac{1}{n} E_j[V_j^{(n)}] = \frac{1}{n}\sum_{k=1}^{n} p_{jj}^{(k)}$ and

$$M_{jj}(n) = \mathbb{E}_j[t_j \mathbb{1}_{\{t_j < n\}}] = \sum_{k=1}^{n} k f_{jj}^{(k)}.$$

Therefore, since $\frac{1}{n}\sum_{k=1}^{n} p_{jj}^{(k)} \to \frac{1}{\overline{T}_j}$, we have

$$\frac{M_{jj}(n) N_{jj}(n)}{n} = \Big(\sum_{k=1}^{n} k f_{jj}^{(k)}\Big)\Big(\frac{1}{n}\sum_{k=1}^{n} p_{jj}^{(k)}\Big) \to \overline{T}_{jj} \frac{1}{\overline{T}_{jj}} = 1.$$

Exercise 5.84 *Let* $\mathbf{P} = (p_{ij})$ *be the transition matrix of a Markov chain with finite state-space S. Prove that* **P** *is irreducible and aperiodic if and only if* **P** *is regular.*

Solution. Assume **P** is irreducibile and aperiodic. Then all the states are pairwise communicating, aperiodic, and, since S is also a minimal closed class, all states are recurrent, $f_{ij} = 1$ $\forall i, j \in S$, see Proposition 5.40. Theorem 5.81 yields $p_{jj}^{(n)} \to w_j := 1/\overline{T}_{jj}$ and $w_j > 0$ by Exercise 5.65. Moreover, Lemma 5.60 yields for any i and $j \in S$

$$p_{ij}^{(n)} \to f_{ij} w_j = w_j > 0.$$

We then conclude that for n large enough, $p_{ij}^{(n)} > 0$ $\forall i, j \in S$.

Conversely, if $\mathbf{P}$ is regular, then trivially $\mathbf{P}$ is irreducible. Moreover, for any large enough n, all the entries of $\mathbf{P}^n$ are nonzero, see Theorem 5.54. $\mathbf{P}$ is therefore aperiodic.

Exercise 5.85 *Let* $\mathbf{P}$ *be a stochastic, irreducible matrix. Show that, if* $\mathbf{P}$ *is* d*-periodic,* $d \geq 2$*, then* $\mathbf{P}^d$ *is not irreducible.*

Solution. Suppose, contrary to our claim, that $\mathbf{P}^d$ is irreducible. Since $\mathbf{P}^d$ is aperiodic, then $\mathbf{P}^d$ is regular, see Exercise 5.84. Therefore $\mathbf{P}$ is regular, hence aperiodic, again by Exercise 5.84, a contradiction, since $d \geq 2$.

Exercise 5.86 *Let* $\mathbf{P}$ *be a stochastic matrix and let* i, j *be two communicating states. Show that either both* i *and* j *are positive recurrent or both are transient.*

Solution. Assume j is positive recurrent. Then $w_j = 1/\overline{T}_{jj} > 0$ and j is recurrent $f_{jj} = 1$. Let d be the period of j. We have $p_{jj}^{(nd)} \to dw_j > 0$, see Theorem 5.81. Since $i \leftrightarrow j$, i is d-periodic, see Proposition 5.76 and recurrent $f_{ii} = 1$, see Proposition 5.40. By Theorem 5.81 $p_{ii}^{(n)} \to w_i$, $w_i = \frac{1}{\overline{T}_{ii}}$. On the other hand, there exist $r, s > 0$ such that $p_{ij}^{(r)} > 0$ and $p_{ji}^{(s)} > 0$. Therefore, $p_{ii}^{(r+n+s)} \geq p_{ij}^{(r)} p_{jj}^{(n)} p_{ji}^{(s)}$. Letting $n \to \infty$, we conclude that

$$w_i \geq p_{ij}^{(r)} w_{jj} p_{ji}^{(s)} > 0,$$

i.e. $\overline{T}_{ii} < \infty$.

6

An introduction to continuous time Markov chains

In this chapter, we give a very short introduction to *continuous time Markov chains*. In Section 6.1 we discuss the *Poisson process* while in Section 6.2 we consider homogeneous continuous time Markov chains with finite state-space and right-continuous trajectories. The convergence to equilibrium of the transition probabilities matrices and the description of the holding times are discussed.

6.1 Poisson process

We go back to Example 3.34 by giving an explicit example: the emission of clicks of a Geiger counter. In this section we follow closely the presentation in [5].

Evidence shows the following features of the distribution of the number of clicks produced by the counter:

(i) The number of clicks produced in pairwise disjoint time intervals are independent.

(ii) The clicks are uniformly distributed with respect to time, i.e. the number of clicks produced in the intervals $[a+c, b+c]$ $a, b, c \geq 0$ does not depend on c.

(iii) In each time interval, the average number of clicks is finite,

(iv) At each moment, the Geiger counter produces at most one click.

When defining a model, the previous features can be formalized as follows. Let $\mathcal{I}$ be the family of time intervals

$$\mathcal{I} := \Big\{ I = [a, b] \,\Big|\, 0 \leq a < b < +\infty \Big\}.$$

A First Course in Probability and Markov Chains, First Edition. Giuseppe Modica and Laura Poggiolini.

Consider the Geiger counter as a probability space $(\Omega, \mathcal{E}, \mathbb{P})$ and, for any $I \in \mathcal{I}$, let $N_I : \Omega \to \mathbb{N}$ be the random variable on $(\Omega, \mathcal{E}, \mathbb{P})$ that counts the number of clicks produced in the interval I. In terms of the family $\{N_I\}_{I\in\mathcal{I}}$, features (i), (ii) and (iii) read:

(P1) $N_{I\cup J} = N_I + N_J$ if $I \cap J = \emptyset$ and $I \cup J \in \mathcal{I}$.

(P2) If $\mathcal{J} \subset \mathcal{I}$ is a family of pairwise disjoint intervals, then $\{N_I\}_{I\in\mathcal{J}}$ is a family of independent random variables.

(P3) For any $c > 0$ and any $I \in \mathcal{I}$, N_I and N_{c+I} follow the same distribution. We recall that $c + I := [c + a, c + b]$ if $I = [a, b]$.

(P4) $\mathbb{E}\big[N_I\big] < +\infty \ \forall I \in \mathcal{I}$.

Properties (P1), (P2), (P3) and (P4) formalize properties (i), (ii) and (iii). We must formulate (iv). Consider the event $F \subset \Omega$ defined by the property 'Two or more clicks are produced at a certain time $t \in [0, 1]$'. Then $x \in F$ if and only if for any positive integer n there exists an interval $I_{k,n} := [k/2^n, (k+1)/2^n]$, $k = k(x, n)$ such that $N_{I_{k,n}}(x) \ge 2$, i.e.

$$F := \bigcap_{n=1}^{\infty} \bigcup_{k=0}^{2^n-1} \Big\{N_{I_{k,n}} \ge 2\Big\}.$$

Consequently, the event $E \subset \Omega$ defined by the property that two or more clicks are produced at a certain time $t \in \mathbb{R}_+$ is

$$E = \bigcup_{q=0}^{\infty} (q + F).$$

and we can formalize feature (iv) as

(P5-1) $\qquad \mathbb{P}(E) = 0 \quad \text{or} \quad \mathbb{P}(F) = 0.$

Lemma 6.1 *Let* $\{N_I, I \in \mathcal{I}\}$ *be a family of random variables satisfying (P1)–(P4). Then* $\mathbb{P}(E) = 0$ *if and only if*

(P5) $$\limsup_{\epsilon\to 0} \frac{\mathbb{P}(N_{[0,\epsilon]} \ge 2)}{\epsilon} = 0.$$

Proof. Using (P1), (P2) and (P3), we get

$$\mathbb{P}(E) = \lim_{n\to\infty} \mathbb{P}\Big(\bigcup_{k=0}^{2^n-1} \Big\{N_{I_{k,n}} \ge 2\Big\}\Big) = 1 - \lim_{n\to\infty} \mathbb{P}\Big(\bigcap_{k=0}^{2^n-1} \Big\{N_{I_{k,n}} \le 1\Big\}\Big)$$

$$= 1 - \lim_{n\to\infty} \prod_{k=0}^{2^n-1} \mathbb{P}\Big(N_{I_{k,n}} \le 1\Big) = 1 - \lim_{n\to\infty} \prod_{k=0}^{2^n-1} \Big(1 - \mathbb{P}(N_{I_{k,n}} \ge 2)\Big)$$

$$= 1 - \lim_{n \to \infty} \left(1 - \mathbb{P}(N_{I_{0,n}} \geq 2)\right)^{2^n}.$$

Let $f(t) := \mathbb{P}(\mathrm{N}_{\{[0,t]\}} \geq 2)$, $t \geq 0$. By the above computation $\mathbb{P}(E) = 0$ if and only if

$$(1 - f(2^{-n}))^{2^n} \to 1. \tag{6.1}$$

Let $t_n := 2^{-n}$ then $(1 - f(t_n))^{1/t_n} \to 1$ if and only if $\frac{\log(1-f(t_n))}{t_n} \to 0$. Since $t_n \to 0$, (6.1) is equivalent to $f(t_n)/t_n \to 0$. Finally, for any $\epsilon \in [0, 1]$, let $n = n(\epsilon)$ such that $t_{n-1} \leq \epsilon < t_n$. Then

$$n(\epsilon) \to +\infty \qquad \text{and} \qquad \frac{f(\epsilon)}{\epsilon} \leq \frac{f(t_n)}{t_{n-1}} = 2\frac{f(t_n)}{t_n} \to 0 \qquad \text{as } \epsilon \to 0.$$

By the above, (P5-1) is equivalent to (P5).

Definition 6.2 (Poisson process) *A family* $\{N_t\}_{t \geq 0}$ *of non-negative integer valued random variables on a probability space* $(\Omega, \mathcal{E}, \mathbb{P})$ *is called a* Poisson process *with intensity* α *if:*

(i) $N_0(x) = 0$ *almost surely.*

(ii) The map $t \to N_t(x)$ *is increasing and right-continuous for a.e.* $x \in \Omega$.

(iii) For any n *and any list* $0 = t_0 < t_1 < \cdots < t_n$ *the random variables* $N_{t_k} - N_{t_{k-1}}$, $k = 1, \ldots, n$, *are pairwise independent.*

(iv) For any $t \geq s \geq 0$ *the random variable* $N_t - N_s$ *follows the Poisson distribution of parameter* $\alpha(t-s)$, *i.e.*

$$\mathbb{P}(N_t - N_s = k) = e^{-\alpha(t-s)} \frac{\alpha^k (t-s)^k}{k!} \qquad \forall k \geq 0. \tag{6.2}$$

In particular, we point out that

$$\mathbb{P}(N_t = k) = e^{-\alpha t} \frac{\alpha^k t^k}{k!} \qquad \forall k \geq 0$$

so that

$$\mathbb{E}\left[N_t\right] = \sum_{k=0}^{\infty} e^{-\alpha t} k \frac{\alpha^k t^k}{k!} = \alpha t.$$

Theorem 6.3 *Let* $\{N_I\}_{I \in \mathcal{I}}$ *be a family of random variables on the probability space* $(\Omega, \mathcal{E}, \mathbb{P})$ *satisfying (P1)–(P5). Then the continuous family of random variables* $\{N_t\}_{t \geq 0}$, $N_t := N_{[0,t]}$, *is a Poisson process with intensity* $\alpha := \mathbb{E}\left[N_{[0,1]}\right]$.

Conversely, let $\{N_t\}_{t \geq 0}$ *be a Poisson process of intensity* $\alpha > 0$ *on* $(\Omega, \mathcal{E}, \mathbb{P})$. *Then the family of random variables* $\{N_I\}_{I \in \mathcal{I}}$ *defined by* $N_I := N_t - N_s$ *if* $I = [s, t]$ *satisfies properties (P1)–(P5) and* $\mathbb{E}\left[N_{[0,1]}\right] = \alpha$.

Proof. (i) Assume $\{N_I\}_{I \in \mathcal{I}}$ satisfies (P1)–(P5). Obviously, (P1)–(P4) imply that the process $\{N_t\}$, $N_t = N_{[0,t]}$ satisfies (i), (ii) and (iii) of Definition 6.2.

Let $\alpha(t) := \mathbb{E}\left[N_{[0,t]} \right]$, $t > 0$. By (P2) and (P3), $\alpha(t)$ is monotone increasing and $\alpha(t+s) = \alpha(t) + \alpha(s)$. Thus, by Lemma 3.50, $\alpha(t) = \alpha\, t$, $\alpha := \alpha(1)$. In particular, $\mathbb{E}\left[N_t \right] = \alpha t$.

In order to prove that N_t follows the Poisson distribution of parameter αt, we define a sequence of Bernoulli processes that approximates N_t. Partition the interval $[0, t]$ in 2^n pairwise intervals of equal length

$$I_{k,n} = \left[\frac{(k-1)t}{2^n}, \frac{kt}{2^n} \right], \qquad k = 1, \dots, 2^n.$$

Let $X_{k,n}(x) := N_{I_{k,n}}(x)$ and $\overline{X}_{k,n}(x) := \mathbb{1}_{\{X_{k,n} \geq 1\}}(x) = \mathbb{1}_{[1,\infty]}(X_{k,n}(x))$. By (P2) and (P3) the random variables $\left\{ X_{k,n} \right\}_k$ are independent and identically distributed and so are the random variables $\left\{ \overline{X}_{k,n} \right\}_k$, see Proposition 4.36. More precisely, the random variables $\left\{ \overline{X}_{k,n} \right\}_k$ are Bernoulli trials of parameter $p_n := \mathbb{P}(X_{k,n} \geq 1) = \mathbb{P}(N_{[0,2^{-n}]} \geq 1)$. Therefore, the random variable $N_t^n := \sum_{k=1}^{2^n} \overline{X}_{k,n}$ follows the binomial distribution $B(2^n, p_n)$. Let us now show that the random variables $\left\{ N_t^n \right\}$ *approximate* the random variable N_t. Clearly, $N_t^{n+1} \geq N_t^n$, i.e. $\left\{ N_t^n \right\}_n$ is a monotone increasing sequence of random variables. Moreover, by (P5) and Lemma 6.1

$$\begin{aligned} \mathbb{P}(N_t^n \neq N_t) &= \sum_{k=1}^{2^n} \mathbb{P}(X_{k,n} - \overline{X}_{k,n}) = \sum_{n=1}^{2^n} \mathbb{P}(X_{k,n} \geq 2) \\ &= 2^n \mathbb{P}(N_{2^{-n}t} \geq 2) \to 0, \end{aligned}$$

i.e. $N_t^n(x) \uparrow N_t(x)$ almost surely. By the Beppo Levi theorem we then get

$$\alpha t = \mathbb{E}\left[N_t \right] = \lim_{n\to\infty} \mathbb{E}\left[N_t^n \right] = \lim_{n\to\infty} 2^n p_n,$$

and

$$\mathbb{P}(N_t^n \geq k) \;\to\; \mathbb{P}(N_t \geq k) \qquad \forall k \in \mathbb{N}.$$

Finally, we recall that limits of binomial distributions are Poisson distributions, see Proposition 3.31. Thus, for any integer k,

$$\mathbb{P}(N_t = k) = \lim_{n\to\infty} \mathbb{P}(N_t^n = k) = \lim_{n\to\infty} B(2^n, p_n)(\{k\}) = P(\alpha t)(\{k\}).$$

Conversely, assume $\left\{ N_t \right\}$ is a Poisson process with intensity α. Let $I = [s, t]$. By properties (i) and (ii) of Poisson processes, the random variables $N_I := N_t - N_s$ are non-negative and (iii) of Definition 6.2 implies (P1). Moreover, $\mathbb{P}(N_I = k) = \mathbb{P}(N_t - N_s = k) = P(\alpha(t-s))(\{k\})$, so that also (P2) and (P3) are satisfied and $\mathbb{E}\left[N_I \right] = \alpha(t-s) < +\infty$, i.e. also (P4) holds. Finally, since

$$\mathbb{P}(N_{[0,\epsilon]} \geq 2) = 1 - \mathbb{P}(N_{[0,\epsilon]} = 0) - \mathbb{P}(N_{[0,\epsilon]} = 1) = 1 - e^{-\alpha\epsilon} - \alpha\epsilon e^{-\alpha\epsilon},$$

dividing by ϵ and letting $\epsilon \to 0$, we get (P5). The claims are then proven.

If there is a click at time $s > 0$, then the probability that there is another click before a further time, i.e. the probability that the time $T(x)$ between two clicks or, more formally, between two singularities of $N_t(x)$ is smaller than t, is given by

$$\mathbb{P}\Big(T < t + s \,\Big|\, T \geq s\Big) := \mathbb{P}(N_{[s,s+t[} = 0) = \mathbb{P}(N_{t+s} - N_s = 0) = e^{-\alpha t}.$$

Thus:

(i) The waiting times between two consecutive clicks are independent.

(ii) The waiting times between two consecutive clicks are exponentially distributed, $\mathbb{P}(T < t + s \mid T \geq s) = e^{-\alpha t}$.

The two properties (i) and (ii) above characterize Poisson processes. Let $\{W_k\}$ be a sequence of positive random variables satisfying (i) and (ii). Let $T_n := \sum_{k=1}^{n} W_k$ and let

$$N_t(x) := \#\Big\{n \,\Big|\, T_n(x) \leq t\Big\}.$$

$\{N_t\}$ is an increasing family of random variables and, moreover, the map $t \mapsto N_t(x)$ is right-continuous. In fact, let $t_k \downarrow t$. Then the sets $E_k := \{n \mid T_{n+1}(x) > t_k\}$ is an increasing family of sets, $E_k \subset E_{k+1}$ $\forall k$ and $E := \cup_k E_k = \{n \mid T_{n+1}(x) > t\}$. Thus, $N_t(x) = \inf E = \lim_k(\inf E_k) = \lim_k N_{t_k}(x)$ and

$$\Big\{x \in \Omega \,\Big|\, N_t(x) = k\Big\} = \Big\{x \in \Omega \,\Big|\, T_k(x) \leq t < T_{k+1}(x)\Big\}.$$

Thus, each N_t, $t \geq 0$, is a random variable and

$$\begin{aligned}\mathbb{P}(N_t = k) &= \mathbb{P}(T_k \leq t < T_{k+1}) = \mathbb{P}(t < T_{k+1}) - \mathbb{P}(t < T_k)\\ &= e^{-\alpha t}\frac{(\alpha t)^k}{k!},\end{aligned}$$

see Exercise 4.65. Then one proves that $\{N_t(x)\}$ is a Poisson process of parameter α. We omit the proof as the proof of independence of the increments, (iii) of Definition 6.2, is technical. We refer the interested reader to, e.g. [5].

Summarizing, we have the following.

Theorem 6.4 *With the previous notation, the following are equivalent:*

(i) $\{N_t\}_{t\geq 0}$ *is a Poisson process with intensity* α.

(ii) The waiting time random variables $\{W_k\}$*, that measure the time between two consecutives jumps of* $N_t(x)$*, are pairwise independent and identically distributed with exponential distribution* $\exp(\alpha)$.

6.2 Continuous time Markov chains

6.2.1 Definitions

6.2.1.1 Transition probabilities matrix

A *continuous time stochastic process* is a one real parameter family $\{X_t\}_{t\geq 0}$ of random variables on a probability space $(\Omega, \mathcal{E}, \mathbb{P})$. We only consider stochastic processes with a finite or denumerable state-space S.

For each $s, t \geq 0$ and each $i, j \in S$, let $\mathbf{P}(s,t) := (\mathbf{P}(s,t)^i_j)$ be the matrix defined by

$$\mathbf{P}^i_j(s,t) := \begin{cases} \mathbb{P}\Big(X_t = j \,\Big|\, X_s = i\Big) & \text{if } \mathbb{P}(X_s = i) > 0, \\ \delta^i_j & \text{if } \mathbb{P}(X_s = i) = 0. \end{cases}$$

We call the matrix $\mathbf{P}(s,t)$ the *transition probabilities matrix* from X_s to X_t. Observe that $\mathbf{P}(s,t)$ is a stochastic matrix.

The law of total probability yields

$$\mathbb{P}(X_t = j) = \sum_{i \in S} \mathbb{P}(X_s = i)\mathbf{P}(s,t)^i_j, \tag{6.3}$$

thus, denoting by $\pi(t) = (\pi_j(t))$, $\pi_j(t) := \mathbf{P}(X_t = j)$, the mass density row vector of X_t, (6.3) reads equivalently as

$$\pi(t) = \pi(s)\mathbf{P}(s,t). \tag{6.4}$$

Of course, $\mathbf{P}^i_j(t,t) = \delta^i_j$ if $\mathbb{P}(X_t = i) > 0$ so $\mathbf{P}(t,t) = \mathrm{Id}\ \forall t \geq 0$.

Finally, by an induction argument, from (6.4) one gets for any $n \in \mathbb{N}$ and any $0 \leq t_0 \leq t_1 \leq \cdots \leq t_n$,

$$\pi(t_n) = \pi(t_0)\mathbf{P}(t_0,t_1)\mathbf{P}(t_1,t_2)\cdots\mathbf{P}(t_{n-1},t_n).$$

6.2.1.2 Homogeneous processes

Definition 6.5 *A continuous time stochastic process* $\{X_t\}_{t\geq 0}$ *with values in a finite or denumerable state-space* S *is called* homogeneous *if for any* $s, t, h \geq 0$ *and for any* $i, j \in S$

$$\mathbb{P}\Big(X_{t+h} = j \,\Big|\, X_{s+h} = i\Big) = \mathbb{P}\Big(X_t = j \,\Big|\, X_s = i\Big).$$

For the transition probabilities matrices $\mathbf{P}(s,t)$ *of the process we then have*

$$\mathbf{P}(s,t) = \mathbf{P}(0, t-s) \qquad \forall 0 \leq s \leq t.$$

The map $t \to \mathbf{P}(t) := \mathbf{P}(0,t)$ *is called the* transition probabilities map *of the process.*

6.2.1.3 Markovian processes

Definition 6.6 (Markov property) *Let $\{X_t\}_{t\geq 0}$ be a continuous time stochastic process with a finite or denumerable state-space S. The process $\{X_t\}_{t\geq 0}$ is called* Markovian *if, for any positive integer n, for any $0 \leq t_0 < t_1 < \cdots < t_n$, and any $i_1, \ldots, i_n \in S$ we have*

$$\begin{aligned}\mathbb{P}\Big(X_{t_n} = i_n \,\Big|\, X_{t_{n-1}} = i_{n-1}, \ldots, X_{t_0} = i_0\Big) \\ = \mathbb{P}\Big(X_{t_n} = i_n \,\Big|\, X_{t_{n-1}} = i_{n-1}\Big)\end{aligned} \tag{6.5}$$

whenever $\mathbb{P}(X_{t_{n-1}} = i_{n-1}, \ldots, X_{t_0} = i_0) > 0$.

With the same reasoning used for discrete time Markov chains, the process $\{X_t\}_{t\geq 0}$ is Markovian if and only if for any $0 \leq r \leq s \leq n$, for any $0 \leq t_0 \leq t_1 \leq \cdots \leq t_n$ and any triplet of events G, F, E detected by $X_{t_0}, \ldots, X_{t_r}$, $X_{t_{r+1}}, \ldots, X_{t_s}$ and $X_{t_{s+1}}, \ldots, X_{t_n}$, respectively, we have

$$\mathbb{P}\Big(E \,\Big|\, F \cap G\Big) = \mathbb{P}\Big(E \,\Big|\, F\Big)$$

or

$$\mathbb{P}\Big(E \cap F \,\Big|\, G\Big) = \mathbb{P}\Big(E \,\Big|\, F\Big)\mathbb{P}\Big(F \,\Big|\, G\Big). \tag{6.6}$$

It is also still possible to compute the joint distribution of a finite number of the random variables X_t's in terms of the transition matrices. Let $k, n \geq 0$, $0 \leq t_0 \leq t_1 \leq \cdots \leq t_{n+k}$. Denote

$$A_k := \Big\{x \in \Omega \,\Big|\, X_{t_k} = i_k\Big\}.$$

By (6.6) one gets

$$\mathbb{P}\Big(A_n \cap \cdots \cap A_1 \,\Big|\, A_0\Big) = \mathbb{P}\Big(A_n \,\Big|\, A_{n-1}\Big)\mathbb{P}\Big(A_{n-1} \cap \cdots \cap A_1 \,\Big|\, A_0\Big).$$

Thus, by induction,

$$\mathbb{P}\Big(A_n \cap \cdots \cap A_1 \,\Big|\, A_0\Big) = \prod_{k=0}^{n-1} \mathbb{P}\Big(A_{k+1} \,\Big|\, A_k\Big)$$

i.e.

$$\mathbb{P}\Big(X_{t_n} = i_n, \ldots, X_{t_0} = i_0\Big) = \mathbb{P}(X_{t_0} = i_0)\mathbf{P}(t_0, t_1)_{i_1}^{i_0} \cdots \mathbf{P}(t_{n-1}, t_n)_{i_n}^{i_{n-1}}. \tag{6.7}$$

Remark 6.7 Since the probabilities of the events detected by the X_t's are actually defined by the value of the probabilities of finite intersections of the generating events, see Theorem B.6, property (6.5) or, equivalently, (6.6) impact on the whole set of the events detected by the X_t's. As in the discrete case, the Markov property says that the *probabilistic prevision of the future state of the chain depends on the past history only through its present state*. Also, formula (6.7) implies that the transiton matrices fix, in principle, the probability of any event detected by the random variables $\{X_t\}$. However, we must be careful, since not every subset of Ω determined by the X_t's is an event, in general. For instance, the set

$$E = \Big\{x \in \Omega \,\Big|\, X_\tau(x) = i \ \forall \tau,\ r \le \tau \le s\Big\}$$

needs not be an event, since E is the intersection of a non denumerable family of events.

6.2.2 Continuous semigroups of stochastic matrices

6.2.2.1 Chapman–Kolmogorov equations

Proposition 6.8 *Let* $\{X_t\}_{t\ge 0}$ *be a Markovian stochastic process with finite or denumerable state-space and transition probabilities matrix* $\mathbf{P}(s,t)$. *Then, the* Chapman–Kolmogorov equations *hold*

$$\begin{cases} \mathbf{P}(s,t) = \mathbf{P}(s,\tau)\mathbf{P}(\tau,t) & \forall s,\tau,t \text{ such that } 0 \le s < \tau < t, \\ \mathbf{P}(s,s) = \mathrm{Id} & \forall s \ge 0. \end{cases}$$

If the process is homogeneous and $\mathbf{P}(t)$ *denotes the transition probabilities map of the process, then the Chapman–Kolmogorov equations reduce to*

$$\begin{cases} \mathbf{P}(t+s) = \mathbf{P}(t)\mathbf{P}(s) & \forall s,t \ge 0, \\ \mathbf{P}(0) = \mathrm{Id}. \end{cases}$$

Proof. By the Markov property,

$$\begin{aligned} \mathbf{P}(s,t)^i_j &= \mathbb{P}\Big(X_t = j \,\Big|\, X_s = i\Big) = \sum_{h\in S} \mathbb{P}\Big(X_t = j,\ X_\tau = h \,\Big|\, X_s = i\Big) \\ &= \sum_{h\in S} \mathbb{P}\Big(X_t = j \,\Big|\, X_\tau = h\Big)\mathbb{P}\Big(X_\tau = h \,\Big|\, X_s = i\Big) \\ &= \sum_{h\in S} \mathbf{P}(\tau,t)^h_j \mathbf{P}(s,\tau)^i_h = (\mathbf{P}(s,\tau)\mathbf{P}(\tau,t))^i_j. \end{aligned}$$

6.2.2.2 Right-continuous Markov processes

Examples show that no regularity can be expected for the transition probabilities map $(s,t) \to \mathbf{P}(s,t)$ of stochastic processes. However, for a subclass of stochastic processes, the situation is better.

Definition 6.9 *A stochastic process* $\{X_t\}_{t\geq 0}$ *on* $(\Omega, \mathcal{E}, \mathbb{P})$ *is said to be* right-continuous *if the trajectories* $t \mapsto X_t(x)$ *are right-continuous for* $\mathbb{P}$*-almost every* $x \in \Omega$.

Proposition 6.10 *Let* $\{X_t\}_{t\geq 0}$ *be a right-continuous stochastic process with a* finite *state-space* S. *Then for any* $s \geq 0$

$$\mathbf{P}(s,t) \to \mathrm{Id} \qquad \text{as } t \to s^+.$$

In particular, the transition probabilities map of a homogeneous process $t \mapsto \mathbf{P}(t)$ *is continuous at* 0,

$$\mathbf{P}(t) \to \mathrm{Id} \qquad \text{as } t \to 0^+.$$

Proof. Let $\{t_n\}$ be such that $t_n \downarrow s$. Observe that since the states are finite, for any $x \in \Omega$ the map $t \to X_t(x)$ is right-continuous at s if and only if there exists $\overline{n} = \overline{n}(x)$ such that $X_{t_n}(x) = i$ for any $n \geq \overline{n}$. In other words, denoting by $E \subset \Omega$ the set of points $x \in \Omega$ such that the map $t \to X_t(x)$ is right-continuous, we have

$$\left\{x \in E \,\middle|\, X_s(x) = i\right\} = \bigcup_{n=1}^{\infty} \bigcap_{k\geq n} \left\{x \in \Omega \,\middle|\, X_{t_k}(x) = i\right\},$$

hence the events

$$\left\{x \in \Omega \,\middle|\, X_s(x) = i\right\} \qquad \text{and} \qquad \bigcup_{k=1}^{\infty} \bigcap_{k\geq n} \left\{x \in \Omega \,\middle|\, X_{t_k}(x) = i\right\}$$

differ by a null set. Therefore, assuming $\mathbb{P}(X_s = i) > 0$,

$$\begin{aligned} 1 = \mathbf{P}(s,s)_i^i &= \mathbb{P}\Big(X_s = i \,\Big|\, X_s = i\Big) \\ &= \lim_{n\to\infty} \mathbb{P}\Big(\bigcap_{k\geq n} \{x \in \Omega \mid X_{t_k}(x) = i\} \,\Big|\, X_s = i\Big) \\ &\leq \lim_{n\to\infty} \mathbb{P}\Big(X_{t_n} = i \,\Big|\, X_s = i\Big) = \lim_{n\to\infty} \mathbf{P}(s,t_n)_i^i. \end{aligned}$$

Since obviously $\mathbf{P}(s,t_n)_i^i \leq 1$, we get $\mathbf{P}(s,t_n)_i^i \to 1$. Thus, since the sequence $\{t_n\}$ is arbitrary, $\mathbf{P}(s,t)_i^i \to 1$ as $t \to s^+$. Furthermore, since for any $i, j \in S$ with $i \neq j$, we have

$$\mathbf{P}(s,t)_j^i = \mathbb{P}\Big(X_t = j \,\Big|\, X_s = i\Big) \leq \mathbb{P}\Big(X_t \neq i \,\Big|\, X_s = i\Big) = 1 - \mathbf{P}(s,t)_i^i,$$

we also deduce that $\mathbf{P}(s,t)_j^i \to 0$ as $t \to s^+$.

Remark 6.11 Let $\{X_t\}_{t\geq 0}$ be a right-continuous stochastic process with finite state-space S and let $\mathcal{D}$ be a denumerable dense set in $\mathbb{R}$. Repeating the argument in the proof of Proposition 6.10 one gets for every $s \in \mathbb{R}_+$, for every $i \in S$ and every sequence $\{t_n\} \subset \mathcal{D}$ such that $t_n \downarrow s$, that

$$\left\{x \in \Omega \,\middle|\, X_s(x) = i\right\} \qquad \text{and} \qquad \bigcup_{k=1}^{\infty} \bigcap_{k\geq n} \left\{x \in \Omega \,\middle|\, X_{t_k}(x) = i\right\}$$

differ by a null set. In particular, the generating events $\{x \in \Omega \mid X_s(x) = i\}$ are generated by the events detected by the random variables $\{X_s, s \in \mathcal{D}\}$. A similar argument shows that any finite intersection of generating events is detected by the random variables $\{X_s, s \in \mathcal{D}\}$. We then infer that the $\mathbb{P}$-completions of the σ-algebra of the events detected by all the X_t's and of the σ-algebra of the events detected by the X_t's with $t \in \mathcal{D}$ coincide, see Theorem B.6.

Definition 6.12 *A homogeneous, Markovian and right-continuous stochastic process* $\{X_t\}_{t\geq 0}$ *is called a* right-continuous Markov chain.

By combining Propositions 6.10 and 6.8 we get the following.

Theorem 6.13 *Let* $\{X_t\}_{t\geq 0}$ *be a right-continuous Markov chain with a* finite *state-space S. Then the associated transition map* $t \mapsto \mathbf{P}(t)$, $t \geq 0$, *is a* continuous semigroup, *i.e.*

$$\begin{cases} \mathbf{P}(t) \to \mathbf{P}(0) & \text{as } t \to 0^+, \\ \mathbf{P}(t+s) = \mathbf{P}(t)\mathbf{P}(s) & \text{for every } t, s \geq 0, \\ \mathbf{P}(0) = \mathrm{Id}. \end{cases} \tag{6.8}$$

The theory of ordinary differential equations then applies. In particular, Theorem C.10 yields the following.

Theorem 6.14 *Let* $\{X_t\}_{t\geq 0}$ *be a right-continuous Markov chain with transition probabilites matrix* $\mathbf{P}(t)$, $t \geq 0$. *Then* $\mathbf{P}(t)$ *is differentiable, and, if*

$$\mathbf{Q} := \mathbf{P}'_+(0) := \lim_{t\to 0^+} \frac{\mathbf{P}(t) - \mathrm{Id}}{t}, \tag{6.9}$$

then

$$\mathbf{P}(t) = e^{\mathbf{Q}t} = \sum_{k=0}^{\infty} \frac{\mathbf{Q}^k t^k}{k!} \qquad \forall t \geq 0. \tag{6.10}$$

Actually, (6.8) are equivalent to (6.10). Also, it is well known that $\mathbf{P}(t) = e^{\mathbf{Q}t}$ is the unique solution of both the systems

$$\begin{cases} \mathbf{P}'(t) = \mathbf{Q}\mathbf{P}(t), \\ \mathbf{P}(0) = \mathrm{Id} \end{cases} \qquad \text{and} \qquad \begin{cases} \mathbf{P}'(t) = \mathbf{P}(t)\mathbf{Q}, \\ \mathbf{P}(0) = \mathrm{Id}. \end{cases}$$

The matrix $\mathbf{Q}$, which is the *infinitesimal generator* of the semigroup $\mathbf{P}(t)$, expresses the rate of change of $\mathbf{P}(t)$. We call $\mathbf{Q}$ the matrix of the *intensity factors* or the *intensities matrix* of the (transition probabilities map of the) chain.

Example 6.15 At the initial stage 0 we wait for a random length of time that follows the exponential distribution with intensity α, then we move to state 1. There again, we wait for a random length of time that follows the same distribution $\exp(\alpha)$ then we move to the state 2 and so on. This is the description of the Poisson process of parameter α, see Theorem 6.4. Its transition matrix and its intensities matrix are

$$\mathbf{P}(t) = \begin{pmatrix} e^{-\alpha t} & 1-e^{-\alpha t} & 0 & \dots \\ 0 & e^{-\alpha t} & 1-e^{-\alpha t} & \dots \\ \vdots & \vdots & \vdots & \ddots \end{pmatrix}$$

and

$$\mathbf{Q} = \begin{pmatrix} -\alpha & \alpha & 0 & 0 & \dots \\ 0 & -\alpha & \alpha & 0 & \dots \\ 0 & 0 & -\alpha & \alpha & \dots \\ \vdots & \vdots & \vdots & \vdots & \ddots \end{pmatrix}.$$

Example 6.16 At the initial stage we are, for certain time in a state 1 and move to states 2 and 3 after a random time following the distribution exp(2) and exp(4), respectively. We move from state 2 to state 3 with a random time following the distribution exp(5). Then the intensities matrix $\mathbf{Q}$ is

$$\mathbf{Q} = \begin{pmatrix} -6 & 2 & 4 \\ 0 & -5 & 5 \\ 0 & 0 & 0 \end{pmatrix}.$$

6.2.2.3 Q-matrices

Definition 6.17 *A matrix* $\mathbf{Q} \in M_{N,N}(\mathbb{R})$ *is called a* Q-matrix *if*

$$\mathbf{Q}^i_j \geq 0 \quad \forall i \neq j, \qquad \sum_{j=1}^{N} \mathbf{Q}^i_j = 0 \quad \forall i.$$

In particular, all the diagonal entries of $\mathbf{Q}$ are nonpositive, $\mathbf{Q}^i_i \leq 0$, and

$$||\mathbf{Q}||_\infty := \max_{i,j=1,\dots,N} |\mathbf{Q}^i_j| = \max_{i=1,\dots,N} (-\mathbf{Q}^i_i).$$

Q-matrices provide an useful tool in order to define stochastic matrices. Let $\mathbf{Q}$ be a $N \times N$ Q-matrix. For $\alpha \geq ||\mathbf{Q}||_\infty$ define

$$\mathbf{R} := \mathbf{R}_\alpha = \mathrm{Id} + \frac{1}{\alpha}\mathbf{Q}.$$

A trivial computation shows that $\mathbf{R}$ is stochastic. Moreover, since $\mathbf{Q} = \alpha(\mathbf{R} - \mathrm{Id})$, we get

$$\mathbf{P}(t) = e^{\mathbf{Q}t} = e^{-\alpha t} e^{\alpha \mathbf{R} t} = e^{-\alpha t} \sum_{k=0}^{\infty} \alpha^k \mathbf{R}^k \frac{t^k}{k!}. \tag{6.11}$$

Proposition 6.18 *Let $\mathbf{Q} \in M_{N,N}(\mathbb{R})$. Then the matrix $\mathbf{P}(t) = e^{\mathbf{Q}t}$ is stochastic for any $t > 0$ if and only if $\mathbf{Q}$ is a Q-matrix. Moreover, for any $i, j \in \{1, \dots, N\}$ the following are equivalent:*

(i) there exists $t_0 > 0$ such that $\mathbf{P}(t_0)^i_j > 0$;

(ii) $\mathbf{P}(t)^i_j > 0$ for any $t > 0$.

In particular, $\mathbf{P}(t)^i_i > 0$ for any state i and any positive t. Finally, if there exists $t_0 > 0$ such that $\mathbf{P}(t_0)$ is irreducible, then $\mathbf{P}(t)^i_j > 0$ for any $i, j \in \{1, \dots, N\}$ and any positive $t > 0$.

Before proving the proposition, let us point out that the proposition asserts that the evolution of a continuous time Markov chain is much more regular than the evolution of a discrete time one. Periodicity issues that could not be neglected in the discrete case do not appear in the continuous time case.

Proof of Proposition 6.18. Assume $\mathbf{P}(t)$ is stochastic for any $t \in \mathbb{R}$. Since $\mathbf{P}(t) = e^{\mathbf{Q}t}$ $\forall t$, then $\mathbf{Q} = \lim_{t \to 0+} \frac{\mathbf{P}(t) - \mathrm{Id}}{t}$ from which we easily infer that $\mathbf{Q}$ is a Q-matrix. Conversely, let $\mathbf{Q}$ be a Q-matrix. Let $\alpha \ge ||\mathbf{Q}||_\infty$ and $\mathbf{R} := \mathrm{Id} + \frac{1}{\alpha}\mathbf{Q}$. By (6.11) $\mathbf{P}(t)$ is the sum of a series with non-negative coefficients

$$\mathbf{P}(t) = e^{-\alpha t} \sum_{k=0}^{\infty} \frac{\alpha^k t^k \mathbf{R}^k}{k!}. \tag{6.12}$$

Since the matrix $\mathbf{R}^k$ is stochastic for any k, we get $\mathbf{P}(t)^i_j \ge 0$ for any $t > 0$ and for any $i, j \in \{1, \dots, N\}$. Moreover,

$$\sum_{j=1}^{N} \mathbf{P}(t)^i_j = e^{-\alpha t} \sum_{k=0}^{\infty} \frac{\alpha^k t^k}{k!} \Big(\sum_{j=1}^{N} (\mathbf{R}^k)^i_j \Big) = e^{-\alpha t} \sum_{k=0}^{\infty} \frac{\alpha^k t^k}{k!} = 1.$$

Let us show the equivalence between claims (i) and (ii). Obviously, (ii) $\Rightarrow$ (i). It remains to prove that (i) $\Rightarrow$ (ii). If, by contradiction, $\mathbf{P}(\bar{t})^i_j = 0$ for some positive $\bar{t}$, then by (6.12) $(\mathbf{R}^k)^i_j = 0$ for any k and $\mathbf{P}(t)^i_j = 0$ $\forall t \ge 0$.

Since $\mathbf{P}(t) = e^{\mathbf{Q}t}$, then for any $t > 0$ and for any positive integer k we have $\mathbf{P}(t)^k = \mathbf{P}(kt)$. If for some $t_0 > 0$ the matrix $\mathbf{P}(t_0)$ is irreducible, then for any $i, j \in \{1, \dots N\}$ there exists $k = k(i, j)$ such that $\mathbf{P}(kt_0)^i_j = (\mathbf{P}(t_0)^k)^i_j > 0$. Therefore equivalence between claims (i) and (ii) yields $\mathbf{P}(t)^i_j > 0$ for any $t > 0$ and for any i, j.

Proposition 6.19 *Let* $\mathbf{Q}$ *be a* Q*-matrix. Then* 0 *is an eigenvalue of* $\mathbf{Q}$*. Any other eigenvalue of* $\mathbf{Q}$ *has strictly negative real part.*

Proof. For any $i = 1, \dots, N$, let $r_i := -\mathbf{Q}_i^i \geq 0$. By Exercise 5.67 the eigenvalues of $\mathbf{Q}$ are in $\cup_{i=1}^N B(-r_i, r_i)$, where $B(x, r)$ denote the *closed* ball of radius r centered at x. Thus, all the eigenvalues but 0 of $\mathbf{Q}$ have strictly negative real part. Moreover, if 0 were not an eigenvalue of $\mathbf{Q}$, then all the eigenvalues of $\mathbf{Q}$ would have strictly negative real part, so that $\mathbf{P}(t) \to 0$ as $t \to \infty$, a contradiction since $\mathbf{P}(t)$ is stochastic for every t.

Example 6.20 Assume the matrix $\mathbf{Q}$ is defined as

$$\mathbf{Q} = \begin{pmatrix} -0.5 & 0.5 & 0 & 0 \\ 0.5 & -1.5 & 1 & 0 \\ 0 & 0 & -2 & 2 \\ 0 & 0 & 3 & -3 \end{pmatrix}$$

and let $\alpha := 4$. Then

$$4\mathbf{R} = 4\mathrm{Id} + \mathbf{Q} = \begin{pmatrix} 3.5 & 0.5 & 0 & 0 \\ 0.5 & 2.5 & 1 & 0 \\ 0 & 0 & 2 & 2 \\ 0 & 0 & 3 & 1 \end{pmatrix}.$$

Figure 6.1 shows a representation of both matrices $\mathbf{Q}$ and $\mathbf{R}$.

6.2.2.4 Asymptotic behaviour

Using the Banach fixed point theorem, it is easy to discuss the asymptotic behaviour of $\mathbf{P}(t) := e^{\mathbf{Q}t}$ as $t \to +\infty$ when $\mathbf{Q}$ is a Q-matrix.

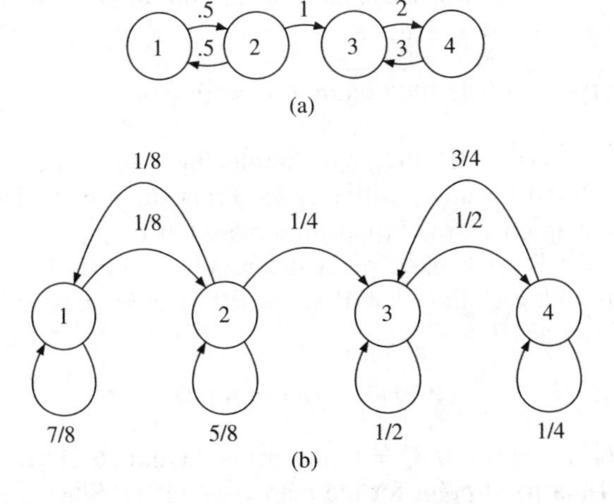

Figure 6.1 Graphs representing the matrices (a) $\mathbf{Q}$ *and (b)* $\mathbf{R}$ *in Example 6.20.*

Lemma 6.21 *Let* $\mathbf{Q}$ *be a* $N \times N$ Q*-matrix, let* $\mathbf{P}(t) := e^{\mathbf{Q}t}$, $t \geq 0$, *and let* w *be a stochastic vector,* $w \in \mathcal{T}$. *The following hold:*

(i) If for some $t_0 > 0$ *the vector* w *is the only solution of* $w\mathbf{P}(t_0) = w$, *then* $w\mathbf{Q} = 0$.

(ii) $w\mathbf{Q} = 0$ *if and only if* $w = w\mathbf{P}(t)$ *for any positive* t.

Proof. (i) If w is the only fixed point of the map $x \mapsto x\mathbf{P}(t_0)$, $x \in \mathcal{T}$, then for any n, w is a fixed point for $x \mapsto x\mathbf{P}(t_0/n)$, since $\mathbf{P}(t_0/n)^n = \mathbf{P}\Big(n\frac{t_0}{n}\Big) = \mathbf{P}(t_0)$. Therefore, for any positive integer n

$$w = we^{\mathbf{Q}t_0/n} = w + \frac{t_0}{n}\, w\mathbf{Q} + o\Big(\frac{1}{n}\Big).$$

Since n is arbitrary, then $w\mathbf{Q} = 0$.
(ii) If $w\mathbf{Q} = 0$, then $w\mathbf{Q}^k = 0$ for any positive integer $k \geq 1$. Thus, for any $t > 0$

$$w\mathbf{P}(t) = w\sum_{k=0}^{\infty} \frac{\mathbf{Q}^k t^k}{k!} = w + \sum_{k=1}^{\infty} \frac{w\mathbf{Q}^k t^k}{k!} = w.$$

Conversely, if $w = w\mathbf{P}(t)$ for any positive t, then

$$w = we^{\mathbf{Q}t} = w + t\, w\mathbf{Q} + o(t) \qquad \text{as } t \to 0^+,$$

so that $w\mathbf{Q} = 0$.

Theorem 6.22 *Let* $\mathbf{P}(t) = e^{\mathbf{Q}t}$ *be stochastic for any positive* t. *The following are equivalent:*

(i) There exists $t_0 > 0$ *such that* $\mathbf{P}(t_0)$ *is irreducible.*

(ii) There exists $w \in \mathcal{T}$ *whose components are all different from zero such that* $x\mathbf{P}(t) \to w$ *for any* $x \in \mathcal{T}$.

In this case, w *is the unique solution in* $\mathcal{T}$ *of* $w\mathbf{Q} = 0$.

Proof. (i) $\Rightarrow$ (ii). If $\mathbf{P}(t_0)$ is irreducible, then $\mathbf{P}(t)^i_j > 0$ for any $i, j \in \{1, \ldots, N\}$ and for any positive t, see Proposition 6.18. Thus $x \mapsto x\mathbf{P}(t)$ is a contraction map on $\mathcal{T}$, see Proposition 5.53. Let $C(t) < 1$ be its contraction coefficient. By the Banach fixed point theorem $x \mapsto x\mathbf{P}(t)$ has a unique fixed point $w(t) \in \mathcal{T}$ and, recalling that $\mathbf{P}(kt) = \mathbf{P}(t)^k$, there exists $A > 0$ such that for any $x \in \mathcal{T}$ and any $k \geq 1$

$$|x\mathbf{P}(kt) - w(t)| \leq A\, C(t)^k. \tag{6.13}$$

(i) of Lemma 6.21 yields $w(t)\mathbf{Q} = 0$ and (ii) of Lemma 6.21 yields that for any $s > 0$ $w(t)$ is also a fixed point for the map $x \mapsto x\mathbf{P}(s)$. Since a contraction has a unique fixed point, we conclude that $w(t) = w(s) =: w$ for any positive s, t.

Let us now show that $x\mathbf{P}(t) \to w$ for any $x \in \mathcal{T}$. From (iv) of Proposition 5.53 we can assume that the contraction coefficient $C(t)$ is continuous with respect to t. Thus we may assume that $C(t) \le C < 1$ for any $t \in [1, 2]$. Let $s \ge 1$. Applying (6.13) with $k := \lfloor s \rfloor$ and $t := s/k$ yields

$$|x\mathbf{P}(s) - w| \le A\, C(t)^k \le A\, C^{\lfloor s \rfloor}.$$

Thus $x\mathbf{P}(s) \to w$ as $s \to \infty$.
(ii) $\Rightarrow$ (i) If $x\mathbf{P}(t) \to w$ for any $x \in \mathcal{T}$, then all the components of w are nonzero, see Exercise 5.65. Thus there exists a large enough t_0 such that $\mathbf{P}(t_0)$ is regular. Consequently, $\mathbf{P}(t)$ is regular for any postive t, by Proposition 6.18.

6.2.2.5 Computing $e^{\mathbf{Q}t}$

Quite often in applications one needs to compute the exponential matrix $\mathbf{P}(t) = e^{\mathbf{Q}t}$, given $\mathbf{Q}$. Various methods can be found in the literature, see e.g. [21], having different degrees of efficiency, accuracy and usefulness either from the analytical, numerical or symbolic viewpoint. Formula (6.11) proves to be quite efficient in computing $\mathbf{P}(t) = e^{\mathbf{Q}t}$ when $\mathbf{Q}$ is a Q-matrix: it permits to compute, for fixed t, good approximations of $\mathbf{P}(t)$ with explicitly bounded errors simply by truncation of the power series on the right-hand side.

Proposition 6.23 (Uniformization method) *Let* $\mathbf{Q}$ *be a* $N \times N$ Q*-matrix and let* $\mathbf{P}(t) = e^{\mathbf{Q}t}$. *Let* $\alpha \ge ||\mathbf{Q}||_\infty$, $\mathbf{R} = \mathrm{Id} + \frac{1}{\alpha}\mathbf{Q}$, *and*

$$\mathbf{P}_M(t) = e^{-\alpha t} \sum_{k=0}^{M} \alpha^k \mathbf{R}^k \frac{t^k}{k!}.$$

Then

$$||\mathbf{P}(t) - \mathbf{P}_M(t)|| \le \frac{(\alpha t)^{M+1}}{(M+1)!}.$$

where $||\mathbf{A}|| := \max_i \sum_{j=1}^N |\mathbf{A}^i_j|$.

Proof. Since $\mathbf{R}$ is stochastic, then so are the matrices $\mathbf{R}^k$'s for every k, hence $||\mathbf{R}^k|| = 1$. Thus the elementary estimate

$$\left| e^x - \sum_{k=0}^{n} \frac{x^k}{k!} \right| \le e^x \frac{x^{n+1}}{(n+1)!} \qquad \forall x \ge 0,$$

gives

$$||\mathbf{P}(t) - \mathbf{P}_M(t)|| \le e^{-\alpha t} \sum_{k=M+1}^{\infty} \alpha^k ||\mathbf{R}^k|| \frac{t^k}{k!} = e^{-\alpha t} \sum_{k=M+1}^{\infty} \frac{(\alpha t)^k}{k!}$$

$$\le e^{-\alpha t} e^{\alpha t} \frac{(\alpha t)^{M+1}}{(M+1)!} = \frac{(\alpha t)^{M+1}}{(M+1)!}.$$

6.2.3 Examples of right-continuous Markov chains

At the initial stage 0 we wait for a random length of time that follows the exponential distribution with intensity α_1, then we move immediately to the state 1. We then wait again independently for a random length of time that follows the exponential distribution $\exp(\alpha_2)$ and we move immediately to the state 2 and so on. This way one describes the *counting process* $\{N_t\}_{t\geq 0}$ associated with a *point process* in time.

If moreover $\{Y_n\}$ is a discrete Markov chain, and the families of random variables $\{Y_n\}$ and $\{N_t\}_{t\geq 0}$ are independent of each other, then it can be shown that $\{X_t\}_{t\geq 0}$, $X_t = Y_{N_t}$ is a right-continuous Markov chain.

We do not discuss the argument in its full generality and refer the interested reader to the specialized literature, see e.g. [22–24]. Here we consider the simplest case in which all the exponential distributions are the same, so that the counting process $\{N_t\}_{t\geq 0}$ is the Poisson process of intensity α, see Theorem 6.4. Thanks to the uniformization method, we have the following.

Theorem 6.24 (Uniformization) *Let* $\mathbf{Q}$ *be a* $N \times N$ *Q-matrix and let* $\mathbf{P}(t) := e^{\mathbf{Q}t}$. *Let* $\alpha \geq ||\mathbf{Q}||_\infty$ *and* $\mathbf{R} := \mathrm{Id} + \mathbf{Q}/\alpha$. *Let* $\{Y_n\}$ *be a discrete time homogeneous Markov chain with finite state-space* S, $|S| = N$, *on a probability space* $(\Omega, \mathcal{E}, \mathbb{P})$ *and transition matrix* $\mathbf{R}$, *and let* $\{N_t\}_{t\geq 0}$ *be a Poisson process with intensity* α *on* $(\Omega, \mathcal{E}, \mathbb{P})$. *If the sets of random variables* $\{Y_n\}$ *and* $\{N_t\}_{t\geq 0}$ *are independent, then the stochastic process* $\{X_t\}_{t\geq 0}$ *defined by* $X_t(x) := Y_{N_t(x)}(x)$ *is a right-continuous Markov chain with transition matrix* $\mathbf{P}(t) := e^{\mathbf{Q}t}$.

Lemma 6.25 *Let* $\{N_t\}_{t\geq 0}$ *be an integer valued right-continuous stochastic process on* $(\Omega, \mathcal{E}, \mathbb{P})$. *Assume that* $N_t : \Omega \to \mathbb{N}$ *has the following properties:*

(i) $N_0(x) = 0$ *and* N_t *is monotone nondecreasing,* $N_s(x) \leq N_t(x)$ *for a.e.* x *for any* s, t *with* $0 \leq s \leq t$.

(ii) $\{N_t\}_{t\geq 0}$ *is time homogeneous, i.e.* $N_t - N_s = N_{t-s}$ *for any* $0 \leq s \leq t$.

(iii) $\{N_t\}_{t\geq 0}$ *has independent increments, i.e. for any* $0 \leq r_n \leq \cdots \leq r_1 \leq s \leq t$ *and for any* $n \leq q \leq p$

$$\mathbb{P}\Big(N_t = p \,\Big|\, N_s = q, N_{r_1} = q_1, \ldots,\ N_{r_n} = q_n\Big) = \mathbb{P}(N_{t-s} = p - q).$$

Let $\{Y_n\}$ *be a homogeneous discrete time Markov chain with a finite or denumerable state-space* S *on* $(\Omega, \mathcal{E}, \mathbb{P})$. *If* $\{N_t\}_{t\geq 0}$ *and* $\{Y_n\}$ *are each other independent, then the stochastic process* $\{X_t\}_{t\geq 0}$,

$$X_t(x) := Y_{N_t(x)}(x)$$

is a right-continuous Markov chain.

Proof. Let us prove that $\{X_t\}_{t\geq 0}$ is a homogeneous process. Let $s < t$ and $i, j \in S$. From the properties (ii) and (iii) of $\{N_t\}_{t\geq 0}$

$$\begin{aligned}\mathbb{P}(N_t = p, N_s = q) &= \mathbb{P}(N_t - N_s = p - q, N_s = q)\\ &= \mathbb{P}(N_{t-s} = p - q)\mathbb{P}(N_s = q).\end{aligned}$$

Moreover, since $\{Y_n\}$ is a homogeneous Markov chain, for non-negative integers p, q with $p \geq q$ we have

$$\mathbb{P}(Y_p = j,\ Y_q = i) = \mathbb{P}\Big(Y_{p-q} = j \,\Big|\, Y_0 = i\Big)\mathbb{P}(Y_q = i).$$

Therefore

$$\begin{aligned}\mathbb{P}(X_t = j,\ X_s = i) &= \sum_{p\geq q}\mathbb{P}(Y_p = j,\ Y_q = i,\ N_t = p,\ N_s = q)\\ &= \sum_{p\geq q}\mathbb{P}(Y_p = j,\ Y_q = i)\mathbb{P}(N_t = p, N_s = q)\\ &= \sum_{p\geq q}\mathbb{P}\Big(Y_{p-q} = j \,\Big|\, Y_0 = i\Big)\mathbb{P}(Y_q = i)\mathbb{P}(N_{t-s} = p - q)\mathbb{P}(N_s = q)\\ &= \Big(\sum_{\ell\geq 0}\mathbb{P}\Big(Y_\ell = j \,\Big|\, Y_0 = i\Big)\mathbb{P}(N_{t-s} = \ell)\Big)\Big(\sum_{q\geq 0}\mathbb{P}(Y_q = i)\mathbb{P}(N_s = q)\Big)\\ &= \mathbb{P}\Big(X_{t-s} = j \,\Big|\, X_0 = i\Big)\mathbb{P}(X_s = i).\end{aligned}$$

i.e.

$$\mathbb{P}\Big(X_t = j \,\Big|\, X_s = i\Big) = \mathbb{P}\Big(X_{t-s} = j \,\Big|\, X_0 = i\Big). \tag{6.14}$$

Let us now prove that $\{X_t\}_{t\geq 0}$ has the Markov property. Let $r < s < t$ and let $h, i, j, \in S$. Then $N_t(x) \geq N_s(x) \geq N_r(x)$, and for any non-negative integers p, q, n such that $p \geq q \geq n$ we get

$$\mathbb{P}(N_t = p,\ N_s = q,\ N_r = n) = \mathbb{P}(N_{t-s} = p - q)\mathbb{P}(N_s = q,\ N_r = n)$$
$$\mathbb{P}(Y_p = j,\ Y_q = i,\ Y_n = h) = \mathbb{P}\Big(Y_{p-q} = j \,\Big|\, Y_0 = i\Big)\mathbb{P}(Y_q = i,\ Y_n = h).$$

We have

$$\begin{aligned}&\mathbb{P}(X_t = j,\ X_s = i,\ X_r = h)\\ &\quad= \sum_{p\geq q\geq n\geq 0}\mathbb{P}(Y_p = j,\ Y_q = i,\ Y_n = h,\ N_t = p,\ N_s = q,\ N_r = n)\\ &\quad= \sum_{p\geq q\geq n\geq 0}\mathbb{P}(Y_p = j,\ Y_q = i,\ Y_n = h)\mathbb{P}(N_t = p,\ N_s = q,\ N_r = n)\end{aligned}$$

$$= \sum_{p \geq q \geq n} \mathbb{P}\Big(Y_{p-q} = j \,\Big|\, Y_0 = i\Big)\mathbb{P}(Y_q = i,\ Y_n = h)$$

$$\mathbb{P}(N_{t-s} = p - q)\mathbb{P}(N_s = q,\, N_r = n)$$

$$= \Big(\sum_{\lambda \geq 0} \mathbb{P}\Big(Y_l = j \,\Big|\, Y_0 = i\Big)\mathbb{P}(N_{t-s} = l)\Big)$$

$$\Big(\sum_{q \geq n \geq 0} \mathbb{P}(Y_q = i,\ Y_n = h)\mathbb{P}(N_s = q,\ N_r = n)\Big)$$

$$= \mathbb{P}\Big(X_{t-s} = j \,\Big|\, X_0 = i\Big)\mathbb{P}(X_t = i,\ X_r = h),$$

therefore, taking also into account (6.14), we finally get

$$\mathbb{P}\Big(X_t = j \,\Big|\, X_s = i,\ X_r = h\Big) = \mathbb{P}\Big(X_{t-s} = j \,\Big|\, X_0 = i\Big)$$
$$= \mathbb{P}\Big(X_t = j \,\Big|\, X_s = i\Big).$$

Similarly one proceeds to get

$$\mathbb{P}\Big(X_t = j \,\Big|\, X_{s_1} = i_1, \ldots,\ X_{i_n} = i_n\Big) = \mathbb{P}\Big(X_t = j \,\Big|\, X_{s_1} = i_1\Big).$$

Proof of Theorem 6.24. By Lemma 6.25 $X_t(x) = Y_{N_t(x)}(x)$ is a right-continuous Markov chain.

Let $\mathbf{S}(t)$ be the transition matrix of $\{Y_{N_t}\}$. Since $N_0(x) = 0$ for a.e. $x \in \Omega$, we have

$$\mathbf{S}(t)^i_j := \mathbb{P}\Big(Y_{N_t(x)}(x) = j \,\Big|\, Y_{N_0(x)}(x) = i\Big)$$

$$= \frac{1}{\mathbb{P}(Y_0(x) = i)} \sum_{k=0}^{\infty} \mathbb{P}\Big(Y_k(x) = j,\, Y_0(x) = i,\, N_t(x) = k\Big)$$

$$= \sum_{k=0}^{\infty} \mathbb{P}\Big(Y_k(x) = j \,\Big|\, Y_0(x) = i\Big)\mathbb{P}\Big(N_t(y) = k\Big)$$

$$= \sum_{k=0}^{\infty} (\mathbf{R}^k)^i_j e^{-\alpha t} \frac{\alpha^k t^k}{k!}$$

$$= e^{-\alpha t} e^{\alpha \mathbf{R} t} = e^{\alpha(-\mathrm{Id}+\mathbf{R})t} = e^{\mathbf{Q}t} = \mathbf{P}(t).$$

Remark 6.26 Notice that Theorem 6.24 reduces the existence of a right-continuous Markov chain with a given Q-matrix $\mathbf{Q}$ to the existence of a Poisson process of suitable intensity $\alpha > ||\mathbf{Q}||_\infty$ and of a discrete time Markov chain with transition matrix $\mathbf{R} := \mathrm{Id} + \mathbf{Q}/\alpha$.

Example 6.27 Poisson processes are right-continuous Markov chains. This is a special case of Lemma 6.25.

6.2.4 Holding times

It can be shown that the holding times between two transitions of a right-continuous Markov chain with a finite state-space are independent and follow exponential distributions of possibly different intensities. We do not enter into the whole argument, which is well-described in the literature, see e.g. [22–24] and restrict ourselves to discuss the first holding time.

Let $\{X_t\}_{t\geq 0}$ be a homogeneous right-continuous stochastic process with a finite or denumerable state-space S. Let $\mathbf{P}(t)$ be the transition matrix of the process. For any state $i \in S$ let SJ_i be the *holding time* in state i, defined for any $x \in \Omega$ as

$$SJ_i(x) := \inf\Big\{t \,\Big|\, X_t(x) \neq i\Big\}.$$

Proposition 6.28 *The holding time in a state $i \in S$ is a random variable. For any $t > 0$ and any $n \geq 0$, let $t_j := jt/2^n$, $j = 0, \ldots, 2^n - 1$. Then*

$$\mathbb{P}(SJ_i(x) > t) = \lim_{n\to\infty} \mathbb{P}\Big(X_{t_j}(x) = i \;\forall j,\; 0 \leq j \leq 2^n - 1\Big).$$

Proof. Let $\mathcal{D}$ be a denumerable dense set of $\mathbb{R}_+$ and let $S_{\mathcal{D}}(x) := \inf\Big\{t \in \mathcal{D} \,\Big|\, X_t(x) \neq i\Big\}$. Observe that

$$SJ_i(x) = S_{\mathcal{D}}(x)$$

if the trajectory $t \mapsto X_t(x)$ is right-continuous. In fact, trivially $SJ_i(x) \leq S_{\mathcal{D}}(x)$. Moreover, let $\{t_n\} \subset \mathcal{D}$ be such that $t_n \downarrow SJ_i(x)$. Since $t \mapsto X_t(x)$ is right-continuous, then $X_{\bar{t}} \neq i$, $\bar{t} := SJ_i(x)$ and $X_{t_n} \neq i$ for large enough n. Therefore $S_{\mathcal{D}}(x) \leq t_n$ for large enough n and passing to the limit as $n \to \infty$, one gets $S_{\mathcal{D}}(x) \leq SJ_i(x)$.

Since the process is right-continuous, by the above the sets

$$\Big\{x \in \Omega \,\Big|\, SJ_i(x) > t\Big\} \quad \text{and} \quad \Big\{x \in \Omega \,\Big|\, S_{\mathcal{D}}(x) > t\Big\}$$

differ by a null set. Observe now that for $t > 0$

$$t < S_{\mathcal{D}}(x) \quad \text{if and only if} \quad X_s(x) = i \;\forall s \in \mathcal{D}, 0 \leq s \leq t$$

for a.e. x. Consequently,

$$\Big\{x \in \Omega \,\Big|\, SJ_i(x) > t\Big\} \quad \text{and} \quad \Big\{x \in \Omega \,\Big|\, X_s(x) = i \;\forall s \in \mathcal{D},\; 0 \leq s \leq t\Big\}$$

differ by a null set and

$$\mathbb{P}(SJ_i(x) > t) = \mathbb{P}\Big(X_s(x) = i \ \forall s \in \mathcal{D},\ 0 \le s \le t\Big).$$

In particular, $SJ_i(x)$ is a random variable on $(\Omega, \mathcal{E}, \mathbb{P})$.

Fix now t and choose $\mathcal{D}$ as $\mathcal{D} := \{jt/2^n\}_{j,n}$. For any n, set $\mathcal{D}_n := \{jt/2^n\}_j$. Trivially, $\mathcal{D}_n \subset \mathcal{D}_{n+1}$ $\forall n$ and $\mathcal{D} = \cup_{n=1}^{\infty} \mathcal{D}_n$. Moreover,

$$\begin{aligned}\mathbb{P}(SJ_i(x) > t) &= \mathbb{P}\Big(X_s(x) = i \ \forall s \in \mathcal{D},\ 0 \le s \le t\Big) \\ &= \mathbb{P}\Big(\bigcap_{n=1}^{\infty} \Big\{x \in \Omega \,\Big|\, X_{t_j}(x) = i \ \forall j,\ 0 \le j \le 2^n - 1\Big\}\Big) \\ &= \lim_{n\to\infty} \mathbb{P}\Big(X_{t_j}(x) = i \ \forall j,\ 0 \le j \le 2^n - 1\Big)\end{aligned}$$

where $t_j = t_{j,n} := tj/2^n$.

Proposition 6.29 *Let* $\{X_t\}_{t\ge 0}$ *be a homogeneous right-continuous Markov chain with finite state-space* S. *Let* $\mathbf{Q} = (q_{ij})$ *be its intensities matrix. Then the holding time in a state* $i \in S$ *follows the exponential distribution of parameter* $-q_{ii}$.

Proof. By Proposition 6.28 SJ_i is a random variable and

$$\mathbb{P}(SJ_i(x) > t) = \lim_{n\to\infty} \mathbb{P}\Big(X_{t_j}(x) = i \ \forall j,\ 0 \le j \le 2^n - 1\Big)$$

where $t_j = t_{j,n} := tj/2^n$. Let $\mathbf{P}(t) = (p_{ij}(t))$ be the transition probabilities matrix of the process. Thus, by the Markov property and the homogeneity of the chain we get

$$\begin{aligned}\mathbb{P}(X_{t_j} = i \ \forall j = 0, \dots, 2^n - 1) &= \prod_{j=0}^{2^n-1} \mathbb{P}\Big(X_{t_{j+1}} = i \,\Big|\, X_{t_j} = i\Big) \\ &= \Big(p_{ii}\Big(\frac{t}{2^n}\Big)\Big)^{2^n}.\end{aligned}$$

Thus, since $\mathbf{P}(t) = (p_{ij}(t))$ is right-differentiable at 0 and $\mathbf{P}'_+(0) = \mathbf{Q}$, we conclude

$$\log \mathbb{P}(SJ_i > t) = \lim_{n\to\infty} 2^n \log p_{ii}\Big(\frac{t}{2^n}\Big) = \lim_{k\to\infty} k\Big(p_{ii}\Big(\frac{t}{k}\Big) - 1\Big) = q_{ii}t.$$

Appendix A

Power series

We recall some results from the theory of power series which should already be known to the reader. Proofs and further results can be found in textbooks on Analysis, see e.g. [25, 26].

A.1 Basic properties

Let $\{a_n\}$ be a sequence of complex numbers. The *power series* centred at zero with coefficients $\{a_n\}$ is

$$\sum_{n=0}^{\infty} a_n z^n := a_0 + \sum_{n=1}^{\infty} a_n z^n, \qquad z \in \mathbb{C},$$

i.e. the sequence of polynomials $\{s_n(z)\}_n$

$$s_n(z) = \sum_{k=0}^{n} a_k z^k := a_0 + \sum_{k=1}^{n} a_k z^k, \qquad z \in \mathbb{C}.$$

Given a power series $\sum_{n=0}^{\infty} a_n z^n$, the non-negative real number ρ defined by

$$\frac{1}{\rho} := \limsup_{n\to\infty} \sqrt[n]{|a_n|},$$

where $1/0^+ = +\infty$ and $1/(+\infty) = 0$, is called the *radius of convergence* of the series $\sum_{n=0}^{\infty} a_n z^n$. In fact, one proves that the sequence $\{s_n(z)\}$ is absolutely convergent if $|z| < \rho$ and it does not converge if $|z| > \rho$. It can be shown that $\rho > 0$ if and only if the sequence $\{|a_n|\}$ is not more than exponentially increasing.

A First Course in Probability and Markov Chains, First Edition. Giuseppe Modica and Laura Poggiolini.

In this case, the *sum* of the series

$$A(z) := \sum_{n=0}^{\infty} a_n z^n$$

is defined in the disc $\{|z| < \rho\}$.

Power series can be integrated and differentiated term by term. More precisely, the following holds.

Theorem A.1 *Let $A(z) = \sum_{n=0}^{\infty} a_n z^n$ be the sum of a power series with radius of convergence ρ. Then the series $\sum_{n=0}^{\infty} n a_n z^{n-1}$ and $\sum_{n=0}^{\infty} a_n \frac{z^{n+1}}{n+1}$ have the same radii of convergence ρ. If $\rho > 0$, then $A(z)$ is holomorphic in $B(0, \rho)$,*

$$A'(z) = \sum_{n=0}^{\infty} n a_n z^{n-1},$$

and, given any piecewise C^1 curve $\gamma : [0, 1] \to B(0, \rho)$,

$$\int_{\gamma} A(z)\, dz = \sum_{n=0}^{\infty} a_n \frac{\gamma(1)^{n+1} - \gamma(0)^{n+1}}{n+1}$$

Then Corollary A.2 easily follows.

Corollary A.2 *Let $A(z) = \sum_{n=0}^{\infty} a_n z^n$ be the sum of a power series of radius of convergence $\rho > 0$. Then*

$$a_n = \frac{D^n A(0)}{n!}.$$

Thus, if $\rho > 0$ the sum $A(z)$, $|z| < \rho$, completely determines the sequence $\{a_n\}$. The function $A(z)$ is called the *generating* function of the sequence $\{a_n\}$. In other words, $\{a_n\}$ and $A(z) := \sum_{n=0}^{\infty} a_n z^n$ contain the same 'information' (we are assuming that $\{a_n\}$ grows at most exponentially or, equivalently that $\rho > 0$): we can say that the function $A(z)$, $|z| < \rho$, is a new 'viewpoint' on the sequence $\{a_n\}$.

By Corollary A.2 if $A(z) = \sum_{n=0}^{\infty} a_n z^n$ and $B(z) = \sum_{n=0}^{\infty} b_n z^n$ are the sums of two power series with positive radii of convergence, that coincide near zero, then $a_n = b_n$ $\forall n$, both series have the same radius ρ, and $A(z) = B(z)$ for any z such that $|z| < \rho$.

Example A.3 (Geometric series) The *geometric series* is the most classical example of power series. It generates the sequence $\{1, 1, 1, \dots\}$

$$\sum_{n=0}^{\infty} z^n$$

and its sum is

$$S(z) := \sum_{n=0}^{\infty} z^n = \frac{1}{1-z}, \qquad |z| < 1.$$

Example A.4 (Exponential) Taking advantage of Taylor expansions, one can prove that

$$\sum_{n=0}^{\infty} \frac{z^n}{n!} = e^z, \qquad z \in \mathbb{C}.$$

Example A.5 (Logarithm) Replacing z with $-z$ in the equality $\frac{1}{1-z} = \sum_{n=0}^{\infty} z^n$ $|z| < 1$, we get

$$\frac{1}{1+z} = \sum_{n=0}^{\infty} (-1)^n z^n, \qquad |z| < 1.$$

Integrating along the interval $[0, x] \subset \mathbb{R}$, we get

$$\begin{aligned}
\log(1+x) &= \int_0^x \frac{1}{1+t}\, dt \\
&= \int_0^x \sum_{n=0}^{\infty} (-1)^n t^n \, dt \\
&= \sum_{n=0}^{\infty} (-1)^n \frac{x^{n+1}}{n+1}, \qquad |x| < 1. \qquad \text{(A.1)}
\end{aligned}$$

A.2 Product of series

Definition A.6 *The* convolution product *of two sequences* $a = \{a_n\}$ *and* $b = \{b_n\}$ *is the sequence* $\{a * b\}_n$ *defined for* $n = 0, 1, \ldots$ *by*

$$(a * b)_n := \sum_{i+j=n} a_i b_j = \sum_{j=0}^{n} a_j b_{n-j}.$$

In the first sum we sum over all couples (i, j) *of non-negative integers such that* $i + j = n$.

The first terms are

$$\begin{aligned}
(a * b)_0 &= a_0 b_0, \\
(a * b)_1 &= a_0 b_1 + a_1 b_0, \\
(a * b)_2 &= a_0 b_2 + a_1 b_1 + a_2 b_0, \\
&\ldots.
\end{aligned}$$

The convolution product is commutative, associative and bilinear. Moreover the following holds.

Theorem A.7 (Cauchy) *Let $a = \{a_n\}$ and $b = \{b_n\}$ be two sequences and let $A(z) = \sum_{n=0}^{\infty} a_n z^n$ and $B(z) = \sum_{n=0}^{\infty} b_n z^n$ be the sums of the associated power series defined for $|z| < \rho_a$ and $|z| < \rho_b$, respectively. Then the power series of their convolution products converges for any z such that $|z| < \min(\rho_a, \rho_b)$ and*

$$\sum_{n=0}^{\infty} (a * b)_n z^n = A(z)B(z) \qquad \forall z, |z| < \min(\rho_a, \rho_b).$$

A.3 Banach space valued power series

Let V be a Banach space with norm $||\ ||$ and let $\{f_n\} \subset V$. Then one may consider the series, with values in V,

$$\sum_{k=0}^{\infty} f_k z^k = \Big\{ \sum_{k=0}^{n} f_k z^k \Big\}_n. \tag{A.2}$$

Let $\rho \geq 0$ be defined by

$$\frac{1}{\rho} := \limsup_{n \to \infty} \sqrt[n]{||f_n||}.$$

As in the case of power series with complex coefficients, the power series in (A.2) absolutely converges in V for any $z \in \mathbb{C}$, $|z| < \rho$. Thus the sum of the series is a well defined function on the disc $|z| < \rho$ of the complex plane with values in V

$$F(z) := \sum_{k=0}^{\infty} f_k z^k \in V, \qquad |z| < \rho.$$

As for complex valued power series, one differentiates term by term Banach valued power series: the sum $F(z)$ is a holomorphic function on the open disc $|z| < \rho$ and

$$F'(z) := \sum_{k=0}^{\infty} k f_k z^k \in V, \qquad |z| < \rho.$$

As a special case, one considers the space $M_{N,N}(\mathbb{C})$ of $N \times N$ complex matrices with norm

$$||\mathbf{F}|| := \sup_{\substack{x \in \mathbb{C}^N \\ x \neq 0}} \frac{|\mathbf{F}x|}{|x|},$$

called the *maximum expansion coefficient* of $\mathbf{F}$. It is easy to prove that

$$|\mathbf{F}z| \leq ||\mathbf{F}||\,|z|, \qquad ||\mathbf{F}^n|| \leq ||\mathbf{F}||^n \quad \forall n,$$

and that $M_{N,N}(\mathbb{C})$ equipped with this norm is a Banach space. Given a sequence $\{\mathbf{F}_n\} \subset M_{N,N}(\mathbb{C})$, the associated power series

$$\sum_{n=0}^{\infty} \mathbf{F}_n z^n \tag{A.3}$$

converges if $|z| < \rho$ where

$$\frac{1}{\rho} := \limsup_{n\to\infty} \sqrt[n]{||\mathbf{F}_n||}.$$

From the above, we have the following:

(i) $\rho > 0$ if and only if the sequence $\{||\mathbf{F}_n||\}$ grows at most exponentially fast.

(ii) If $|z| < \rho$, then the series (A.3) converges, both pointwise and absolutely, to a matrix $\mathbf{F}(z) \in M_{N,N}(\mathbb{C})$,

$$\mathbf{F}(z) := \sum_{n=0}^{\infty} \mathbf{F}_n z^n, \qquad |z| < \rho,$$

i.e. we have $\forall i, j \in \{1, \dots, N\}$ the complex valued limits

$$\mathbf{F}^i_j(z) = \sum_{n=0}^{\infty} (\mathbf{F}_n)^i_j z^n, \qquad |z| < \rho.$$

A.3.1.1 Power expansion of the inverse of a matrix

Let $\mathbf{P} \in M_{N,N}(\mathbb{C})$ be a $N \times N$ square matrix with complex entries. If $||\mathbf{P}||\,|z| < 1$, then $\sum_{n=0}^{\infty} ||\mathbf{P}||^n |z|^n = \frac{1}{1-||\mathbf{P}||\,|z|} < \infty$. Thus the series $\sum_{n=0}^{\infty} \mathbf{P}^n z^n$ converges, both pointwisely and absolutely, to a matrix $\mathbf{S}(z) := \sum_{n=0}^{\infty} \mathbf{P}^n z^n$. For any positive n we can write

$$(\mathrm{Id} - z\mathbf{P}) \sum_{k=0}^{n} \mathbf{P}^k z^k = \sum_{k=0}^{n} (\mathbf{P}^k z^k - \mathbf{P}^{k+1} z^{k+1}) = \mathrm{Id} - \mathbf{P}^{n+1} z^{n+1}$$

so that

$$\left|\Big(\mathrm{Id} - z\mathbf{P}\Big)\Big(\sum_{k=0}^{n} \mathbf{P}^k z^k\Big) - \mathrm{Id}\right| = ||\mathbf{P}^{n+1}||\,|z|^{n+1} \le ||\mathbf{P}||^{n+1}|z|^{n+1}.$$

Since $||\mathbf{P}||\,|z| < 1$, as $n \to \infty$ we get

$$\Big(\mathrm{Id} - z\mathbf{P}\Big)\mathbf{S}(z) = \mathrm{Id}.$$

Therefore, we conclude that, if $|z| < \frac{1}{||\mathbf{P}||}$, then $\mathrm{Id} - z\mathbf{P}$ is invertible and

$$(\mathrm{Id} - z\mathbf{P})^{-1} = \mathbf{S}(z) = \sum_{k=0}^{\infty} \mathbf{P}^k z^k.$$

A.3.1.2 Exponential of a matrix

Let $\mathbf{Q} \in M_{N,N}(\mathbb{C})$. The radius of convergence of the power series

$$\sum_{n=0}^{\infty} \frac{\mathbf{Q}^n}{n!} z^n$$

is $+\infty$, so that, for any $z \in \mathbb{C}$ the power series $\sum_{n=0}^{\infty} \frac{\mathbf{Q}^n}{n!} z^n$ converges, both absolutely and pointwisely, to a matrix denoted by $e^{\mathbf{Q}z}$,

$$e^{\mathbf{Q}z} := \sum_{n=0}^{\infty} \frac{\mathbf{Q}^n}{n!} z^n \qquad \forall z \in \mathbb{C}. \tag{A.4}$$

Proposition A.8 *The following hold:*

(i) $e^{\mathbf{Q}0} = \mathrm{Id}$.

(ii) $e^{\mathbf{Q}z}$ *and* $\mathbf{Q}$ *commute.*

(iii) $\left(e^{\mathbf{Q}z}\right)' = \mathbf{Q}e^{\mathbf{Q}z}$ *for any* $z \in \mathbb{C}$.

(iv) $e^{\mathbf{Q}(z+w)} = e^{\mathbf{Q}z}e^{\mathbf{Q}w}$ *for any* $z, w \in \mathbb{C}$.

(v) $e^{\mathbf{Q}z}$ *is invertible and its inverse is* $(e^{\mathbf{Q}z})^{-1} = e^{-\mathbf{Q}z}$.

(vi) $(e^{\mathbf{Q}z})^n = e^{\mathbf{Q}nz}$ *for any* $z \in \mathbb{C}$ *and any* $n \in \mathbb{Z}$.

Proof. Properties (i) and (ii) are a direct consequence of the definition of $e^{\mathbf{Q}z}$. Property (iii) follows differentiating term by term the series $\sum_{n=0}^{\infty} \frac{\mathbf{Q}^n}{n!} z^n$. Property (iv) is a consequence of the formula for the product of power series: in fact,

$$\begin{aligned} e^{\mathbf{Q}z}e^{\mathbf{Q}w} &= \sum_{k=0}^{\infty} \frac{\mathbf{Q}^k}{k!} z^k \sum_{k=0}^{\infty} \frac{\mathbf{Q}^k}{k!} w^k = \sum_{n=0}^{\infty} \sum_{k=0}^{n} \frac{\mathbf{Q}^k \mathbf{Q}^{n-k} z^k w^{n-k}}{k!(n-k)!} \\ &= \sum_{n=0}^{\infty} \frac{\mathbf{Q}^n}{n!} \sum_{k=0}^{n} \binom{n}{k} z^k w^{n-k} = \sum_{n=0}^{\infty} \frac{\mathbf{Q}^n}{n!} (z+w)^n = e^{\mathbf{Q}(z+w)}. \end{aligned}$$

Finally, properties (v) and (vi) are particular cases of (iv).

A.3.2 Exercises

Exercise A.9 *Prove the Newton binomial theorem:*

(i) directly, with an induction argument on n;

(ii) taking advantage of the Taylor expansions;

(iii) starting from the formula $D((1+z)^n) = n(1+z)^{n-1}$;

(iv) starting from the identity $e^{t(x+y)} = e^{tx}e^{ty}$.

Exercise A.10 *Prove the following equalities:*

$$\sum_{j=0}^{n} 2^j \binom{n}{j} = 3^n,$$

$$\sum_{j=0}^{n} \binom{n}{2j} = \sum_{j=0}^{n} \binom{n}{2j+1} = 2^{n-1},$$

$$\sum_{j=0}^{n} (-1)^j \binom{n}{j}^2 = \begin{cases} (-1)^{n/2}\binom{n}{n/2} & \text{if } n \text{ is even,} \\ 0 & \text{if } n \text{ is odd,} \end{cases}$$

$$\sum_{j=a}^{n} \binom{j}{a} = \binom{n+1}{a+1},$$

$$\sum_{j=0}^{\infty} \binom{\alpha+j}{j} x^j = \frac{1}{(1-x)^{\alpha+1}},$$

$$\binom{2n}{n} = (-4)^n \binom{-1/2}{n}.$$

Exercise A.11 *Show that*

$$\sum_{j=0}^{n} (-1)^j \binom{n}{j} \binom{n+m-j}{k-j} = \begin{cases} \binom{n}{k} & \text{if } m \geq k, \\ 0 & \text{if } m < k. \end{cases}$$

Exercise A.12 *Let $\{a_n\}$ be a complex valued sequence such that $\{|a_n|\}$ grows at most exponentially fast. Let $A(z) = \sum_{n=0}^{\infty} a_n z^n$, $|z| < \rho$ be its generating function. Compute the generating function of the following sequences:*

- $\{\alpha a_0, \alpha a_1, \alpha a_2, \alpha a_3, \dots\}$, $\alpha \in \mathbb{C}$,
- $\{a_0, 0, a_1, 0, a_2, 0, \dots\}$,

- $\{a_0, 0, a_2, 0, a_4, 0, \dots\}$,
- $\{a_1, 0, a_3, 0, a_5, 0, \dots\}$,
- $\{0, 0, 0, a_0, a_1, a_2, a_3, \dots\}$,
- $\{a_3, a_4, a_5, a_6, \dots\}$,
- $\{a_0, 2a_1, 3a_2, 4a_3, 5a_4, \dots\}$,
- $\{a_0, a_1/2, a_2/3, a_3/4, a_4/5, \dots\}$,
- $\{a_0, a_0 + a_1, a_1 + a_2, a_2 + a_3, \dots\}$,
- $\{a_0 + a_1, a_1 + a_2, a_2 + a_3, a_3 + a_4, \dots\}$,
- $\{a_0, a_0 + a_1, a_0 + a_1 + a_2, a_0 + a_1 + a_2 + a_3, \dots\}$.

Exercise A.13 *Compute* $\sum_{j=0}^{p}(-1)^j\binom{n}{j}$. $\left[(-1)^p\binom{n-1}{p}\right]$.

Solution.

We prove this equality directly. Since for any $p = 1, \dots, n-1$ we have $\binom{n}{p} = \binom{n-1}{p} + \binom{n-1}{p-1}$, one gets

$$\begin{aligned}\sum_{j=0}^{p}(-1)^j\binom{n}{j} &= 1 - n + \sum_{j=2}^{p}(-1)^j\binom{n-1}{j} + \sum_{j=1}^{p-1}(-1)^{j+1}\binom{n-1}{j} \\ &= 1 - n + \sum_{j=2}^{p-1}(-1)^j\left(\binom{n-1}{j} - \binom{n-1}{j}\right) \\ &\qquad + \binom{n-1}{1} + (-1)^p\binom{n-1}{p} \\ &= 1 - n + 0 + (n-1) + (-1)^p\binom{n-1}{p} = (-1)^p\binom{n-1}{p}.\end{aligned}$$

The same result can be obtained via generating functions. From the formula for the product of two power series, we get

$$\begin{aligned}\sum_{p=0}^{\infty}\left(\sum_{j=0}^{p}(-1)^{p-j}\binom{n}{j}\right)x^p &= \left(\sum_{j=0}^{\infty}(-x)^j\right)\left(\sum_{j=0}^{\infty}\binom{n}{j}x^j\right) \\ &= \frac{1}{1+x}(1+x)^n = (1+x)^{n-1} = \sum_{p=0}^{n-1}\binom{n-1}{p}x^p.\end{aligned}$$

The claim follows by the identity principle for polynomials.

Exercise A.14 *For any $p, q, k \geq 0$, prove* Vandermonde formula

$$\sum_{j=0}^{k} \binom{p}{j}\binom{q}{k-j} = \binom{p+q}{k}. \tag{A.5}$$

Solution. We prove the equality (A.5) by using generating functions. From the formula for the product of power series, we get

$$\sum_{k=0}^{\infty} \Big(\sum_{j=0}^{k} \binom{p}{j}\binom{q}{k-j}\Big) z^k = \Big(\sum_{k=0}^{\infty} \binom{p}{k} z^k\Big)\Big(\sum_{k=0}^{\infty} \binom{q}{k} z^k\Big)$$

$$= (1+z)^{p+q} = \sum_{k=0}^{\infty} \binom{p+q}{k} z^k,$$

hence the claim.

Exercise A.15 *Show that* $\sum_{j=0}^{\infty} \binom{p}{j}\binom{q}{j-1} = \binom{p+q}{p-1}$ $\forall p, q \geq 0$.

Solution. Applying the Vandermonde formula (A.5), we get

$$\sum_{j=0}^{\infty} \binom{p}{j}\binom{q}{j-1} = \sum_{j=0}^{\infty} \binom{p}{j}\binom{q}{q-j+1} = \binom{p+q}{q+1} = \binom{p+q}{p-1}.$$

Exercise A.16 *Show that* $\sum_{k=0}^{n} \binom{n}{k}^2 = \binom{2n}{n}$.

Solution. Applying Vandermonde formula we get

$$\sum_{k=0}^{n} \binom{n}{k}^2 = \sum_{k=0}^{n} \binom{n}{k}\binom{n}{n-k} = \binom{2n}{n}.$$

Appendix B

Measure and integration

The axiomatic approach to probability by Andrey Kolmogorov (1903–1987) makes essential use of the measure theory. In this appendix we review the aspects of the theory that are relevant to us. We do not prove everything and refer the interested reader for proofs and further study to one of the many volumes on this now classic subject, see e.g. [7, 27].

B.1 Measures

B.1.1 Basic properties

Here Ω shall denote a generic set. For a generic subset E of Ω, $E^c := \Omega \setminus E$ denotes the *complement* of E in Ω and $\mathcal{P}(\Omega)$ denotes the family of all subsets of Ω. A family $\mathcal{E}$ of subsets of Ω is then a subset of $\mathcal{P}(\Omega)$, $\mathcal{E} \subset \mathcal{P}(\Omega)$. We say that a family $\mathcal{E} \subset \mathcal{P}(\Omega)$ of subsets of a set Ω is an *algebra* if $\emptyset, \Omega \in \mathcal{E}$ and $E \cup F$, $E \cap F$ and $E^c \in \mathcal{E}$ whenever $E, F \in \mathcal{E}$.

Definition B.1 *We say that $\mathcal{E}$ is a* σ-algebra *if $\mathcal{E}$ is an algebra and for every sequence of subsets $\{E_k\} \subset \mathcal{E}$ we also have $\cup_k E_k$ and $\cap_k F_k \in \mathcal{E}$.*

In other words, if we operate on sets of a σ-algebra with differences, countable unions or intersections, we get sets of the same σ-algebra: we also say that a σ-algebra is *closed* with respect to differences, countable unions and intersections.

Let $\mathcal{D} \subset \mathcal{P}(\Omega)$ be a family of subsets of Ω. It is readily seen that the class

$$\mathcal{S} := \bigcap \Big\{ \mathcal{E} \,\Big|\, \mathcal{E} \text{ is a } \sigma\text{-algebra, } \mathcal{E} \supset \mathcal{D} \Big\}$$

is again a σ-algebra, hence *the smallest σ-algebra containing* $\mathcal{D}$. We say that $\mathcal{S}$ is the σ-algebra *generated* by $\mathcal{D}$.

A First Course in Probability and Markov Chains, First Edition. Giuseppe Modica and Laura Poggiolini.

Definition B.2 *The smallest σ-algebra $\mathcal{B} \subset \mathcal{P}(\mathbb{R}^n)$ containing the open sets of $\mathbb{R}^n$ is called the σ-algebra of* Borel sets.

Definition B.3 *A* measure *on Ω is a couple $(\mathcal{E}, \mathbb{P})$ of a σ-algebra $\mathcal{E} \subset \mathcal{P}(\Omega)$ and of a map $\mathbb{P} : \mathcal{E} \to \overline{\mathbb{R}}_+$ with the following properties:*

(i) $\mathbb{P}(\emptyset) = 0$.

(ii) (Monotonicity) *If $A, B \in \mathcal{E}$ with $A \subset B$, then $\mathbb{P}(A) \leq \mathbb{P}(B)$.*

(iii) (σ-additivity) *For any disjoint sequence $\{A_i\} \subset \mathcal{E}$ we have*

$$\mathbb{P}\Big(\bigcup_{i=1}^{\infty} A_i\Big) = \sum_{i=1}^{\infty} \mathbb{P}(A_i).$$

Obviously (iii) reduces to finite additivity for pairwise disjoint subsets if $\mathcal{E}$ is finite. When $\mathcal{E}$ is infinite, the infinite sum on the right-hand side is understood as the sum of a series of non-negative terms. From Definition B.3 we easily get the following.

Proposition B.4 *Let $(\mathcal{E}, \mathbb{P})$ be a measure on Ω. We have:*

(i) If $A \in \mathcal{E}$, then $0 \leq \mathbb{P}(A)$.

(ii) If $A, B \in \mathcal{E}$ with $A \subset B$, then $\mathbb{P}(B \setminus A) + \mathbb{P}(A) = \mathbb{P}(B)$.

(iii) If $A, B \in \mathcal{E}$ then $\mathbb{P}(A \cup B) + \mathbb{P}(A \cap B) = \mathbb{P}(A) + \mathbb{P}(B)$.

(iv) If $A, B \in \mathcal{E}$ then $\mathbb{P}(A \cup B) \leq \mathbb{P}(A) + \mathbb{P}(B)$.

(v) (σ-subadditivity) *For any sequence $\{A_i\} \subset \mathcal{E}$ we have*

$$\mathbb{P}\Big(\bigcup_{i=1}^{\infty} A_i\Big) \leq \sum_{i=1}^{\infty} \mathbb{P}(A_i). \tag{B.1}$$

(vi) (Disintegration formula) *If $\{D_i\} \subset \mathcal{E}$ is a partition of Ω, then for every $A \subset \Omega$,*

$$\mathbb{P}(A) = \sum_{i=1}^{\infty} \mathbb{P}(A \cap D_i). \tag{B.2}$$

(vii) (Continuity)

(a) If $\{E_i\} \subset \mathcal{E}$ with $E_i \subset E_{i+1}$ $\forall i$, then $\cup_{i=1}^{\infty} E_i \in \mathcal{E}$ and

$$\mathbb{P}(\cup_{i=1}^{\infty} E_i) = \lim_{i \to \infty} \mathbb{P}(E_i). \tag{B.3}$$

(b) Let $\{E_i\} \subset \mathcal{E}$ be such that $E_i \supset E_{i+1}$ $\forall i$. Then $\cap_{i=1}^{\infty} E_i \in \mathcal{E}$ and moreover, if $\mathbb{P}(E_1) < +\infty$, then

$$\mathbb{P}(\cap_{i=1}^{\infty} E_i) = \lim_{i \to \infty} \mathbb{P}(E_i). \tag{B.4}$$

Proof. (i)–(vi) follow trivially from the definition of measure. Let us prove claim (a) of (vii). Since $\mathbb{P}(E_k) \le \mathbb{P}(\cup_k E_k)$ for every k, the claim is trivial if for some k $\mathbb{P}(E_k) = +\infty$. We may therefore assume $\mathbb{P}(E_k) < \infty$ for all k. We set $E := \cup_k E_k$ and decompose E as

$$E = E_1 \bigcup \Big(\bigcup_{k=2}^{\infty} (E_k \setminus E_{k-1}) \Big).$$

The sets E_1 and $E_k \setminus E_{k-1}$, $k \ge 1$, are of course in $\mathcal{E}$ and pairwise disjoint. Because of the σ-additivity of $\mathbb{P}$ we then have

$$\begin{aligned}\mathbb{P}(E) &= \mathbb{P}(E_1) + \sum_{k=2}^{\infty} \mathbb{P}(E_k \setminus E_{k-1}) \\ &= \mathbb{P}(E_1) + \sum_{k=2}^{\infty} (\mathbb{P}(E_k) - \mathbb{P}(E_{k-1})) = \lim_{k\to\infty} \mathbb{P}(E_k).\end{aligned}$$

Claim (b) of (vii) easily follows. In fact, since $\mathbb{P}(E_1) < +\infty$ and $E_k \subset E_1$, we have $\mathbb{P}(E_k) = \mathbb{P}(E_1) - \mathbb{P}(E_1 \setminus E_k)$ for all k. Since $\{E_1 \setminus E_k\}$ is an increasing sequence of sets, we deduce from (a) that

$$\begin{aligned}\mathbb{P}(E_1) - \lim_{k\to\infty} \mathbb{P}(E_k) &= \lim_{k\to\infty} \mathbb{P}(E_1 \setminus E_k) \\ &= \mathbb{P}\Big(\bigcup_k (E_1 \setminus E_k) \Big) = \mathbb{P}(E_1) - \mathbb{P}\Big(\bigcap_k E_k \Big).\end{aligned}$$

Let $(\mathcal{E}, \mathbb{P})$ be a measure on Ω. We say that $N \subset \Omega$ is $\mathbb{P}$-*negligible*, or simply a *null set*, if there exists $E \in \mathcal{E}$ such that $N \subset E$ and $\mathbb{P}(E) = 0$. Let $\widetilde{\mathcal{E}}$ be the collection of all the subsets of Ω of the form $F = E \cup N$ where $E \in \mathcal{E}$ and N is $\mathbb{P}$-negligible. It is easy to check that $\widetilde{\mathcal{E}}$ is a σ-algebra which is called the $\mathbb{P}$-*completion* of $\mathcal{E}$. Moreover, setting $\mathbb{P}(F) := \mathbb{P}(E)$ if $F = E \cup N \in \widetilde{\mathcal{E}}$, then $(\widetilde{\mathcal{E}}, \mathbb{P})$ is also a measure on Ω called the $\mathbb{P}$-*completion* of $(\mathcal{E}, \mathbb{P})$. It is often customary to consider measures as $\mathbb{P}$-complete measures.

B.1.2 Construction of measures

B.1.2.1 Uniqueness

Let $(\mathcal{E}, \mathbb{P})$ be a measure on Ω. The σ-additivity property of $\mathbb{P}$ suggests that the values of $\mathbb{P}$ on $\mathcal{E}$ are in fact determined by the values of $\mathbb{P}$ on a restricted class of subsets of $\mathcal{E}$.

Definition B.5 *A family $\mathcal{D} \subset \mathcal{P}(\Omega)$ of subsets of Ω is said to be* closed under finite intersections *if $A, B \in \mathcal{D}$ implies $A \cap B \in \mathcal{D}$.*

A set function $\alpha : \mathcal{I} \subset \mathcal{P}(\Omega) \to \overline{\mathbb{R}}_+$ is σ-finite *if there exists a sequence $\{I_k\} \subset \mathcal{I}$ such that $\Omega = \cup_k I_k$ and $\alpha(I_k) < \infty$ $\forall k$.*

We have the following coincidence criterion. A proof can be found in, e.g. [7].

Theorem B.6 (Coincidence criterion) *Let* $(\mathcal{E}_1, \mathbb{P}_1)$ *and* $(\mathcal{E}_2, \mathbb{P}_2)$ *be two measures on* Ω *and let* $\mathcal{D} \subset \mathcal{E}_1 \cap \mathcal{E}_2$ *be a family that is closed under finite intersections. Assume that* $\mathbb{P}_1(A) = \mathbb{P}_2(A)$ $\forall A \in \mathcal{D}$ *and that there exists a sequence* $\{D_h\} \in \mathcal{D}$ *such that* $\Omega = \cup_h D_h$ *and* $\mathbb{P}_1(D_h) = \mathbb{P}_2(D_h) < +\infty$ *for any* h*. Then* $\mathbb{P}_1$ *and* $\mathbb{P}_2$ *coincide on the* σ*-algebra generated by* $\mathcal{D}$*.*

Corollary B.7 (Uniqueness of extension) *Let* Ω *be an open set, let* $\mathcal{I}$ *be a family of subsets of* Ω *closed under finite intersections and let* $\alpha : \mathcal{I} \to \overline{\mathbb{R}}_+$ *be* σ*-finite. Then* α *has at most one extension* $\mu : \mathcal{E} \to \overline{\mathbb{R}}_+$ *to the* σ*-algebra* $\mathcal{E}$ *generated by* $\mathcal{I}$ *such that* $(\mathcal{E}, \mu)$ *is a measure.*

B.1.2.2 Carathéodory Method I

We now present the so-called Method I for constructing measures.

Let $\mathcal{I}$ be a family of subsets of Ω containing the empty set, and let $\alpha : \mathcal{I} \to \overline{\mathbb{R}}_+$ be a set function such that $\alpha(\emptyset) = 0$. For any $E \subset \Omega$ set

$$\mu^*(E) := \inf\left\{ \sum_{i=1}^{\infty} \alpha(I_i) \;\middle|\; \cup_i I_i \supset E,\ I_i \in \mathcal{I} \right\}. \tag{B.5}$$

Of course, we set $\mu^*(E) = +\infty$ if no covering of E by subsets in $\mathcal{I}$ exists. It is easy to check that $\mu^*(\emptyset) = 0$, that μ^* is monotone increasing and that μ^* is σ-subadditive, i.e.

$$\mu^*\Big(\bigcup_{i=1}^{\infty} E_i\Big) \le \sum_{i=1}^{\infty} \mu^*(E_i)$$

for every denumerable family $\{E_i\}$ of subsets of Ω.

We now define a σ-algebra $\mathcal{M}$ on which μ^* is σ-additive. A first attempt is to choose the class of sets on which μ^* is σ-additive, i.e. the class of sets E such that

$$\mu^*(B \cup E) = \mu^*(E) + \mu^*(B)$$

for every subset B disjoint from E, or, equivalently such that

$$\mu^*(A \cap E) + \mu^*(A \cap E^c) \le \mu^*(A) \qquad \forall A,\ A \supset E. \tag{B.6}$$

However, in general, this class is not a σ-algebra. Following Carathéodory, a localization of (B.6) suffices. A set $E \subset \Omega$ is said to be μ^**-measurable* if

$$\mu^*(A \cap E) + \mu^*(A \cap E^c) = \mu^*(A) \qquad \forall A \in \mathcal{P}(\Omega)$$

and the class of μ^*-measurable sets will be denoted by $\mathcal{M}$.

Theorem B.8 (Carathéodory) $\mathcal{M}$ *is a* σ*-algebra and* μ^* *is* σ*-additive on* $\mathcal{M}$*. In other words,* $(\mathcal{M}, \mu^*)$ *is a measure on* Ω*.*

Without additional hypotheses both on $\mathcal{I}$ and α, we might end up with $\mathcal{I}$ not included in $\mathcal{M}$ or with a μ^* that is not an extension of α.

Definition B.9 *A family $\mathcal{I} \subset \mathcal{P}(\Omega)$ of subsets of Ω is a* semiring *if:*

(i) $\emptyset \in \mathcal{I}$.

(ii) For any $E, F \in \mathcal{I}$ we have $E \cap F \in \mathcal{I}$.

(iii) If $E, F \in \mathcal{I}$, then $E \setminus F = \cup_{j=1}^{N} I_j$ where the I_j's are pairwise disjoint elements in $\mathcal{I}$.

Notice that, if $E, F \in \mathcal{I}$, then $E \cup F$ decomposes as $E \cup F = \cup_{j=1}^{n} I_j$ where $I_1, \dots, I_n$ belong to $\mathcal{I}$ and are pairwise disjoint.

Theorem B.10 (Carathéodory) *Let $\mathcal{I} \subset \mathcal{P}(\Omega)$ be a semiring of subsets of Ω, let $\alpha : \mathcal{I} \to \overline{\mathbb{R}}_+$ be a σ-additive set function such that $\alpha(\emptyset) = 0$ and let $(\mathcal{M}, \mu^*)$ be the measure constructed by the above starting from $\mathcal{I}$ and α. Then:*

(i) $\mathcal{I} \subset \mathcal{M}$.

(ii) μ^* extends α to $\mathcal{M}$.

(iii) Let $E \subset \Omega$ with $\mu^*(E) < \infty$. Then $E \in \mathcal{M}$ if and only if $E = \cap_k F_k \setminus N$ where $\mu^*(N) = 0$, $\{F_k\}$ is a decreasing sequence of sets F_k and, for $k \geq 1$, F_k is a disjoint union $F_k = \cup_j I_{k,j}$ of sets $I_{k,j} \in \mathcal{I}$.

Assume $\alpha : \mathcal{I} \to \overline{\mathbb{R}}_+$ be such that $\mathcal{I}$ is a semiring, α is σ-additive and σ-finite and let $\mathcal{E}$ be the σ-algebra generated by $\mathcal{I}$. From Theorems B.6 and B.10 the following easily follows:

- (iii) of Theorem B.10 implies that $\mathcal{M}$ is the μ^*-completion of $\mathcal{E}$: every set $A \in \mathcal{M}$ has the form $A = E \cup N$ where $E \in \mathcal{E}$ and $\mu^*(N) = 0$.
- Corollary B.7 imples that $(\mathcal{E}, \mu^*)$ is the unique measure that extends α to $\mathcal{E}$.

B.1.2.3 Lebesgue measure in $\mathbb{R}^n$

A *right-closed interval* I of $\mathbb{R}^n$, $n \geq 1$, is the product of n intervals closed to the right and open to the left, $I = \prod_{i=1}^{n}]a_i, b_i]$. The elementary volume of this interval is $|I| := \prod_{i=1}^{n} (b_i - a_i)$.

An induction argument on the dimension n shows that $\mathcal{I}$ is a semiring. For instance, let $n = 2$ and let A, B, C, D be right-closed intervals on $\mathbb{R}$. Then

$$(A \times B) \setminus (C \times D) = \Big((A \setminus C) \times (B \setminus D)\Big) \cup \Big((A \setminus C) \times (B \cap D)\Big)$$
$$\cup \Big((A \cap C) \times (B \setminus D)\Big).$$

The family of right-closed intervals of $\mathbb{R}^n$ will be denoted by $\mathcal{I}$. We know that $\mathcal{B}(\mathbb{R}^n)$ is the σ-algebra generated by $\mathcal{I}$, see Exercise B.16. Since $\mathcal{I}$ is trivially closed under finite intersections, we infer from Theorem B.6 that *two measures that coincide on* $\mathcal{I}$ *and that are finite on bounded open sets coincide on every Borel set* $E \in \mathcal{B}(\mathbb{R}^n)$.

Proposition B.11 *The volume map* $|\ | : \mathcal{I} \to \mathbb{R}_+$ *is a* σ*-additive set function.*

Proof. It is easily seen that the elementary measure $|\ |$ is finitely additive on intervals. Let us prove that it is σ-subadditive. For that, let I, I_k be intervals with $I = \cup_k I_k$ and, for $\epsilon > 0$ and any k denote by J_k an interval centred as I_k that contains strictly I_k with $|J_k| \le |I_k| + \epsilon\, 2^{-k}$. The family of open sets $\{\text{int}(J_k)\}_k$ covers the compact set $\overline{I}$, hence we can select $k_1, k_2, \dots k_N$ such that $I \subset \cup_{i=1}^N \text{int}(J_{k_i})$ concluding

$$|I| \le \sum_{i=1}^N |J_{k_i}| \le \sum_{k=1}^\infty |J_k| \le \sum_{k=1}^\infty (|I_k| + \epsilon 2^{-k}) \le \sum_{k=1}^\infty |I_k| + 2\epsilon,$$

i.e. that $||$ is σ-subadditive on $\mathcal{I}$.

Suppose now that $I = \cup_k I_k$ where the $\{I_k\}$'s are pairwise disjoint. Of course, by the σ-subadditivity property of $||$, $|I| \le \sum_{k=1}^\infty |I_k|$. On the other hand, $\cup_{k=0}^N I_k \subset I$ for any integer N. Finite additivity then yields

$$\sum_{k=0}^N |I_k| = \left| \bigcup_{k=0}^N I_k \right| \le |I| \qquad \forall N$$

and, as $N \to \infty$, also the opposite inequality $\sum_{k=0}^\infty |I_k| \le |I|$.

Taking advantage of Proposition B.11, Theorem B.10 applies. We get the existence of a unique measure $(\mathcal{B}(\mathbb{R}^n), \mathcal{L}^n)$ that is finite on bounded open sets, called the *Lebesgue measure* on $\mathbb{R}^n$, that extends to Borel sets the elementary measure of intervals. From (B.5) we also get a formula for the measure of a Borel set $E \in \mathcal{B}(\mathbb{R}^n)$,

$$\mathcal{L}^n(E) = \inf \left\{ \sum_{k=1}^\infty |I_k| \,\middle|\, I_k \text{ intervals, } E \subset \bigcup_{k=1}^\infty I_k \right\}. \tag{B.7}$$

B.1.2.4 Stieltjes–Lebesgue measure

Proposition B.12 *Let* $F : \mathbb{R} \to \mathbb{R}$ *be a right-continuous and monotone nondecreasing function. Then the set function* $\zeta : \mathcal{I} \to \mathbb{R}$ *defined by* $\zeta(]a, b]) := F(b) - F(a)$ *on the class* $\mathcal{I}$ *of right-closed intervals is* σ*-additive.*

Proof. Obviously, ζ is additive and monotone increasing, hence finitely subadditive, on $\mathcal{I}$. We now prove that ζ is σ-additive. Let $\{I_i\} \subset \mathcal{I}$, $I_i :=]x_i, y_i]$, be a disjoint partition of $I :=]a, b]$. Since ζ is additive, we get

$$\sum_{i=1}^\infty \zeta(I_i) \le \zeta(I).$$

Let us prove the opposite inequality. For $\epsilon > 0$, let $\{\delta_i\}$ be such that $F(y_i + \delta_i) \leq F(y_i) + \epsilon\, 2^{-i}$. The open intervals $]x_i, y_i + \delta_i[$ form an open covering of $[a+\epsilon, b]$, hence finitely many among them cover again $[a+\epsilon, b]$. Therefore, by the finite subadditivity of ζ,

$$F(b) - F(a+\epsilon) \leq \sum_{k=1}^{N} (F(y_{i_k} + \delta_{i_k}) - F(x_{i_k}))$$

$$= \sum_{k=1}^{N} (\zeta(I_{i_k}) + \epsilon\, 2^{-i_k}) \leq \sum_{i=1}^{\infty} \zeta(I_i) + \epsilon.$$

Letting ϵ go to zero, we conclude

$$\zeta(I) = F(b) - F(a) \leq \sum_{i=1}^{\infty} \zeta(I_i).$$

Example B.13 If F is not right-continuous, the set function $\zeta : \mathcal{I} \to \mathbb{R}$, $\zeta(]a, b]) := F(b) - F(a)$ is not in general subadditive. For instance, for $0 \leq a \leq 1$, set

$$F(t) = F_a(t) = \begin{cases} 0 & \text{if } -1 \leq t < 0, \\ a & \text{if } t = 0, \\ 1 & \text{if } 0 < t \leq 1. \end{cases}$$

Let $\mathcal{I} = \{]-1, 0]\} \cup \left\{]\frac{1}{j+1}, \frac{1}{j}]\right\}_j$, clearly $\cup_{I \in \mathcal{I}} I =]-1, 1]$, but

$$1 = F(1) - F(-1) = \zeta\Big(\bigcup_{I \in \mathcal{I}} I\Big) \neq \sum_{I \in \mathcal{I}} \zeta(I) = F(0) - F(-1) = a$$

as soon as $a < 1$.

Theorem B.14 (Lebesgue) *The following hold:*

(i) Let $(\mathcal{B}(\mathbb{R}), \mathbb{P})$ be a finite measure on $\mathbb{R}$. Then the law $F(t) := \mathbb{P}(]-\infty, t])$, $t \in \mathbb{R}$, is real valued, monotone nondecreasing and right-continuous.

(ii) Let $F : \mathbb{R} \to \mathbb{R}$ be a real valued, monotone nonderecreasing and right-continuous function. Then there exists a unique measure $(\mathcal{B}(\mathbb{R}), \mathbb{P})$ finite on bounded sets of $\mathbb{R}$ such that

$$\mathbb{P}(]a, b]) = F(b) - F(a) \qquad \forall a, b,\ a < b.$$

Proof. (i) F is real valued since $(\mathcal{B}(\mathbb{R}), \mathbb{P})$ is finite on bounded Borel sets. Moreover, monotonicity property of measures implies that F is monotone nondecreasing. Let us prove that F is right-continuous. Let $t \in \mathbb{R}$ and let $\{t_n\}$

be a monotone decreasing sequence such that $t_n \downarrow t$. Since $]-\infty, t] = \bigcap_{n=1}^{\infty}]-\infty, t_n]$ and $\mathbb{P}$ is finite, one gets $F(t_n) = \mathbb{P}(]-\infty, t_n]) \to \mathbb{P}(]-\infty, t]) = F(t)$ by the continuity property of measures.

(ii) Assume $F(t)$ is right-continuous and monotone nondecreasing. Let $\mathcal{I}$ be the semiring of right-closed intervals. The set function $\zeta : \mathcal{I} \to \mathbb{R}_+$, $\zeta(]a, b]) := F(b) - F(a)$ is σ-additive, see Proposition B.12. Therefore Theorem B.6 and B.10 apply and ζ extends in a unique way to a measure on the σ-algebra generated by $\mathcal{I}$, i.e. on $\mathcal{B}(\mathbb{R})$, that is finite on bounded open sets.

The measure $(\mathcal{B}(\mathbb{R}), \mathbb{P})$ in Theorem B.14 is called the *Stieltjes–Lebesgue measure* associated with the right-continuous monotone nondecreasing function F.

B.1.2.5 Approximation of Borel sets

Borel sets are quite complicated if compared with open sets that are simply denumerable unions of closed cubes with disjoint interiors. However, the following holds.

Theorem B.15 *Let $(\mathcal{B}(\mathbb{R}^n), \mathbb{P})$ be a measure on $\mathbb{R}^n$ that is finite on bounded open sets. Then for any $E \in \mathcal{B}(\mathbb{R}^n)$*

$$\mathbb{P}(E) = \inf\left\{\mathbb{P}(A) \,\middle|\, A \supset E,\ A \text{ open}\right\}, \tag{B.8}$$

$$\mathbb{P}(E) = \sup\left\{\mathbb{P}(F) \,\middle|\, F \subset E,\ F \text{ closed}\right\}. \tag{B.9}$$

In particular, if $E \in \mathcal{B}(\mathbb{R}^n)$ has finite measure, $\mathbb{P}(E) < +\infty$, then for every $\epsilon > 0$ there exists an open set Ω and a compact set K such that $K \subset E \subset \Omega$ and $\mathbb{P}(\Omega \setminus K) < \epsilon$.

Although the result can be derived from (iii) of Theorem B.10, it is actually independent of it. We give here a proof that does not use Theorem B.10.

Proof. Step 1. Let us prove the claim assuming $\mathbb{P}$ is finite. Consider the family

$$\mathcal{A} := \left\{E \text{ Borel} \,\middle|\, (B.8) \text{ holds true for } E\right\}.$$

Of course, $\mathcal{A}$ contains the family of open sets. We prove that $\mathcal{A}$ *is closed under denumerable unions and intersections*. Let $\{E_j\} \subset \mathcal{A}$ and, for $\epsilon > 0$ and $j = 1, 2, \ldots$, let A_j be open sets with $A_j \supset E_j$ and $\mathbb{P}(A_j) \le \mathbb{P}(E_j) + \epsilon\, 2^{-j}$, that we rewrite as $\mathbb{P}(A_j \setminus E_j) < \epsilon\, 2^{-j}$ since E_j and A_j are measurable with finite measure. Since

$$\Big(\bigcup_j A_j\Big) \setminus \Big(\bigcup_j E_j\Big) \subset \bigcup_j (A_j \setminus E_j), \quad \Big(\bigcap_j A_j\Big) \setminus \Big(\bigcap_j E_j\Big) \subset \bigcup_j (A_j \setminus E_j),$$

we infer

$$\mathbb{P}(A \setminus \cup_j E_j) \le \epsilon, \qquad \mathbb{P}(B \setminus \cap_j E_j) \le \epsilon, \tag{B.10}$$

where $A := \cup_j A_j$ and $B := \cap_j A_j$. Since A is open and $A \supset \cup_j E_j$, the first inequality of (B.10) yields $\cup_j E_j \in \mathcal{A}$. On the other hand, $C_N := \cap_{j=1}^N A_j$ is open, contains $\cap_j E_j$ and, by the second inequality of (B.10), $\mathbb{P}(C_N \setminus \cap_j E_j) \leq 2\epsilon$ for sufficiently large N. Therefore $\cap_j E_j \in \mathcal{A}$.

Moreover, since every closed set is the intersection of a denumerable family of open sets, $\mathcal{A}$ also contains all closed sets. In particular, the family

$$\tilde{\mathcal{A}} := \{A \in \mathcal{A},\ A^c \in \mathcal{A}\}$$

is a σ-algebra that contains the family of open sets. Consequently, $\mathcal{A} \supset \tilde{\mathcal{A}} \supset \mathcal{B}(\mathbb{R}^n)$ and (B.8) holds for all Borel sets of Ω.

Since (B.9) for E is (B.8) for E^c, we have also proved (B.9).

Step 2. Let us prove (B.8) and (B.9) for measures that are finite on bounded open sets. Let us introduce the following notation: given a Borel set $A \subset \mathbb{R}^n$, define the *restriction* of $\mathbb{P}$ to A as the set function

$$\mathbb{P} \llcorner A(E) := \mathbb{P}(A \cap E) \qquad \forall E \in \mathcal{B}(\mathbb{R}^n). \tag{B.11}$$

It is easily seen that $(\mathcal{B}(\mathbb{R}^n), \mathbb{P} \llcorner A)$ is a measure on $\mathbb{R}^n$ that is finite if $\mathbb{P}(A) < \infty$.

Let us prove (B.8). We may assume $\mathbb{P}(E) < +\infty$ since otherwise (B.8) is trivial. Let $V_j := B(0, j)$ be the open ball centred at 0 or radius j. The measures $\mathbb{P} \llcorner V_j$ are Borel and $\mathbb{P} \llcorner V_j(\mathbb{R}^n) = \mathbb{P}(V_j) < +\infty$. Step 1 then yields that for any $\epsilon > 0$ there are open sets A_j with $A_j \supset E$ and $\mathbb{P} \llcorner V_j(A_j \setminus E) < \epsilon 2^{-j}$. The set $A := \cup_j (A_j \cap V_j)$ is open, $A \supset E$ and, by the subadditivity of $\mathbb{P}$

$$\begin{aligned}\mathbb{P}(A \setminus E) &= \mathbb{P}(\cup_j((A_j \cap V_j) \setminus E)) \\ &\leq \sum_{j=1}^{\infty} \mathbb{P}((A_j \cap V_j) \setminus E) \leq \sum_{j=1}^{\infty} \mathbb{P} \llcorner V_j(A_j \setminus E) \leq \epsilon.\end{aligned}$$

Let us prove (B.9). The claim easily follows applying Step 1 to the measure $\mathbb{P} \llcorner E$ if $\mathbb{P}(E) < +\infty$. If $\mathbb{P}(E) = +\infty$, then $E = \cup_j E_j$ with E_j measurable and $\mathbb{P}(E_j) < +\infty$, then for every $\epsilon > 0$ and every j there exists a closed set F_j with $F_j \subset E_j$ and $\mathbb{P}(E_j \setminus F_j) < \epsilon 2^{-j}$. The set $F := \cup_j F_j$ is contained in E and

$$\mathbb{P}(E \setminus F) \leq \mathbb{P}(\cup_j(E_j \setminus F_j)) \leq \sum_{j=i}^{\infty} \mathbb{P}(E_j \setminus F_j) \leq \epsilon,$$

hence, for sufficiently large N, $G_N := \cup_{i=1}^N F_i$ is closed and $\mathbb{P}(E \setminus G_N) < 2\epsilon$.

Step 3. By assumption, $E \in \mathcal{B}(\mathbb{R}^n)$ and $\mathbb{P}(E) < +\infty$. By Step 2 for each $\epsilon > 0$ there exists an open set Ω and a closed set F such that $F \subset E \subset \Omega$ and

$\mathbb{P}(\Omega) \leq \mathbb{P}(E) + \epsilon$, $\mathbb{P}(E) \leq \mathbb{P}(F) + \epsilon$, so that $\mathbb{P}(\Omega) - \mathbb{P}(F) \leq 2\epsilon$. Setting $K := F \cap \overline{B(0, n)}$ with large enough n, we still get

$$\mathbb{P}(E) \leq \mathbb{P}(K) + 2\epsilon,$$

thus concluding that $A \supset E \supset K$ and $\mathbb{P}(A \setminus K) < 3\epsilon$.

B.1.3 Exercises

Exercise B.16 *Show that $\mathcal{B}(\mathbb{R})$ is the smallest σ-algebra generated by one of the following families of sets:*

- *the closed sets;*
- *the open intervals;*
- *the closed intervals;*
- *the intervals $]a, b]$, $a, b \in \mathbb{R}$, $a < b$;*
- *the intervals $[a, b[$, $a, b \in \mathbb{R}$, $a < b$;*
- *the closed half-lines $]-\infty, t]$, $t \in \mathbb{R}$.*

Solution. [*Hint*. Show that any open set can be written as the union of an at most denumerable family of intervals.]

Exercise B.17 *The* law *of a finite measure $(\mathcal{B}(\mathbb{R}^n), \mathbb{P})$ on $\mathbb{R}^n$ is defined by*

$$F(t) = F(t_1, \dots, t_n) := \mathbb{P}(]-\infty, t_1] \times \cdots \times]-\infty, t_n]).$$

Show that two finite measures $(\mathcal{B}(\mathbb{R}^n), \mathbb{P})$ and $(\mathcal{B}(\mathbb{R}^n), \mathbb{Q})$ on $\mathbb{R}^n$ coincide if and only if the corresponding laws agree.

B.2 Measurable functions and integration

Characterizing the class of Riemann integrable functions and understanding the range of applicability of the fundamental theorem of calculus were the problems that led to measures and to a more general notion of integral due to Henri Lebesgue (1875–1941). The approach we follow here, which is very well adapted to calculus of probability, is to start with a measure and define the associated notion of integral.

The basic idea is the following. Suppose one wants to compute the area of the subgraph of a non-negative function $f : \mathbb{R} \to \mathbb{R}$. One can do it by approximating the subgraph in two different ways, see Figure B.1. One can take partitions of the

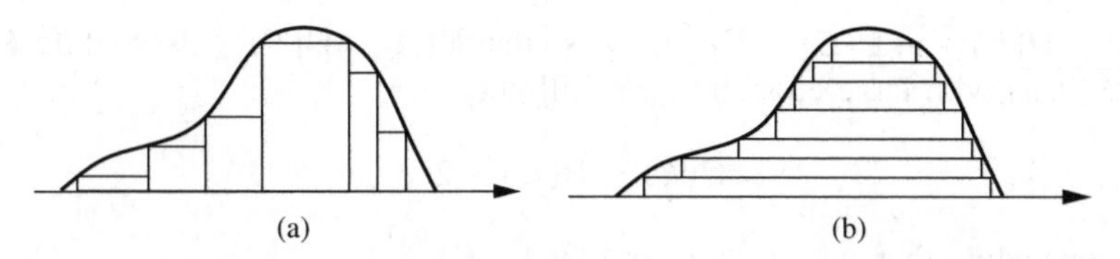

Figure B.1 The computation of an integral according to (a) Riemann and (b) Lebesgue.

x axis, and approximate the integral by the area of a piecewise constant function as we do when defining the Riemann integral, or one can take a partition of the y axis, and approximate the area of the subgraph by the areas of the strips. The latter defines the area of the subgraph as

$$\int_a^b f(x)\,dx = \lim_{N\to\infty} \frac{1}{2^N} \sum_{k=1}^{\infty} |E_{f,k\,2^{-N}}|, \tag{B.12}$$

where

$$E_{f,t} := \Big\{x \in \mathbb{R} \,\Big|\, f(x) > t\Big\}, \qquad t \in \mathbb{R},$$

is the t upper level of f and $|E_{f,t}|$ denotes its 'measure'. Since $t \to |E_{f,t}|$ is monotone nonincreasing, hence Riemann integrable, (B.12) suggests defining the integral by means of the *Cavalieri formula*

$$\text{Lebesgue} \int_a^b f(x)\,dx := \text{Riemann} \int_0^{\infty} |E_{f,t}|\,dt \tag{B.13}$$

From this point of view, it is clear that the notion of integral makes essential use of a *measure* on the domain, that must be able to measure even irregular sets, since the upper levels can be very irregular, for instance if the function is oscillating.

In the following, instead of defining the integral by means of (B.13), we adopt a slightly more direct approach to the integral and then prove (B.13).

B.2.1 Measurable functions

Definition B.18 *Let $\mathcal{E}$ be a σ-algebra of subsets of a set Ω. We say that $f : \Omega \to \overline{\mathbb{R}}$ is* $\mathcal{E}$-measurable *if for any $t \in \mathbb{R}$ we have*

$$\Big\{x \in \Omega \,\Big|\, f(x) > t\Big\} \in \mathcal{E}.$$

There are several equivalent ways to say that a function is $\mathcal{E}$-measurable. Taking advantage of the fact that $\mathcal{E}$ is a σ-algebra, one proves that the following are equivalent:

(i) $\{x \in \Omega \,|\, f(x) > t\} \in \mathcal{E}$ for any t.

(ii) $\{x \in \Omega \,|\, f(x) \geq t\} \in \mathcal{E}$ for any t.

(iii) $\{x \in \Omega \mid f(x) \leq t\} \in \mathcal{E}$ for any t.

(iv) $\{x \in \Omega \mid f(x) < t\} \in \mathcal{E}$ for any t.

Moreover, in the previous statements one can substitute 'for any t' with 'for any t in a dense subset of $\mathbb{R}$', in particular, with 'for any $t \in \mathbb{Q}$'.

Since any open set of $\mathbb{R}$ is an at most denumerable union of intervals, the following are also equivalent:

(i) $\{x \in \Omega \mid f(x) > t\} \in \mathcal{E}$ for any t.

(ii) For any open set $A \subset \mathbb{R}$ we have $f^{-1}(A) \in \mathcal{E}$.

(iii) For any closed $F \subset \mathbb{R}$ we have $f^{-1}(F) \in \mathcal{E}$.

(iv) For any Borel set $B \subset \mathbb{R}$ we have $f^{-1}(B) \in \mathcal{E}$.

The three last statements are independent of the ordering relation of $\mathbb{R}$. They suggest the following extension.

Definition B.19 *Let $\mathcal{E}$ be a σ-algebra of subsets of a set Ω. A vector valued function $f : \Omega \to \mathbb{R}^N$, $N \geq 1$, is* $\mathcal{E}$-measurable *if one of the following holds:*

(i) For any open set $A \subset \mathbb{R}^N$ we have $f^{-1}(A) \in \mathcal{E}$.

(ii) For any closed set $F \subset \mathbb{R}^N$ we have $f^{-1}(F) \in \mathcal{E}$.

(iii) For any Borel set $B \subset \mathbb{R}^N$ we have $f^{-1}(B) \in \mathcal{E}$.

In general, not every function is $\mathcal{E}$-measurable. However, since $\mathcal{E}$ is a σ-algebra, one can prove that the algebraic manipulations as well as the pointwise limits of $\mathcal{E}$-measurable functions always result in $\mathcal{E}$-measurable functions. For instance, if f and g are $\mathcal{E}$-measurable and $\alpha \in \mathbb{R}$, then the functions

$$f + g, \quad fg, \quad \frac{1}{f}, \quad \alpha f, \quad |f|, \quad \max(f, g), \quad \min(f, g)$$

are $\mathcal{E}$-measurable. Moreover, let $\{f_n\}$ be a sequence of $\mathcal{E}$-measurable functions. Then:

- The functions

$$\sup_n f_n(x), \qquad \inf_n f_n(x), \qquad \liminf_{n\to\infty} f_n(x), \qquad \limsup_{n\to\infty} f_n(x)$$

  are $\mathcal{E}$-measurable.

- Let $E \subset \Omega$ be the set

$$E := \left\{x \in \Omega \,\middle|\, \exists \lim_{n\to\infty} f_n(x)\right\}$$

and let $f(x) := \lim_{n\to\infty} f_n(x)$, $x \in E$. Then $E \in \mathcal{E}$ and for any $t \in \mathbb{R}$ we have $\{x \in E \mid f(x) > t\} \in \mathcal{E}$.

Recalling that a function $\phi : X \to Y$ between metric spaces is continuous if and only if for any open set $A \subset Y$ the set $\phi^{-1}(A) \subset X$ is open, we get immediately the following:

- Continuous functions $g : \mathbb{R}^N \to \mathbb{R}^m$ are $\mathcal{B}(\mathbb{R}^N)$-measurable,
- Let $f : \Omega \to \mathbb{R}^N$ be $\mathcal{E}$-measurable and let $\phi : \mathbb{R}^N \to \mathbb{R}^m$ be $\mathcal{B}(\mathbb{R}^N)$-measurable, then $\phi \circ f$ is $\mathcal{E}$-measurable.

In particular, (ii) implies that $|f|^p, \log|f|, \ldots$ are $\mathcal{E}$-measurable functions if f is $\mathcal{E}$-measurable and that $f : \Omega \to \mathbb{R}^N$, $f = (f^1, \ldots, f^N)$ is $\mathcal{E}$-measurable if and only if its components $f^1, \ldots, f^N$ are $\mathcal{E}$-measurable.

Let $(\mathcal{E}, \mathbb{P})$ be a measure on Ω. A *simple function* $\varphi : \Omega \to \mathbb{R}$ is a function with a finite range and $\mathcal{E}$-measurable level sets, that is,

$$\varphi(x) = \sum_{j=1}^{k} a_j \chi_{E_j}(x), \qquad E_j := \{x \mid \varphi(x) = a_j\} \in \mathcal{E}.$$

The class of simple functions will be denoted by $\mathcal{S}$. Simple functions being linear combinations of $\mathcal{E}$-measurable functions are $\mathcal{E}$-measurable.

Proposition B.20 (Sampling) *Let $\mathcal{E}$ be a σ-algebra of subsets of a set Ω. A non-negative function $f : \Omega \to \overline{\mathbb{R}}_+$ is $\mathcal{E}$-measurable if and only if there exists a nondecreasing sequence $\{\varphi_k\}$ of non-negative simple functions such that $\varphi_k(x) \uparrow f(x)$ $\forall x \in \Omega$.*

Proof. Let f be the pointwise limit of a sequence $\{\varphi_k\}$ of simple functions. Since every φ_k is $\mathcal{E}$-measurable, then f is $\mathcal{E}$-measurable.

Conversely, let $f : \Omega \to \mathbb{R}$ be a function. By sampling f, we then construct a sequence $\{\varphi_k\}$ of functions with finite range approaching f, see Figure B.1. More precisely, let $E_k := \{x \mid f(x) > 2^k\}$ and for $h = 0, 1, , \ldots 4^k - 1$, let

$$E_{k,h} := \{x \mid h/2^k < f(x) \leq (h+1)/2^k\}.$$

Define $\varphi_k : \Omega \to \mathbb{R}$ as

$$\varphi_k(x) = \begin{cases} 2^k & \text{if } x \in E_k, \\ \frac{h}{2^k} & \text{if } x \in E_{k,h}. \end{cases} \tag{B.14}$$

By definition, $\varphi_k(x) \leq f(x)$, moreover, $\varphi_k(x) \leq \varphi_{k+1}(x)$ $\forall x \in \Omega$, since passing from k to $k+1$ we half the sampling intervals. Let us prove that $\varphi_k(x) \to f(x)$ $\forall x$. If $f(x) = +\infty$, then $\varphi_k(x) = 2^k$ $\forall k$, hence $\varphi_k(x) \to +\infty =$

$f(x)$. If $f(x) < +\infty$, then for sufficiently large k, $f(x) \leq 2^k$, hence there exists $h \in \{0, \dots, 4^k - 1\}$ such that $x \in E_{k,h}$. Therefore,

$$f(x) - \varphi_k(x) \leq \frac{h+1}{2^k} - \frac{h}{2^k} = \frac{1}{2^k}.$$

Passing to the limit as $k \to \infty$ we get again $\varphi_k(x) \to f(x)$.

The previous construction applies to any non-negative function $f : \Omega \to \overline{\mathbb{R}}_+$. To conclude, notice that if f is $\mathcal{E}$-measurable, then the sets E_k and $E_{k,h}$ are $\mathcal{E}$-measurable for every k, h. Since

$$\varphi_k(x) = \sum_{h=0}^{4^k-1} \frac{h}{2^k} \chi_{E_{k,h}}(x) + 2^k \chi_{E_k}(x) \qquad \forall x \in \Omega,$$

φ_k is a simple function.

B.2.2 The integral

Let $(\mathcal{E}, \mathbb{P})$ be a measure on Ω. For any simple function $\varphi : \Omega \to \mathbb{R}$, one defines the *integral* of φ with respect to the measure $(\mathcal{E}, \mathbb{P})$ as

$$I(\varphi) := \sum_{j=1}^{k} a_j \mathbb{P}(\varphi = a_j), \tag{B.15}$$

as intuition suggests. Since a priori $\mathbb{P}(\varphi = a_j)$ may be infinite, we adopt the convention that $a_j \mathbb{P}(\varphi = a_j) := 0$ if $a_j = 0$. Notice that the integral may be infinite.

We then define the integral of a non-negative $\mathcal{E}$-measurable function with respect to $(\mathcal{E}, \mathbb{P})$ as

$$\int_\Omega f(x)\, \mathbb{P}(dx) := \sup \left\{ I(\varphi) \,\middle|\, \varphi \text{ simple}, \varphi(x) \leq f(x)\ \forall x \in \Omega \right\}. \tag{B.16}$$

For a generic $\mathcal{E}$-measurable function $f : \Omega \to \overline{\mathbb{R}}$, decompose f as $f(x) = f_+(x) - f_-(x)$ where

$$f_+(x) := \max(f(x), 0), \qquad f_-(x) := \max(-f(x), 0),$$

and define

$$\int_\Omega f(x)\, \mathbb{P}(dx) := \int_\Omega f_+(x)\, \mathbb{P}(dx) - \int_\Omega f_-(x)\, \mathbb{P}(dx) \tag{B.17}$$

provided that at least one of the integrals on the right-hand side of (B.17) is finite. In this case one says that f is *integrable* with respect to $(\mathcal{E}, \mathbb{P})$. If both the integrals on the right-hand side of (B.17) are finite, then one says that f is *summable*. Notice that for functions that do not change sign, integrability is equivalent to measurability.

Since $|f(x)| = f_+(x) + f_-(x)$ and $f_+(x), f_-(x) \le |f(x)|$, it is easy to check that if f is $\mathcal{E}$-measurable then so is $|f|$, and f is summable if and only if f is $\mathcal{E}$-measurable and $\int |f(x)|\,\mathbb{P}(dx) < +\infty$. Moreover,

$$\left|\int_\Omega f(x)\,\mathbb{P}(dx)\right| \le \int_\Omega |f(x)|\,\mathbb{P}(dx).$$

The class of summable functions will be denoted by $\mathcal{L}^1(\Omega, \mathcal{E}, \mathbb{P})$ or simply by $\mathcal{L}^1(\Omega)$ when the measure is understood.

When $\Omega = \mathbb{R}^n$, one refers to the integral with respect to the Lebesgue measure $(\mathcal{B}(\mathbb{R}^n), \mathcal{L}^n)$ in (B.17) as the *Lebesgue integral*.

Finally, let $f : \Omega \to \mathbb{R}$ be a function and let $E \in \mathcal{E}$. One says that f is *measurable on* E, *integrable on* E, and *summable on* E if $f(x)\chi_E(x)$ is $\mathcal{E}$-measurable, integrable, and summable, respectively. If f is integrable on E, one sets

$$\int_E f(x)\,\mathbb{P}(dx) := \int_\Omega f(x)\chi_E(x)\,\mathbb{P}(dx). \tag{B.18}$$

In particular,

$$\int_E 1\,\mathbb{P}(dx) = \int_\Omega \chi_E(x)dx = \mathbb{P}(E).$$

B.2.3 Properties of the integral

From the definition of integral with respect to the measure $(\mathcal{E}, \mathbb{P})$ and taking also advantage of the σ-additivity of the measure one gets the following.

Theorem B.21 *Let $(\mathcal{E}, \mathbb{P})$ be a measure on Ω.*

(i) For any $c \in \mathbb{R}$ and f integrable on E, we have $\int_E c\,f(x)\,\mathbb{P}(dx) = c\int_E f(x)\,\mathbb{P}(dx)$.

(ii) (Monotonicity) *Let f, g be two integrable functions such that $f(x) \le g(x)$ $\forall x \in \Omega$. Then*

$$\int_\Omega f(x)\,\mathbb{P}(dx) \le \int_\Omega g(x)\,\mathbb{P}(dx).$$

(iii) (Continuity, the Beppo Levi theorem) *Let $\{f_k\}$ be a nondecreasing sequence of non-negative $\mathcal{E}$-measurable functions $f_k : \Omega \to \overline{\mathbb{R}}_+$ and let $f(x) := \lim_{k\to\infty} f_k(x)$ be the pointwise limit of the f_k's. Then f is integrable and*

$$\int_\Omega f(x)\,\mathbb{P}(dx) = \lim_{k\to\infty}\int_\Omega f_k(x)\,\mathbb{P}(dx). \tag{B.19}$$

(iv) (Linearity) *$\mathcal{L}^1$ is a vector space and the integral is a linear operator on it: for $f, g \in \mathcal{L}^1$ and $\alpha, \beta \in \mathbb{R}$ we have*

$$\int_\Omega (\alpha f(x) + \beta g(x))\,\mathbb{P}(dx) = \alpha\int_\Omega f(x)\,\mathbb{P}(dx) + \beta\int_\Omega g(x)\,\mathbb{P}(dx).$$

A few comments on the Beppo Levi theorem are appropriate. Notice that the measurability assumption is on the sequence $\{f_k\}$. The measurability of the limit f is for free, thanks to the fact that $\mathcal{E}$ is a σ-algebra. Moreover, the integrals in (B.19) may be infinite, and the equality is in both directions: we can compute one side of the equality in (B.19) and conclude that the other side has the same value. The Beppo Levi theorem is of course strictly related to the continuity property of measures, and at the end, to the σ-additivity of measures.

Proof of theorem B.21. (i) and (ii) are trivial.

(iii) Let us prove the Beppo Levi theorem. Since f is the pointwise limit of $\mathcal{E}$-measurable functions, f is $\mathcal{E}$-measurable. Moreover, since $f_k(x) \le f(x)$ for any $x \in \Omega$ and every k, from (i) we infer $\lim_{k\to\infty} \int_\Omega f_k \, \mathbb{P}(dx) \le \int_\Omega f \, \mathbb{P}(dx)$. We now prove the opposite inequality

$$\int_\Omega f(x)\,\mathbb{P}(dx) \le \alpha := \lim_{k\to\infty} \int_\Omega f_k(x)\,\mathbb{P}(dx).$$

Assume without loss of generality that $\alpha < +\infty$. Let ϕ be a simple function, $\phi = \sum_{i=1}^N a_i \chi_{B_i}$, $B_i = \{x \mid \phi(x) = a_i\} \in \mathcal{E}$, such that $\phi \le f$ and let β be a real number, $0 < \beta < 1$. For $k = 1, 2, \dots$ set

$$A_k := \left\{x \in \Omega \,\middle|\, f_k(x) \ge \beta\,\phi(x)\right\}.$$

$\{A_k\}$ is a nondecreasing sequence of measurable sets such that $\cup_k A_k = \Omega$. Hence, from (B.15) and the continuity property of measures

$$\int_{A_k} \phi(x)\,\mathbb{P}(dx) = \sum_{i=1}^N a_i \mathbb{P}(B_i \cap A_k) \to \sum_{i=1}^N a_i \mathbb{P}(B_i) = \int \phi(x)\,\mathbb{P}(dx)$$

as $k \to \infty$. On the other hand, for any k we have

$$\begin{aligned} \beta \int_{A_k} \phi(x)\,\mathbb{P}(dx) &= \int_{A_k} \beta\,\phi(x)\,\mathbb{P}(dx) \le \int_{A_k} f_k(x)\,\mathbb{P}(dx) \\ &\le \int_\Omega f_k(x)\,\mathbb{P}(dx) \le \alpha. \end{aligned} \tag{B.20}$$

Therefore, passing to the limit first as $k \to \infty$ in (B.20) and then letting $\beta \to 1^-$ we get

$$\int_\Omega \phi(x)\,\mathbb{P}(dx) \le \alpha.$$

Since the previous inequality holds for any simple function ϕ below f, the definition of integral yields $\int_\Omega f(x)\,\mathbb{P}(dx) \le \alpha$, as required.

(iv) We have already proved the linearity of the integral on the class of simple functions, see Proposition 2.28. To prove (iv), it suffices to approximate f and g by simple functions, see Proposition B.20, and then pass to the limit using (iii).

We conclude with a few simple consequences.

Proposition B.22 *Let $(\mathcal{E}, \mathbb{P})$ be a measure on Ω.*

(i) Let $E \in \mathcal{E}$ have finite measure and let $f : E \to \mathbb{R}$ be an integrable function on $E \in \mathcal{E}$ such that $|f(x)| \leq M$ for any $x \in E$. Then f is summable on E and $\int_E |f| \, \mathbb{P}(dx) \leq M \, \mathbb{P}(E)$.

(ii) Let $E, F \in \mathcal{E}$ and let $f : E \cup F \to \overline{\mathbb{R}}$ be an integrable function on $E \cup F$. Then f is integrable both on E and F and

$$\int_E f(x) \, \mathbb{P}(dx) + \int_F f(x) \, \mathbb{P}(dx) = \int_{E\cup F} f(x) \, \mathbb{P}(dx) + \int_{E\cap F} f(x) \, \mathbb{P}(dx). \tag{B.21}$$

B.2.4 Cavalieri formula

Theorem B.23 (Cavalieri formula) *Let $(\mathcal{E}, \mathbb{P})$ be a measure on Ω. For any non-negative $\mathcal{E}$-measurable function $f : \Omega \to \overline{\mathbb{R}}_+$ we have*

$$\int_\Omega f(x) \, \mathbb{P}(dx) = (Riemann) \int_0^\infty \mathbb{P}(f > t) \, dt. \tag{B.22}$$

As usual, we shorten the notation $\mathbb{P}(\{x \in \Omega \mid f(x) > t\})$ to $\mathbb{P}(f > t)$.

Proof. Let us prove the claim for non-negative simple functions. Assume $\varphi(x) = \sum_{i=1}^N a_i \chi_{E_i}(x)$, where the sets $\{E_i\}$ are measurable and pairwise disjoint. For $i = 1, \ldots, N$ let $\alpha_i(t) := \mathbb{1}_{[0,a_i]}(t)$. For the piecewise (hence simple) function $t \mapsto \mathbb{P}(\varphi > t)$ we have

$$\mathbb{P}(\varphi > t) = \sum_{i=1}^N \mathbb{P}(E_i \cap \{\varphi > t\}) = \sum_{i=1}^N \alpha_i(t)\mathbb{P}(E_i)$$

hence, integrating with respect to t

$$\int_0^\infty \mathbb{P}(\varphi > t) \, dt = \sum_{i=1}^N \left(\int_0^\infty \alpha_i(t) \, dt \right) \mathbb{P}(E_i)$$
$$= \sum_{i=1}^N a_i \mathbb{P}(E_i) = \int \varphi(x) \, \mathbb{P}(dx).$$

Assume now $f : \Omega \to \overline{\mathbb{R}}$ is non-negative and $\mathcal{E}$-measurable. Proposition B.20 yields a nondecreasing sequence $\{\varphi_k\}$ of non-negative simple functions such that

$\varphi_k(x) \uparrow f(x)$ pointwisely. As shown before, for each $k = 1, 2, \dots$

$$\int_\Omega \varphi_k(x)\,\mathbb{P}(dx) = \int_0^\infty \mathbb{P}(\varphi_k > t)dt.$$

Since $\varphi_k \uparrow f(x)$ and $\mathbb{P}(\varphi_k > t) \uparrow \mathbb{P}(f > t)$ as k goes to ∞, we can pass to the limit in the previous equality using the Beppo Levi theorem to get

$$\text{Lebesgue} \int_\Omega f(x)\,\mathbb{P}(dx) = \text{Lebesgue} \int_0^\infty \mathbb{P}(f > t)\,d\mathcal{L}^1(t).$$

The claim then follows, since $t \to \mathbb{P}(f > t)$, being nondecreasing, is Riemann integrable and Riemann and Lebesgue integrals of Riemann integrable functions coincide.

Corollary B.24 *Let $f : \Omega \to \mathbb{R}$ be integrable and for any $t \in \mathbb{R}$, let $F(t) := \mathbb{P}(f \le t)$. Then*

$$\int_\Omega f(x)\,\mathbb{P}(dx) = \int_0^{+\infty} (1 - F(t))\,dt - \int_{-\infty}^0 F(t)\,dt.$$

Proof. Apply (B.22) to the positive and negative parts of f and sum the resulting equalities.

B.2.5 Markov inequality

Let $(\mathcal{E}, \mathbb{P})$ be a measure on Ω. From the monotonicity of the integral one deduces that for any non-negative measurable function $f : \Omega \to \mathbb{R}$ we have the inequality

$$t\,\mathbb{P}(f > t) = \int_{\{f > t\}} t\,\mathbb{P}(dx) \le \int_{\{f > t\}} f(x)\,\mathbb{P}(dx) \qquad \forall t \ge 0$$

i.e.

$$\mathbb{P}(f > t) \le \frac{1}{t} \int_{\{f > t\}} f(x)\,\mathbb{P}(dx) \qquad \forall t > 0. \tag{B.23}$$

This last inequality has different names: *Markov inequality*, *weak estimate* or *Chebyshev inequality*.

B.2.6 Null sets and the integral

Let $(\mathcal{E}, \mathbb{P})$ be a measure on Ω.

Definition B.25 *We say that a set $N \subset \Omega$ is a* null set *if there exists $F \in \mathcal{E}$ such that $N \subset F$ and $\mathbb{P}(F) = 0$. We say that a predicate $p(x)$, $x \in \Omega$, is true* for

$\mathbb{P}$-almost every x or $\mathbb{P}$-a.e., and we write '$p(x)$ is true a.e.' if the set

$$N := \Big\{x \in \Omega \,\Big|\, p(x) \text{ is false}\Big\}$$

is a null set.

In particular, given an $\mathcal{E}$-measurable function $f : \Omega \to \mathbb{R}$, we say that '$f = 0$ $\mathbb{P}$-a.e.' or that '$f(x) = 0$ for $\mathbb{P}$-almost every $x \in \Omega$' if the set $\{x \in \Omega \,|\, f(x) \neq 0\}$ has zero measure,

$$\mathbb{P}\Big(\Big\{x \in \Omega \,\Big|\, f(x) \neq 0\Big\}\Big) = 0.$$

Similarly, one says that '$|f| \leq M$ $\mathbb{P}$-a.e.' or that '$|f(x)| \leq M$ for $\mathbb{P}$-almost every x', if $\mathbb{P}(\{x \in \Omega \,|\, |f(x)| > M\}) = 0$. From the σ-additivity of the measure, we immediately get the following.

Proposition B.26 *Let $(\mathcal{E}, \mathbb{P})$ be a measure on Ω and let $f : \Omega \to \mathbb{R}$ be a $\mathcal{E}$-measurable function.*

(i) If $\int_\Omega |f(x)|\, \mathbb{P}(dx) < \infty$, then $|f(x)| < +\infty$ $\mathbb{P}$-a.e.

(ii) $\int_\Omega |f(x)|\, \mathbb{P}(dx) = 0$ if and only if $f(x) = 0$ for $\mathbb{P}$-almost every $x \in \Omega$.

Proof. (i) Let $C := \int_\Omega |f(x)|\, \mathbb{P}(dx)$. Markov inequality yields for any positive integer k

$$\mathbb{P}\Big(\Big\{x \in \Omega \,\Big|\, f(x) = +\infty\Big\}\Big) \leq \mathbb{P}\Big(\Big\{x \in \Omega \,\Big|\, f(x) > k\Big\}\Big) \leq \frac{C}{k}$$

Hence, passing to the limit as $k \to \infty$ we infer that $\mathbb{P}(\{x \in \Omega \,|\, f(x) = +\infty\}) = 0$.

(ii) If $f(x) = 0$ for almost every $x \in \Omega$, then every simple function φ such that $\varphi \leq |f|$, is nonzero on at most a null set. Thus $\int_\Omega \varphi(x)\, \mathbb{P}(dx) = 0$ and, by the definition of the integral of $|f|$, $\int_\Omega |f(x)|\, \mathbb{P}(dx) = 0$.

Conversely, from the Markov inequality we get for any positive integer k

$$k\, \mathbb{P}\Big(\Big\{x \in \Omega \,\Big|\, |f(x)| > 1/k\Big\}\Big) \leq \int_\Omega |f(x)|\, \mathbb{P}(dx) = 0$$

so that $\mathbb{P}(\{x \in \Omega \,|\, |f(x)| > 1/k\}) = 0$. Since

$$\Big\{x \in \Omega \,\Big|\, |f(x)| > 0\Big\} = \bigcup_k \Big\{x \in \Omega \,\Big|\, |f(x)| > 1/k\Big\},$$

passing to the limit as $k \to \infty$ thanks to the continuity property of the measure, we conclude that $\mathbb{P}(\{x \in \Omega \,|\, |f(x)| > 0\}) = 0$, i.e. $|f(x)| = 0$ $\mathbb{P}$-a.e.

B.2.7 Push forward of a measure

Let $(\mathcal{E}, \mathbb{P})$ be a measure on Ω and let $f : \Omega \to \mathbb{R}^N$ be an $\mathcal{E}$-measurable function. Since inverse images of Borel sets are $\mathcal{E}$-measurable, we define a set function $\mathbb{P}_f : \mathcal{B}(\mathbb{R}^N) \to [0, 1]$ on $\mathbb{R}^N$, also denoted by $f_\#\mathbb{P}$, by

$$\mathbb{P}_f(A) = f_\#\mathbb{P}(A) := \mathbb{P}(f^{-1}(A)) \qquad \forall A \in \mathcal{B}(\mathbb{R}^N), \tag{B.24}$$

called the *pushforward* or *image* of the measure $\mathbb{P}$. It is easy to check the following.

Proposition B.27 *$(\mathcal{B}(\mathbb{R}^N), f_\#\mathbb{P})$ is a measure on $\mathbb{R}^N$ and for every non-negative Borel function φ on $\mathbb{R}^N$ we have*

$$\int_{\mathbb{R}^N} \varphi(t)\, f_\#\mathbb{P}(dt) = \int_\Omega \varphi(f(x))\, \mathbb{P}(dx). \tag{B.25}$$

Proof. We essentially repeat the proof of Theorem 3.9. For the reader's convenience, we outline it again.

The σ-additivity of $f_\#\mathbb{P}$ follows from the σ-additivity of $\mathbb{P}$ using the De Morgan formulas and the relations

$$f^{-1}(\cup_i A_i) = \cup_i f^{-1}(A_i), \qquad f^{-1}(\cap_i A_i) = \cap_i f^{-1}(A_i),$$

which are true for any family of subsets $\{A_i\}$ of $\mathbb{R}^N$.

In order to prove (B.25), we first consider the case in which φ is a simple function, $\varphi(t) = \sum_{i=1}^n c_i \mathbb{1}_{E_i}(t)$ where $c_1, \dots, c_n$ are distinct constants and the level sets $E_i := \{t \mid \varphi(t) = c_i\}$, $i = 1, \dots, n$, are measurable. Then

$$\varphi \circ f(x) = \sum_{i=1}^n c_i \mathbb{1}_{f^{-1}(E_i)}(x),$$
$$f^{-1}(E_i) = \{x \in \Omega \mid \varphi \circ f(x) = c_i\}$$

so that

$$\begin{aligned}\int_{\mathbb{R}^N} g(t)\, f_\#\mathbb{P}(dt) &= \sum_{i=1}^n c_i\, f_\#\mathbb{P}(E_i)\\ &= \sum_{i=1}^n c_i\, \mathbb{P}(\varphi \circ f = c_i) = \int_\Omega \varphi(f(x))\, \mathbb{P}(dx),\end{aligned}$$

i.e. (B.25) holds when φ is simple.

Let now φ be a non-negative measurable function. Proposition B.20 yields an increasing sequence $\{\varphi_k\}$ of simple functions pointwisely converging to φ.

Since for every k we have already proved that

$$\int_{\mathbb{R}^N} \varphi_k(t)\, f_{\#}\mathbb{P}(dt) = \int_{\Omega} \varphi_k(f(x))\, \mathbb{P}(dx)$$

we can pass to the limit as $k \to \infty$ and take advantage of the Beppo Levi theorem to get (B.25).

Pushforward of measures can be composed. Let $(\mathcal{E}, \mathbb{P})$ be a measure on Ω, let $f : \Omega \to \mathbb{R}^N$ be $\mathcal{E}$-measurable and let $g : \mathbb{R}^N \to \mathbb{R}^M$ be $\mathcal{B}(\mathbb{R}^N)$-measurable. Then from (B.24) we infer

$$(g \circ f)_{\#}\mathbb{P}(A) = g_{\#}(f_{\#}\mathbb{P}(A)) =: (g_{\#} \circ f_{\#})\mathbb{P}(A) \qquad \forall A \in \mathcal{B}(\mathbb{R}^M). \tag{B.26}$$

From (B.25) we infer the following relations for the associated integrals

$$\int_{\mathbb{R}^M} \varphi(t)\, (g \circ f)_{\#}\mathbb{P}(dt) = \int_{\Omega} \varphi(g(f(x)))\, \mathbb{P}(dx) = \int_{\mathbb{R}^N} \varphi(g(s))\, f_{\#}\mathbb{P}(ds) \tag{B.27}$$

for every non-negative, $\mathcal{B}(\mathbb{R}^M)$-measurable function $\varphi : \mathbb{R}^M \to \mathbb{R}$, see Theorem 4.6.

B.2.8 Exercises

Exercise B.28 *Let $\mathcal{E}$ be a σ-algebra of subsets of a set Ω and let $f, g : \Omega \to \overline{\mathbb{R}}$ be $\mathcal{E}$-measurable. Then $\{x \in \Omega \,|\, f(x) > g(x)\} \in \mathcal{E}$.*

Solution. For any rational number $r \in \mathbb{Q}$, the set $A_r := \{x \in \Omega \,|\, f(x) > r, g(x) < r\}$ belongs to $\mathcal{E}$. Moreover,

$$\Big\{x \in E \,\Big|\, f(x) > g(x)\Big\} = \bigcup_{r \in \mathbb{Q}} A_r.$$

Thus $\{x \in E \,|\, f(x) > g(x)\}$ is a denumerable union of sets in $\mathcal{E}$.

Exercise B.29 *Let $\mathcal{E}$ be a σ-algebra of subsets of a set Ω, let $E \in \mathcal{E}$ and let $f, g : \Omega \to \overline{\mathbb{R}}$ be two $\mathcal{E}$-measurable functions. Then the function*

$$h(x) := \begin{cases} f(x) & \text{if } x \in E, \\ g(x) & \text{if } x \in E^c \end{cases}$$

is $\mathcal{E}$-measurable.

Exercise B.30 *Let $\mathcal{E}$ be a σ-algebra of subsets of a set Ω, let $f : \Omega \to \mathbb{R}$ be $\mathcal{E}$-measurable and let $E \in \mathcal{E}$ be such that $\mathbb{P}(E) < \infty$. Then $\mathbb{P}(E \cap \{f = t\}) \neq 0$ for at most a denumerable set of t's.*

Exercise B.31 *Show that if φ is a simple function, then $\int_{\Omega} \varphi(x)\, \mathbb{P}(dx) = I(\varphi)$.*

Exercise B.32 (Discrete value functions) *Let $(\mathcal{E}, \mathbb{P})$ be a measure on a set Ω. Let $X : \Omega \to \mathbb{R}$ be an $\mathcal{E}$-measurable non-negative function with* discrete values, *i.e. $X(\Omega)$ is a countable set $\{a_k\}$. Give an explicit formula for $\int_\Omega X(x)\,\mathbb{P}(dx)$.*

Solution. Let $E_k := \{x \in \Omega \,|\, X(x) = a_k\}$. Then

$$X(x) = \sum_{j=1}^{\infty} a_j \mathbb{1}_{E_j}(x) \qquad \forall x \in \Omega.$$

Given x, the series has only one addendum since only one set E_j contains x.

If X has a finite range, then X is a simple function so that by definition

$$\int_\Omega X(x)\,\mathbb{P}(dx) = \sum_{j=1}^{k} a_j \mathbb{P}(X = a_j). \tag{B.28}$$

If $X(\Omega)$ is denumerable, then for any non-negative integer k we have

$$\int_\Omega \sum_{j=1}^{k} a_j \mathbb{1}_{E_j}(x)\,\mathbb{P}(dx) = \sum_{j=1}^{k} a_j \mathbb{P}(X = a_j).$$

Since X is non-negative, we can apply the Beppo Levi theorem and, as $k \to \infty$, we get

$$\begin{aligned}\int_\Omega X(x)\,\mathbb{P}(dx) &= \lim_{k\to\infty} \int_\Omega \Big(\sum_{j=1}^{k} a_j \mathbb{1}_{E_j}(x)\Big)\,\mathbb{P}(dx) \\ &= \lim_{k\to\infty} \sum_{j=1}^{k} a_j \mathbb{P}(X = a_j) = \sum_{j=1}^{\infty} a_j \mathbb{P}(X = a_j).\end{aligned} \tag{B.29}$$

Formula (B.29) can also be written as

$$\int_\Omega X(x)\,\mathbb{P}(dx) = \sum_{t\in\mathbb{R}} t\mathbb{P}(X = t) \tag{B.30}$$

since $\mathbb{P}(X = t) = 0$ if $t \notin \{a_k\}$.

Exercise B.33 *Let $(\mathcal{E}, \mathbb{P})$ be a measure on a set Ω and let $X : \Omega \to \overline{\mathbb{R}}$ be an $\mathcal{E}$-measurable non-negative function with* discrete values *and such that $+\infty \in X(\Omega)$. Give an explicit formula for $\int_\Omega X(x)\,\mathbb{P}(dx)$.*

Solution. Let $\{a_k\} \cup \{+\infty\}$ be the range of X. For $k \geq 1$, let $E_k := \{x \,|\, X(x) = a_k\}$, so that

$$\bigcup_{k=1}^{\infty} \{x \in \Omega \,|\, X = a_k\} = \{x \in \Omega \,|\, X < +\infty\}$$

and $X(x) = \sum_{j=1}^{\infty} a_j \mathbb{1}_{E_j}(x)$ if $X(x) < +\infty$. From Exercise B.32,

$$\sum_{i=1}^{\infty} a_i \mathbb{P}(X = a_i) = \int_{\{X<+\infty\}} X(x)\, \mathbb{P}(dx).$$

Moreover,

$$\begin{aligned}\int_{\{X=+\infty\}} X(x)\, \mathbb{P}(dx) &= \lim_{k\to+\infty} k\, \mathbb{P}(X = +\infty) \\ &= \begin{cases} 0 & \text{if } \mathbb{P}(X = +\infty) = 0, \\ +\infty & \text{if } \mathbb{P}(X = +\infty) > 0. \end{cases}\end{aligned}$$

Thus

$$\int_{\Omega} X(x)\, \mathbb{P}(dx) = \begin{cases} \sum_{i=1}^{\infty} a_i \mathbb{P}(X = a_i) & \text{if } \mathbb{P}(X = +\infty) = 0, \\ +\infty & \text{if } \mathbb{P}(X = +\infty) > 0. \end{cases}$$

Exercise B.34 (Integral on countable sets) *Let* $(\mathcal{E}, \mathbb{P})$ *be a measure on a finite or denumerable set* Ω. *Denote by* $p : \Omega \to \mathbb{R}$, $p(x) := \mathbb{P}(\{x\})$, *its mass density. Let* $X : \Omega \to \mathbb{R}$ *be a non-negative function. Give an explicit formula for* $\int_{\Omega} X(x)\, \mathbb{P}(dx)$.

Solution. Let $\{a_j\}$ be the range of X. By (B.29)

$$\begin{aligned}\int_{\Omega} X(x)\, \mathbb{P}(dx) &= \sum_{j=0}^{\infty} a_j \mathbb{P}(X = a_j) = \sum_{j=0}^{\infty} a_j \Big(\sum_{X(x)=a_j} p(x) \Big) \\ &= \sum_{j=0}^{\infty} \sum_{X(x)=a_j} X(x) p(x) = \sum_{x\in\Omega} X(x) p(x).\end{aligned}$$

Exercise B.35 (Dirac delta) *Let* Ω *be a set and let* $x_0 \in \Omega$. *The set function* $\delta_{x_0} : \mathcal{P}(\Omega) \to \mathbb{R}$ *such that*

$$\delta_{x_0}(A) = \begin{cases} 1 & \textit{if } x_0 \in A, \\ 0 & \textit{otherwise}, \end{cases}$$

is called the Dirac delta *[named after Paul Dirac (1902–1984)] at* x_0, *and is a probability measure on* Ω. *Prove that for any* $X : \Omega \to \mathbb{R}$,

$$\int_{\Omega} X(x)\, \delta_{x_0}(dx) = X(x_0). \tag{B.31}$$

Exercise B.36 (Sum of measures) *Let* $(\mathcal{E}, \alpha)$ *and* $(\mathcal{E}, \beta)$ *be two measures on* Ω *and let* $\lambda, \mu \in \mathbb{R}_+$.

(i) *Show that* $\lambda\alpha + \mu\beta : \mathcal{E} \to \overline{\mathbb{R}}_+$ *defined by* $(\lambda\alpha + \mu\beta)(E) := \lambda\alpha(E) + \mu\beta(E)\ \forall E \in \mathcal{E}$ *is such that* $(\mathcal{E}, \lambda\alpha + \mu\beta)$ *is a measure on* Ω.

(ii) *Show that for amy* $\mathcal{E}$*-measurable non-negative function* $f : \Omega \to \overline{\mathbb{R}}_+$

$$\int_\Omega f(x)\,(\lambda\alpha + \mu\beta)(dx) = \lambda \int_\Omega f(x)\,\alpha(dx) + \mu \int_\Omega f(x)\,\beta(dx). \quad \text{(B.32)}$$

Solution. We first consider the case when f is a non-negative simple function: $f = \sum_{i=1}^N c_i \chi_{E_i}$ where $E_i = \{x \in \Omega \mid f = c_i\}$, $c_i \geq 0$. Thus

$$\begin{aligned}\int_\Omega f(x)\,(\lambda\alpha + \mu\beta)(dx) &= \sum_{i=1}^n c_i(\lambda\alpha + \mu\beta)(f = c_i)\\ &= \lambda \sum_{i=1}^n c_i \alpha(f = c_i) + \mu \sum_{i=1}^n c_i \beta(f = c_i)\\ &= \lambda \int_\Omega f(x)\,\alpha(dx) + \mu \int_\Omega f(x)\,\beta(dx).\end{aligned}$$

When f is an $\mathcal{E}$-measurable non-negative function, we approximate it from below with an increasing sequence $\{\varphi_k\}$ of simple functions that pointwise converges to $f(x)$. Since any φ_k is simple, we have

$$\int_\Omega \varphi_k(x)\,(\lambda\alpha + \mu\beta)(dx) = \lambda \int_\Omega \varphi_k(x)\,\alpha(dx) + \mu \int_\Omega \varphi_k(x)\,\beta(dx).$$

Letting $k \to +\infty$ and taking advantage of the Beppo Levi theorem we get (B.32).

Example B.37 (Counting measure) Let Ω be a set. Given a subset $A \subset \Omega$ let $\mathcal{H}^0(A) := |A|$ be the cardinality of A. It is easy to see that $(\mathcal{P}(\Omega), \mathcal{H}^0)$ is a measure on Ω, called *counting measure*. Clearly,

$$\mathcal{H}^0(A) = \sum_{x \in A} 1,$$

where the sum on the right-hand side is $+\infty$ if A has infinite many points. The corresponding integral is

$$\int_\Omega f(x)\,\mathcal{H}^0(dx) = \sum_{x \in \Omega} f(x).$$

The formula above is obvious if f is nonzero on a finite set only and can be proven by passing to the limit and taking advantage of the Beppo Levi theorem in the general case.

Exercise B.38 (Absolutely continuous measures) *Let* $(\mathcal{B}(\mathbb{R}), \mathbb{P})$ *be an absolutely continuous measure with respect to the Lebesgue measure, i.e. assume there exists a summable function* $\rho : \mathbb{R} \to \overline{\mathbb{R}}_+$ *such that*

$$\mathbb{P}(A) = \int_A \rho(t)\,dt \qquad \forall A \in \mathcal{B}(\mathbb{R}).$$

Show that, for any non-negative $\mathcal{B}(\mathbb{R})$*-measurable function* f*,*

$$\int_{\mathbb{R}} f(x)\,\mathbb{P}(dx) = \int_{\mathbb{R}} f(x)\rho(x)\,dx. \tag{B.33}$$

Solution. Assume f is simple, i.e. $f(x) = \sum_{i=1}^n c_i \mathbb{1}_{E_i}(x)$, $E_i = \{x \mid f(x) = c_i\} \in \mathcal{E}$. Then

$$\begin{aligned}\int_{\mathbb{R}} f(x)\,\mathbb{P}(dx) &= \sum_{i=1}^n c_i \mathbb{P}(f = c_i) = \sum_{i=1}^n c_i \int_{\mathbb{R}} \mathbb{1}_{E_i}(x)\rho(x)\,dx \\ &= \int_{\mathbb{R}} \Big(\sum_{i=1}^n c_i \mathbb{1}_{E_i}(x)\Big)\rho(x)\,dx = \int_{\mathbb{R}} f(x)\rho(x)\,dx.\end{aligned}$$

The general case can be proven by an approximation argument, using Proposition B.20 and the Beppo Levi theorem.

B.3 Product measures and iterated integrals

B.3.1 Product measures

Let $(\mathcal{E}, \mathbb{P})$ and $(\mathcal{F}, \mathbb{Q})$ be measures on two sets X and Y, respectively. Denote by $\mathcal{I}$ the family of all 'rectangles' in the Cartesian product $X \times Y$

$$\mathcal{I} := \Big\{A = E \times F \,\Big|\, E \in \mathcal{E},\ F \in \mathcal{F}\Big\}$$

and let $\zeta : \mathcal{I} \to \overline{\mathbb{R}}_+$ be the set function that maps any rectangle $A \times B \in \mathcal{I}$ into $\zeta(A \times B) := \mathbb{P}(A)\mathbb{Q}(B)$. The following can be easily shown.

Proposition B.39 $\mathcal{I}$ *is a semiring and* $\zeta : \mathcal{I} \to \overline{\mathbb{R}}$ *is a* σ*-additive set function.*

Proof. It is quite trivial to show that $\mathcal{I}$ is a semiring. In fact, if $E := A \times B$ and $F := C \times D \in \mathcal{I}$, then $E \cap F = (A \cap C) \times (B \cap D)$ and

$$\begin{aligned}E \setminus F = \Big((A \setminus C) \times (B \setminus D)\Big) &\cup \Big((A \cap C) \times (B \setminus D)\Big) \\ &\cup \Big((A \setminus C) \times (B \cap D)\Big).\end{aligned}$$

Let us prove that ζ is σ-additive. Let $E \times F = \cup_k (E_k \times F_k)$, $E, E_k \in \mathcal{E}$, $F, F_k \in \mathcal{F}$ be such that the sets $\{E_k \times F_k\}$ are pairwise disjoint so that

$$\chi_E(x)\,\chi_F(y) = \sum_{k=1}^{\infty} \chi_{E_k}(x)\chi_{F_k}(y) \qquad \forall x \in X,\ y \in Y.$$

Integrating with respect to $\mathbb{Q}$ on Y and applying the Beppo Levi theorem, we obtain

$$\chi_E(x)\,\mathbb{Q}(F) = \sum_{k=1}^{\infty} \mathbb{Q}(F_k)\chi_{E_k}(x) \qquad \forall x \in X.$$

Moreover, integrating with respect to $\mathbb{P}$ on X and again by the Beppo Levi theorem we get

$$\zeta(E \times F) := \mathbb{P}(E)\mathbb{Q}(F) = \sum_{k=1}^{\infty} \mathbb{P}(E_k)\mathbb{Q}(F_k).$$

Thus, see Theorem B.10, ζ extends to a measure denoted $(\mathcal{G}, \mathbb{P} \times \mathbb{Q})$ on the smallest σ-algebra $\mathcal{G}$ containing $\mathcal{I}$. This measure is called the *product measure* of $(\mathcal{E}, \mathbb{P})$ and $(\mathcal{F}, \mathbb{Q})$. Moreover, such an extension is unique, provided $\zeta : \mathcal{I} \to \overline{\mathbb{R}}_+$ is σ-finite, see Theorem B.6. This happens in particular, if both $(\mathcal{E}, \mathbb{P})$ and $(\mathcal{F}, \mathbb{Q})$ are σ-finite.

Of course one can consider the product of finitely many measures. Taking for instance the product of n Bernoulli trials, one obtains the Bernoulli distribution on $\{0, 1\}^n$

$$Ber(n, p) = B(1, p) \times \cdots \times B(1, p).$$

B.3.1.1 Infinite Bernoulli process

Let $(\Omega, \mathcal{E}, \mathbb{P})$ be a *probability measure* on Ω. Consider the set $\Omega^\infty = \prod_{i=1}^\infty \Omega$ of Ω-valued sequences, and consider the family $\mathcal{C} \subset \mathcal{P}(\Omega^\infty)$ of sets $E \subset \Omega^\infty$ of the form

$$E = \prod_{i=1}^{\infty} E_i$$

where $E_i \in \mathcal{E}\ \forall i$ and $E_i = \Omega$ except for a finite number of indexes, i.e. the family of 'cylinders' with the terminology of Section 2.2.7. Define also $\alpha : \mathcal{C} \to [0, 1]$ by setting for $E = \prod_{i=1}^\infty E_i \in \mathcal{C}$,

$$\alpha(E) = \prod_{i=1}^{\infty} \mathbb{P}(E_i).$$

Notice that the product is actually a finite product, since $\mathbb{P}(E_i) = 1$ except for a finite number of indexes.

The following theorem holds. The interested reader may find a proof in, e.g. [7].

Theorem B.40 (Kolmogorov) *$\mathcal{C}$ is a semiring and α is σ-additive on $\mathcal{C}$.*

Therefore, Theorem B.6 and B.10 apply so that there exists a unique probability measure $(\mathcal{E}, \mathbb{P}^\infty)$ on Ω^∞ that extends α to the σ-algebra generated by $\mathcal{C}$.

The existence and uniqueness of the Bernoulli distribution of parameter p introduced in Section 2.2.7 is a particular case of the previous statement. One obtains it by choosing the Bernoulli trial distribution $B(1, p)$ on $\{0, 1\}$ as starting probability space $(\Omega, \mathcal{E}, \mathbb{P})$.

B.3.2 Reduction formulas

Let $A \subset X \times Y$. For any point $x \in A$ let A_x be the subset of Y defined as

$$A_x := \left\{y \in Y \,\middle|\, (x, y) \in A\right\}.$$

A_x is called the *section of A at x*.

Theorem B.41 (Fubini) *Let X, Y be two sets and let $(\mathcal{G}, \mathbb{P} \times \mathbb{Q})$ be the product measure on $X \times Y$ of the two σ-finite measures $(\mathcal{E}, \mathbb{P})$ and $(\mathcal{F}, \mathbb{Q})$ on X and Y, respectively. Then, for any $A \in \mathcal{G}$ the following hold:*

(i) $A_x \in \mathcal{F}$ *$\mathbb{P}$-a.e.* $x \in X$.

(ii) $x \mapsto \mathbb{Q}(A_x)$ *is an $\mathcal{E}$-measurable function.*

(iii) $(\mathbb{P} \times \mathbb{Q})(A) = \int_X \mathbb{Q}(A_x)\, \mathbb{P}(dx)$.

Changing the roles of the two variables, one also has:

(iv) $A_y \in \mathcal{E}$ *$\mathbb{Q}$-a.e.* $y \in Y$.

(v) $y \mapsto \mathbb{P}(A_y)$ *is an $\mathcal{F}$-measurable function.*

(vi) $(\mathbb{P} \times \mathbb{Q})(A) = \int_Y \mathbb{P}(A_y)\, \mathbb{Q}(dy)$.

From the Fubini theorem, Theorem B.41, one obtains the following *reduction formulas*.

Theorem B.42 (Fubini–Tonelli) *Let $(\mathcal{E}, \mathbb{P})$ and $(\mathcal{F}, \mathbb{Q})$ be two σ-finite measures on the sets X and Y, respectively, and let $(\mathcal{G}, \mathbb{P} \times \mathbb{Q})$ be the product measure on $X \times Y$. Let $f : X \times Y \to \overline{\mathbb{R}}$ be $\mathcal{G}$-measurable and non-negative (respectively, $\mathbb{P} \times \mathbb{Q}$ summable). Then the following hold:*

(i) $y \mapsto f(x, y)$ *is $\mathcal{F}$-measurable (respectively, $\mathbb{Q}$-summable) $\mathbb{P}$-a.e.* $x \in X$.

(ii) $x \to \int_Y f(x, y)\, \mathbb{Q}(dy)$ *is $\mathcal{E}$-measurable (respectively, $\mathbb{P}$-summable).*

(iii) We have

$$\int_{X\times Y} f(x,y)\,\mathbb{P}\times\mathbb{Q}(dx\,dy) = \int_X \mathbb{P}(dx)\int_Y f(x,y)\,\mathbb{Q}(dy).$$

Of course, the two variables can be interchanged, so under the same assumption of Theorem B.42 we also have:

(i) $x \mapsto f(x,y)$ is $\mathcal{E}$-measurable (respectively, $\mathbb{P}$-summable) $\mathbb{Q}$-a.e. $y \in Y$.

(ii) $y \to \int_X f(x,y)\,\mathbb{P}(dx)$ is $\mathcal{F}$-measurable (respectively, $\mathbb{Q}$-summable).

(iii) We have

$$\int_{X\times Y} f(x,y)\,\mathbb{P}\times\mathbb{Q}(dx\,dy) = \int_Y \mathbb{Q}(dy)\int_X f(x,y)\,\mathbb{P}(dx).$$

Proof. The proof is done in three steps.

(i) If f is the characteristic function of a $\mathbb{P}\times\mathbb{Q}$ measurable set, then we apply the Fubini theorem, Theorem B.41. Because of additivity, the result still holds true for any $\mathcal{G}$-measurable simple function f.

(ii) If f is non-negative, then f can be approximated from below by an increasing sequence of simple functions. Applying the Beppo Levi theorem and the continuity of measures, the result holds true for f.

(iii) If f is $\mathbb{P}\times\mathbb{Q}$ summable, then one applies the result of Step (ii) to the positive and negative parts f_+ and f_- of f.

Notice that the finiteness assumption on the two measures $(\mathcal{E}, \mathbb{P})$ and $(\mathcal{F}, \mathbb{Q})$ in Theorems B.41 and B.42 cannot be dropped as the following example shows.

Example B.43 Let $X = Y = \mathbb{R}$, $\mathbb{P} = \mathcal{L}^1$, and let $\mathbb{Q}$ be the measure that *counts the points*: $\mathbb{Q}(A) = |A|$. Let $S := \{(x,x)\,|\,x \in [0,1]\}$ and let $f(x,y) = \chi_S(x,y)$ be its characteristic function. S is closed, hence S belongs to the smallest σ-algebra generated by 'intervals', i.e. $\mathcal{B}(\mathbb{R}^2)$. Clearly $(\mathbb{P}\times\mathbb{Q})(S) = \infty$, but

$$\int_{\mathbb{R}} \mathbb{P}(dx)\int_{\mathbb{R}} f(x,y)\,\mathbb{Q}(dy) = \int_0^1 1\,dx = 1,$$

$$\int_{\mathbb{R}} \mathbb{Q}(dy)\int_{\mathbb{R}} f(x,y)\,\mathbb{P}(dx) = \int_0^1 0\,d\mathbb{Q} = 0.$$

B.3.3 Exercises

Exercise B.44 *Show that* $\mathcal{L}^n \times \mathcal{L}^k = \mathcal{L}^{n+k}$ *on the Borel sets of* $\mathbb{R}^n \times \mathbb{R}^k$.

Exercise B.45 *Let $(\mathcal{E}, \mathbb{P})$ be a measure on Ω and let $f : \Omega \to \overline{\mathbb{R}}_+$ be a $\mathbb{P}$-measurable function. Show that the subgraph of f*

$$SG_f := \left\{(x, t) \in \Omega \times \mathbb{R} \,\middle|\, 0 < t < f(x)\right\}$$

is $\mathbb{P} \times \mathcal{L}^1$-measurable and

$$\mathbb{P} \times \mathcal{L}^1(SG_f) = \int_\Omega f(x)\,\mathbb{P}(dx).$$

[*Hint*. Prove the claim for simple functions and use an approximation argument for the general case.]

B.4 Convergence theorems

B.4.1 Almost everywhere convergence

Definition B.46 *Let $(\mathcal{E}, \mathbb{P})$ be a measure on Ω and let $\{X_n\}$ and X be $\mathcal{E}$-measurable functions.*

(i) We say that $\{X_n\}$ converges in measure *to X, if for any $\delta > 0$*

$$\mathbb{P}\Big(\Big\{x \,\Big|\, |X_n - X| > \delta\Big\}\Big) \to 0 \qquad \text{as } n \to \infty.$$

(ii) We say that $\{X_n\}$ converges to X almost everywhere, *and we write $X_n \to X$ $\mathbb{P}$-a.e., if the measure of the set*

$$\begin{aligned} E :&= \left\{x \,\middle|\, X(x) = \pm\infty \text{ o } X_n(x) \not\to X(x)\right\} \\ &= \left\{x \,\middle|\, \limsup_{n\to\infty} |X_n(x) - X(x)| > 0\right\} \end{aligned}$$

is null, $\mathbb{P}(E) = 0$.

The difference between the above defined convergences becomes clear if one first considers the following sets, which can be constructed starting from a given sequence of sets $\{A_n\}$; namely, the sets

$$\limsup_n A_n := \bigcap_{m=1}^{\infty} \bigcup_{n=m}^{\infty} A_n \qquad \text{and} \qquad \liminf_n A_n := \bigcup_{m=1}^{\infty} \bigcap_{n=m}^{\infty} A_n$$

In the following proposition we collect the elementary properties of such sets.

Proposition B.47 *We have the following:*

(i) $x \in \liminf_n A_n$ if and only if there exists $\overline{n}$ such that $x \in A_n$ $\forall n \geq \overline{n}$.

(ii) $x \in \limsup_n A_n$ *if and only if there exists infinite values of n such that* $x \in A_n$.

(iii) $x \in (\limsup_n A_n)^c$ *if and only if* $\{n \mid x \in A_n\}$ *is finite.*

(iv) $(\limsup_n A_n)^c = \liminf_n A_n^c$.

(v) Let $\mathcal{E}$ *be a* σ*-algebra of subsets of* Ω*. If* $\{A_n\} \subset \mathcal{E}$*, then both* $\liminf_n A_n$ *and* $\limsup_n A_n$ *are* $\mathcal{E}$*-measurable. Moreover,*

$$\mathbb{P}(\liminf_n A_n) \le \liminf_{n\to\infty} \mathbb{P}(A_n) \le \limsup_{n\to\infty} \mathbb{P}(A_n) \le \mathbb{P}(\limsup_n A_n).$$

Proof. (i) and (ii) agree with the definitions of $\liminf_n A_n$ and $\limsup_n A_n$, respectively. (iii) is a rewrite of (ii) and (iv) is a consequence of De Moivre formulas. To prove (v) it suffices to observe that the $\mathcal{E}$-measurability of $\liminf_n A_n$ and $\limsup_n A_n$ comes from the properties of σ-algebras and that the inequality in (v) is a consequence of the continuity of measures.

Let $(\mathcal{E}, \mathbb{P})$ be a measure on Ω and let $\{X_n\}$ and X be $\mathcal{E}$-measurable functions. Given any $\delta \ge 0$, define

$$A_{n,\delta} := \Big\{x \in \Omega \,\Big|\, |X_n(x) - X(x)| > \delta\Big\},$$

$$E_\delta := \Big\{x \in \Omega \,\Big|\, \limsup_{n\to\infty} |X_n(x) - X(x)| > \delta\Big\}.$$

Since $x \in E_\delta$ if and only if there exists a sequence $\{k_n\}$ such that $|X_{k_n}(x) - X(x)| > \delta$, then

$$E_\delta = \limsup_n A_{n,\delta} \qquad \forall \delta \ge 0. \tag{B.34}$$

Proposition B.48 *Let* $(\mathcal{E}, \mathbb{P})$ *be a measure on* Ω *and let* $\{X_n\}$ *and* X *be* $\mathcal{E}$*-measurable functions. With the notation above,* $\{X_n\}$ *converges to* X *in measure if and only if* $\mathbb{P}(A_{n,\delta}) \to 0$ *as* $n \to \infty$ *for any positive* δ*. Moreover, the following are equivalent:*

(i) $X_n \to X$ $\mathbb{P}$*-a.e.*

(ii) $\mathbb{P}(\limsup_n A_{n,0}) = 0$.

(iii) $\mathbb{P}(\limsup_n A_{n,\delta}) = 0$ *for any* $\delta > 0$.

Proof. By definition, $X_n \to X$ $\mathbb{P}$-a.e. if and only if $\mathbb{P}(E_0) = 0$. For any $\delta \ge 0$, $E_\delta \subset E_0 = \cup_{\delta>0} E_\delta$, hence $\mathbb{P}(E_0) = 0$ if and only if $\mathbb{P}(E_\delta) = 0$ for any $\delta > 0$. The claim follows from (B.34).

Convergence in measure and almost everywhere convergence are not equivalent, see Example 4.76. Nevertheless, the two convergences are related, as the following proposition shows.

Proposition B.49 *Let $(\mathcal{E}, \mathbb{P})$ be a measure on Ω and let $\{X_n\}$ and X be $\mathcal{E}$-measurable functions on Ω. Then:*

(i) *If $X_n \to X$ $\mathbb{P}$-a.e., then $X_n \to X$ in measure.*

(ii) *If $X_n \to X$ in measure, then there exists a subsequence $\{X_{k_n}\}$ of $\{X_n\}$ such that $X_{k_n} \to X$ $\mathbb{P}$-a.e.*

Proof. (i) Let $\delta > 0$. For any n let $A_{n,\delta} := \{x \in \Omega \mid |X_n - X| > \delta\}$. By Proposition B.48, $\mathbb{P}(\limsup_n A_{n,\delta}) = 0$ for any $\delta > 0$. Let $m \geq 1$ and define $B_m := \cup_{n=m}^{\infty} A_{n,\delta}$. Then $A_{n,\delta} \subset B_m$ $\forall n \geq m$ hence

$$\mathbb{P}(A_{n,\delta}) \leq \mathbb{P}(B_m) \to \mathbb{P}\Big(\bigcap_{m=1}^{\infty} B_m\Big) = \mathbb{P}(\limsup_n A_{n,\delta}) = 0.$$

(ii) Let $A_{n,0} := \{x \in \Omega \mid |X_n - X| > 0\}$. We must show that there exists a sequence n_j such that

$$\mathbb{P}(\limsup_j A_{n_j,0}) = 0,$$

see Definition 4.75 and (B.34). Let $A_{n,\delta} = \{x \in \Omega \mid |X_n - X| > \delta\}$. By assumption $\mathbb{P}(A_{n,\delta}) \to 0$ for any $\delta > 0$. Let n_1 be the smallest integer such that $\mathbb{P}(A_{n_1,1}) < 1/2$, and for any $k \geq 2$, let n_{k+1} be the smallest integer greater than n_k such that $\mathbb{P}(A_{n_{k+1},1/(k+1)}) \leq 1/2^{k+1}$. Let $B_m := \cup_{j \geq m} A_{n_j,1/j}$. Since $B_m \downarrow \cap_m B_m = \limsup_j (A_{n_j,1/j})$ we obtain

$$\mathbb{P}(\limsup_j A_{n_j,1/j}) = \lim_{m\to\infty} \mathbb{P}(B_m) \leq \lim_{m\to+\infty} \sum_{j=m}^{\infty} 1/2^{j+1} = 0.$$

Since

$$\mathbb{P}(\limsup_j A_{n_j,0}) \leq \mathbb{P}(\limsup_j A_{n_j,1/j}),$$

the claim follows.

B.4.2 Strong convergence

We see here some different results related to the Beppo Levi theorem and the convergence of integrals.

The first result is about the convergence of integrals of series of non-negative functions.

Proposition B.50 (Series of non-negative functions) *Let $(\mathcal{E}, \mathbb{P})$ be a measure on Ω. Let $\{f_k\}$ be a sequence of $\mathcal{E}$-measurable non-negative functions. Then*

$$\int_{\Omega} \sum_{k=1}^{\infty} f_k(x)\, \mathbb{P}(dx) = \sum_{k=1}^{\infty} \int_{\Omega} f_k(x)\, \mathbb{P}(dx).$$

Proof. The partial sums $\{\sum_{k=1}^{n} f_k(x)\}$ are a nondecreasing sequence of $\mathcal{E}$-measurable non-negative functions. Applying the Beppo Levi theorem to this sequence yields the result.

B.4.3 Fatou lemma

In the following lemma, the monotonicity assumption in the Beppo Levi theorem is removed.

Lemma B.51 (Fatou) *Let $(\mathcal{E}, \mathbb{P})$ be a measure on Ω and let $\{f_k\}$ be a sequence of $\mathcal{E}$-measurable non-negative functions. Then*

$$\int_\Omega \liminf_{k\to\infty} f_k(x)\,\mathbb{P}(dx) \le \liminf_{k\to\infty} \int_\Omega f_k(x)\,\mathbb{P}(dx).$$

Proof. Let $g_n(x) := \inf_{k\ge n} f_k(x)$. $\{g_n(x)\}$ is an increasing sequence of $\mathcal{E}$-measurable non-negative functions. Moreover,

$$0 \le g_n(x) \le f_k(x),\ \forall k \ge n, \qquad \liminf_{\kappa\to\infty} f_k(x) = \lim_{n\to\infty} g_n(x).$$

Thus $\int_\Omega g_n(x)\,\mathbb{P}(dx) \le \inf_{k\ge n} \int_\Omega f_k(x)\,\mathbb{P}(dx)$ and, applying the Beppo Levi theorem, we get

$$\begin{aligned}\int_\Omega \liminf_{k\to\infty} f_k(x)\,\mathbb{P}(dx) &= \lim_{n\to\infty} \int_\Omega g_n(x)\,\mathbb{P}(dx) \le \lim_{n\to\infty} \inf_{k\ge n} \int_\Omega f_k(x)\,\mathbb{P}(dx)\\ &= \liminf_{k\to\infty} \int_\Omega f_k(x)\,\mathbb{P}(dx).\end{aligned}$$

Remark B.52 The Fatou lemma implies the Beppo Levi theorem. In fact, let $\{f_n\}$ be an increasing sequence of functions that converges to $f(x)$. Then $f(x) = \lim_{k\to\infty} f_k(x) = \liminf_{k\to\infty} f_k(x)$. Since the sequence $\{f_k\}$ is monotone, we get

$$\liminf_{k\to\infty} \int_\Omega f_k(x)\,\mathbb{P}(dx) = \lim_{k\to\infty} \int_\Omega f_k(x)\,\mathbb{P}(dx) \le \int_\Omega f(x)\,\mathbb{P}(dx),$$

and, by the Fatou lemma, we get the opposite inequality:

$$\int_\Omega f(x)\,\mathbb{P}(dx) = \int_\Omega \liminf_{k\to\infty} f_k(x)\,\mathbb{P}(dx) \le \liminf_{k\to\infty} \int_\Omega f_k(x)\,\mathbb{P}(dx).$$

Corollary B.53 (Fatou lemma) *Let $(\mathcal{E}, \mathbb{P})$ be a measure on Ω. Let $\{f_k\}$ be a sequence of $\mathcal{E}$-measurable functions and let $\phi : \Omega \to \overline{\mathbb{R}}$ be a $\mathbb{P}$-summable function.*

(i) If $f_k(x) \ge \phi(x)$ $\forall k$ and $\mathbb{P}$-a.e. $x \in \Omega$, then

$$\int_\Omega \liminf_{k\to\infty} f_k(x)\,\mathbb{P}(dx) \le \liminf_{k\to\infty} \int_\Omega f_k(x)\,\mathbb{P}(dx).$$

(ii) If $f_k(x) \le \phi(x)$ $\forall k$ *and* $\mathbb{P}$*-a.e., then*

$$\limsup_{k\to\infty} \int_\Omega f_k(x)\,\mathbb{P}(dx) \le \int_\Omega \limsup_{k\to\infty} f_k(x)\,\mathbb{P}(dx).$$

Proof. Let $E_k := \{x \in \Omega \mid f_k(x) \ge \phi(x)\}$ and let $E := \cap_k E_k$. Since $\mathbb{P}(E^c) = 0$, we can assume without loss of generality that $f_k(x) \ge \phi(x)$ $\forall k$ and $\forall x \in \Omega$. To prove (i) it suffices to apply the Fatou lemma, Lemma B.51, to the sequence $\{f_k - \phi\}$. (ii) is proven similarly.

B.4.4 Dominated convergence theorem

Theorem B.54 (Lebesgue dominated convergence) *Let* $(\mathcal{E}, \mathbb{P})$ *be a measure on* Ω *and let* $\{f_k\}$ *be a sequence of* $\mathcal{E}$*-measurable functions. Assume:*

(i) $f_k(x) \to f(x)$ $\mathbb{P}$*-a.e.* $x \in \Omega$.

(ii) There exists a $\mathbb{P}$*-summable function* ϕ *such that* $|f_k(x)| \le \phi(x)$ $\forall k$ *and for* $\mathbb{P}$*-a.e.* x.

Then

$$\int_\Omega |f_k(x) - f(x)|\,\mathbb{P}(dx) \to 0$$

and, in particular,

$$\int_\Omega f_k(x)\,\mathbb{P}(dx) \to \int_\Omega f(x)\,\mathbb{P}(dx).$$

Proof. By assumption $|f_k(x) - f(x)| \le 2\phi(x)$ for $\mathbb{P}$-a.e. x and for any k. Moreover, $|f_k(x) - f(x)| \to 0$ $\forall k$ and for $\mathbb{P}$-a.e. x. Thus, by the Fatou lemma, Corollary B.53, we get

$$\limsup_{k\to\infty} \int_\Omega |f_k(x) - f(x)|\,\mathbb{P}(dx) \le \int_\Omega \limsup_{k\to\infty} |f_k(x) - f(x)|\,dx$$
$$= \int_\Omega 0\,\mathbb{P}(dx) = 0.$$

The last claim is proven by the following inequality:

$$\left|\int_\Omega f_k(x)\,\mathbb{P}(dx) - \int_\Omega f(x)\,\mathbb{P}(dx)\right| = \left|\int_\Omega (f_k(x) - f(x))\,\mathbb{P}(dx)\right|$$
$$\le \int_\Omega |f_k(x) - f(x)|\,\mathbb{P}(dx).$$

Remark B.55 Notice that in Theorem B.54:

- Assumption (ii) is equivalent to the $\mathbb{P}$-summability of the *envelope* $\phi(x) := \sup_k |f_k(x)|$ of the functions $|f_k|$.
- Assumption (ii) cannot be dropped as the following sequence $\{f_k\}$ shows:

$$f_k(x) = \begin{cases} k & \text{if } 0 < x < 1/k, \\ 0 & \text{otherwise.} \end{cases}$$

Example B.56 The dominated convergence theorem extends to arbitrary measures a classical dominated convergence theorem for series.

Theorem (Dominated convergence for series) *Let $\{a_{j,n}\}$ be a double sequence such that:*

(i) For any j, $a_{j,n} \to a_j$ as $n \to \infty$.

(ii) There exists a non-negative sequence $\{c_j\}$ such that $|a_{j,n}| \le c_j$ for any n and any j and $\sum_{j=1}^{\infty} |c_j| < \infty$.

Then the series $\sum_{j=1}^{\infty} a_j$ is absolutely convergent and

$$\sum_{j=1}^{\infty} a_{j,n} \to \sum_{j=1}^{\infty} a_j \qquad \text{as } n \to \infty.$$

Proof. Consider the counting measure $(\mathcal{P}(\mathbb{N}), \mathcal{H}^0)$ on $\Omega = \mathbb{N}$ and apply the Lebesgue dominated convergence theorem to the sequence $\{f_n\}$ defined by $f_n(j) = a_{j,n}$.

For the reader's convenience, we include a direct proof. Since $a_{j,n} \to a_j$ and $|a_{n,j}| \le c_j$ $\forall n, j$, we get $|a_j| \le c_j$ $\forall j$ so that $\sum_{j=0}^{\infty} a_j$ is absolutely convergent.

Let $\epsilon > 0$. Choose $p = p(\epsilon)$ such that $2\sum_{j=p+1}^{\infty} c_j < \epsilon$. Then

$$\left|\sum_{j=0}^{\infty} a_{j,n} - \sum_{j=0}^{\infty} a_j\right| \le \sum_{j=0}^{\infty} |a_{j,n} - a_j| = \sum_{j=0}^{p} |a_{j,n} - a_j| + \sum_{j=p+1}^{\infty} |a_{j,n} - a_j|$$

$$\le \sum_{j=0}^{p} |a_{j,n} - a_j| + 2\sum_{j=p+1}^{\infty} c_j \le \sum_{j=0}^{p} |a_{j,n} - a_j| + \epsilon.$$

Thus, as $n \to \infty$, we obtain

$$\limsup_{n\to\infty} \left|\sum_{j=0}^{\infty} a_{j,n} - \sum_{j=0}^{\infty} a_j\right| \le \epsilon.$$

Since ϵ is arbitrary, the claim follows.

The next theorem is an important consequence on the convergence of integrals of series of functions.

Theorem B.57 (Lebesgue) *Let $(\mathcal{E}, \mathbb{P})$ be a measure on Ω and let $\{f_k\}$ be a sequence of $\mathcal{E}$-measurable functions such that*

$$\sum_{k=0}^{\infty} \int_{\Omega} |f_k(x)| \, \mathbb{P}(dx) < +\infty.$$

Then for $\mathbb{P}$-a.e. x the series $\sum_{k=0}^{\infty} f_k(x)$ is absolutely convergent to a $\mathbb{P}$-summable function $f(x)$. Moreover,

$$\int_{\Omega} \Big| f(x) - \sum_{k=0}^{p} f_k(x) \Big| \, \mathbb{P}(dx) \to 0 \qquad \text{as } p \to \infty, \tag{B.35}$$

and

$$\int_{\Omega} f(x) \, \mathbb{P}(dx) = \sum_{k=0}^{\infty} \int_{\Omega} f_k(x) \, \mathbb{P}(dx).$$

Notice that the assumptions are on the integrals only, while the claim is about the $\mathbb{P}$-a.e. convergence of the series $\sum_{k=0}^{\infty} |f_k(x)|$.

Proof. For any $x \in \Omega$ let $g(x) \in \overline{\mathbb{R}}_+$ be the sum of the non-negative addenda series $\sum_{k=0}^{\infty} |f_k(x)|$. Applying the Beppo Levi theorem, the assumption gives

$$\int_{\Omega} g(x) \, \mathbb{P}(dx) = \sum_{k=0}^{\infty} \int_{\Omega} |f_k(x)| \, \mathbb{P}(dx) < +\infty,$$

i.e. g is $\mathbb{P}$-summable. Thus, by Proposition B.26, $g(x) < +\infty$ for $\mathbb{P}$-a.e. x, i.e. the series $\sum_{k=0}^{\infty} f_k(x)$ absolutely converges to $f(x) := \sum_{k=0}^{\infty} f_k(x)$ and, for any integer $p \geq 1$ we have

$$\Big| \sum_{k=p}^{\infty} f_k(x) \Big| \leq \sum_{k=p}^{\infty} |f_k(x)|. \tag{B.36}$$

In particular,

$$|f(x)| \leq \sum_{k=0}^{\infty} |f_k(x)| = g(x) \qquad \mathbb{P}\text{-a.e. } x \in \Omega,$$

so that f is summable. Integrating (B.36) we get

$$\int_{\Omega} \Big| f(x) - \sum_{k=0}^{p-1} f_k(x) \Big| \, \mathbb{P}(dx) = \int_{\Omega} \Big| \sum_{k=p}^{\infty} f_k(x) \Big| \, \mathbb{P}(dx) \leq \int_{\Omega} \sum_{k=p}^{\infty} |f_k(x)| \, \mathbb{P}(dx)$$
$$= \sum_{k=p}^{\infty} \int_{\Omega} |f_k(x)| \, \mathbb{P}(dx).$$

As $p \to \infty$ we get the first part of the claim. The second part of the claim easily follows since

$$\left| \int_\Omega f(x)\,\mathbb{P}(dx) - \sum_{k=0}^{p-1} \int_\Omega f_k(x)\,\mathbb{P}(dx) \right|$$

$$\leq \int_\Omega \left| f(x) - \sum_{k=0}^{p-1} f_k(x) \right| \mathbb{P}(dx) \to 0 \qquad \text{as } p \to \infty.$$

B.4.5 Absolute continuity of integrals

Theorem B.58 (Absolute continuity of integrals) *Let $(\mathcal{E}, \mathbb{P})$ be a measure on Ω and let f be a $\mathbb{P}$-summable function. For any $\epsilon > 0$ there exists $\delta > 0$ such that $\int_E |f|\,\mathbb{P}(dx) < \epsilon$ for any $E \in \mathcal{E}$ such that $\mathbb{P}(E) < \delta$. Equivalently,*

$$\int_E f(x)\,\mathbb{P}(dx) \to 0 \qquad \textit{as } \mathbb{P}(E) \to 0.$$

Proof. Let

$$f_k(x) = \begin{cases} k & \text{if } f(x) > k, \\ f(x) & \text{if } -k \leq f(x) \leq k, \\ -k & \text{if } f(x) < -k. \end{cases}$$

Then $|f_k(x) - f(x)| \to 0$ for $\mathbb{P}$-a.e. x and $|f_k(x) - f(x)| \leq 2|f(x)|$, Since $|f|$ is $\mathbb{P}$-summable, the dominated convergence theorem, Theorem B.54, applies for any $\epsilon > 0$ there exists $N = N_\epsilon$ such that

$$\int_\Omega |f(x) - f_N(x)|\,\mathbb{P}(dx) < \epsilon/2.$$

Let $\delta := \epsilon/(2N)$. Then for any $E \in \mathcal{E}$ such that $\mathbb{P}(E) \leq \delta$ we get

$$\int_E |f_N(x)|\,\mathbb{P}(dx) \leq N\,\mathbb{P}(E) \leq N\,\frac{\epsilon}{2N} = \epsilon/2$$

so that

$$\int_E |f(x)|\,\mathbb{P}(dx) \leq \int_E |f_N(x)|\,\mathbb{P}(dx) + \int_\Omega |f - f_N|\,\mathbb{P}(dx) \leq \frac{\epsilon}{2} + \frac{\epsilon}{2} = \epsilon.$$

B.4.6 Differentiation of the integral

Let $f, g : \mathbb{R}^n \to \overline{\mathbb{R}}$ be Lebesgue-summable non-negative functions. Clearly, $\int_A f(x)\,dx = \int_A g(x)\,dx \quad \forall A \subset \mathbb{R}^n$ if and only if $f(x) = g(x)$ almost everywhere. Thus one would like to characterize $f(x)$ in terms of its integral, i.e. of the map $A \to \int_A f(x)\,dx$. *Differentiation theory* provides such a characterization. Obviously, if f is continuous, then the mean value theorem gives

$$\lim_{r \to 0} \frac{1}{|B(x,r)|} \int_{B(x,r)} f(y)\,dy = f(x) \quad \forall x.$$

More generally, the following theorem holds.

Theorem B.59 (Lebesgue) *Let $E \subset \mathbb{R}^n$ be a $\mathcal{B}(\mathbb{R}^n)$-measurable set and let $f : E \to \mathbb{R}$ be $\mathcal{B}(\mathbb{R}^n)$-measurable such that $\int_E |f|^p\,dx < +\infty$ for some $1 \le p < +\infty$. Then for almost every $x \in E$,*

$$\frac{1}{|B(x,r)|} \int_{E \cap B(x,r)} |f(y) - f(x)|^p\,dy \to 0 \quad \text{as } r \to 0^+.$$

In particular, for almost every $x \in E$, the limit

$$\lim_{r \to 0^+} \frac{1}{|B(x,r)|} \int_{E \cap B(x,r)} f(y)\,dy$$

exists, is finite and equal to $f(x)$.

Example B.60 Let f be $\mathcal{L}^1$-summable on $]-1, 1[$. Show that

$$\lim_{r \to 0^+} \frac{1}{2r} \int_{x-r}^{x+r} f(y)\,dy = f(x) \qquad \text{a.e. } x \in]-1, 1[.$$

Definition B.61 *Let $f : E \subset \mathbb{R}^n \to \mathbb{R}$ be $\mathcal{L}^n$-summable on E. The points of the set*

$$\mathcal{L}_f := \left\{ x \in E \,\middle|\, \exists \lambda \in \mathbb{R} \text{ such that } \frac{1}{|B(x,r)|} \int_{E \cap B(x,r)} |f(y) - \lambda|\,dy \to 0 \right\}$$

are called Lebesgue points *of f. For any $x \in \mathcal{L}_f$ the limit*

$$\lambda_f(x) := \lim_{r \to 0^+} \frac{1}{|B(x,r)|} \int_{B(x,r)} f(y)\,dy$$

exists and is finite thus it defines a function $\lambda_f : \mathcal{L}_f \to \mathbb{R}$ called the Lebesgue representative *of f.*

From the Lebesgue differentiation theorem, Theorem B.59, we get

Theorem B.62 *Let f be a $\mathcal{L}^n$-summable function on $\mathbb{R}^n$. Then $\mathcal{L}^n(\mathbb{R}^n \setminus \mathcal{L}_f) = 0$ and $f = \lambda_f$ $\mathcal{L}^n$-a.e.*

The differentiation theorem can be extended to more general sets than balls centred at x. One may use cubes centred at x, cubes containing x or even different objects. For example, let A be a bounded Borel set such that $\mathcal{L}^n(A) > 0$. Assume, e.g.

$$A \subset B(0, 100) \subset \mathbb{R}^n, \qquad |A| = c\,|B_1|.$$

For any $x \in \mathbb{R}^n$ and any $r > 0$, let $A_{x,r} := x + r\,A$. Obviously, $A_{x,r} \subset B(x, 100\,r)$ and $|A_{x,r}| = r^n\,|A| = c\,r^n\,|B_1| = c\,|B(x,r)|$. Theorem B.59 implies the following.

Theorem B.63 *Let $E \subset \mathbb{R}^n$ be a Borel-measurable set and let $f : E \to \mathbb{R}$ be $\mathcal{B}(\mathbb{R}^n)$-measurable with $\int_E |f|^p\,dx < +\infty$ for some $1 \leq p < \infty$. Then for $\mathcal{L}^n$-a.e. $x \in E$*

$$\frac{1}{|A_{x,r}|}\int_{E\cap A_{x,r}} |f(y) - f(x)|^p\,dy \to 0 \qquad \text{as } r \to 0^+.$$

We now collect some results due to Giuseppe Vitali (1875–1932) on the differentiation of integrals and of monotone functions.

Theorem B.64 (Vitali) *Let $h : \mathbb{R} \to \mathbb{R}$ be monotone nondecreasing. Then h is differentiable at $\mathcal{L}^1$-a.e. $x \in \mathbb{R}$ and the derivative $h'(x)$ is non-negative at $\mathcal{L}^1$-a.e. $x \in \mathbb{R}$. Moreover, h' is $\mathcal{L}^1$-summable on any bounded interval $]a, b[\in \mathbb{R}$ and*

$$\int_x^y h'(t)\,dt \leq h(y) - h(x) \qquad \forall x < y. \tag{B.37}$$

Remark B.65 Equality may not hold in (B.37). Take, e.g. $h(x) := \operatorname{sgn}(x)$ so that $h'(x) = 0\ \forall x \neq 0$, and, of course, $0 = \int_{-1}^1 h'(t)\,dt < h(1) - h(-1) = 2$. Although surprising, one may construct examples of continuous and strictly increasing functions whose derivative is zero almost everywhere: one somewhat simpler example of a continuous, nonconstant and nondecreasing function with zero derivative almost everywhere is the famous *Cantor–Vitali function*. Obviously, for such functions the inequality in (B.37) may be strict.

Definition B.66 *A function $f : \mathbb{R} \to \mathbb{R}$ is said to be* absolutely continuous *if for any $\epsilon > 0$ there exists $\delta > 0$ such that for any pair of sequences $\{x_k\}$, $\{y_k\}$ such that $\sum_{k=1}^\infty |x_k - y_k| < \delta$ we have $\sum_{k=1}^\infty |f(x_k) - f(y_k)| < \epsilon$.*

Let $f \in L^1([a, b])$, Theorem B.58 implies that the integral function

$$F(x) := \int_a^x f(t)\,dt, \qquad x \in [a, b],$$

is absolutely continuous. The next theorem shows that integral functions are the only absolutely continuous functions.

Theorem B.67 (Vitali) *A function $h : [a, b] \to \mathbb{R}$ is absolutely continuous if and only if there exists a $\mathcal{L}^1$-summable function f on $[a, b]$ such that*

$$h(y) - h(x) = \int_x^y f(s)\,ds \qquad \forall x, y \in [a, b],\ x < y.$$

Moreover, h is differentiable at almost every $x \in [a, b]$ and $h'(x) = f(x)$ for a.e. $x \in [a, b]$.

Lipschitz continuous functions $f : \mathbb{R} \to \mathbb{R}$ are absolutely continuous; thus by Theorem B.67 they are differentiable $\mathcal{L}^1$-a.e., the derivative $f'(x)$ is $\mathcal{L}^1$-summable and

$$f(y) - f(x) = \int_x^y f'(s)\,ds \qquad \forall x, y \in \mathbb{R},\ x < y.$$

Moreover, the following holds in $\mathbb{R}^n$.

Theorem B.68 (Rademacher) *Let $f : \mathbb{R}^n \to \mathbb{R}$ be Lipschitz continuous. Then f is differentiable at $\mathcal{L}^n$-almost every $x \in \mathbb{R}^n$, the map $x \to \mathbf{D}f(x)$ is $\mathcal{B}(\mathbb{R}^n)$-measurable and $|\mathbf{D}f(x)| \leq Lip(f)$ $\mathcal{L}^n$-a.e. $x \in \mathbb{R}^n$.*

B.4.7 Weak convergence of measures

In this section we consider *Borel measures* on $\mathbb{R}$, that is measures $(\mathcal{B}(\mathbb{R}), \mu)$ on $\mathbb{R}$. Since the σ algebra is understood, we simply write μ to denote the measure $(\mathcal{B}(\mathbb{R}), \mu)$.

Recall that the *law* of a finite Borel measure μ on $\mathbb{R}$ is the function $F : \mathbb{R} \to \mathbb{R}$, $F(t) := \mu(]-\infty, t])$. We recall that F is monotone nondecreasing, in particular, is right-continuous on $\mathbb{R}$ and the set of its discontinuity points is at most denumerable. Moreover, $F(t) \to F(-\infty)$ as $t \to -\infty$ and $F(t) \to F(+\infty)$ as $t \to +\infty$ and the measure μ is completely determined by F.

Definition B.69 *Let $\{\mu_n\}$, μ be finite Borel measures on $\mathbb{R}$. We say that $\{\mu_n\}$* weakly *converges to μ, and we write $\mu_n \rightharpoonup \mu$, if for any continuous bounded function $\varphi : \mathbb{R} \to \mathbb{R}$ one has*

$$\int_{\mathbb{R}} \varphi(t)\,\mu_n(dt) \to \int_{\mathbb{R}} \varphi(t)\,\mu(dt).$$

Proposition B.70 *If the weak limit of a sequence of measures exists, then it is unique.*

Proof. Assume $\mu_n \rightharpoonup \mu_1$ and $\mu_n \rightharpoonup \mu_2$, and let $\mathbb{P} := \mu_1 - \mu_2$. Then, for every continuous bounded function $\varphi : \mathbb{R} \to \mathbb{R}$

$$\int_{\mathbb{R}} \varphi(t)\,\mathbb{P}(dt) = 0.$$

The characteristic function of $]-\infty, a]$ can be approximated from below by an increasing sequence of continuous non-negative functions, thus obtaining $\mathbb{P}([-\infty, a]) = 0\ \forall \in \mathbb{R}$, hence $\mathbb{P}(A) = 0\ \forall A \in \mathcal{B}(\mathbb{R})$.

Theorem B.71 *Let $\{\mu_n\}$, μ be finite Borel measures on $\mathbb{R}$. Assume $\mu_n(\mathbb{R}) = \mu(\mathbb{R})$ $\forall n$ and let F_n and F be their laws, respectively. The following are equivalent:*

(i) If F is continous at t, then $F_n(t) \to F(t)$.

(ii) μ_n weakly converges to μ.

Proof. (i) $\Rightarrow$ (ii) Without any loss of generality, we can assume that for any $n \in \mathbb{N}$, $F_n(-\infty) = F(-\infty) = 0$ and $F_n(+\infty) = \mu_n(\mathbb{R}) = \mu(\mathbb{R}) = F(+\infty) = 1$. Fix $\delta > 0$ and let $a, b \in \mathbb{R}$, $a < b$, such that F is continuous at a and b, and such that $F(a) \le \delta$ and $1 - F(b) \le \delta$. $F_n(a)$ and $F_n(b)$ converge to $F(a)$ and $F(b)$, respectively, hence for large enough n's, $F_n(a) \le 2\delta$, $1 - F_n(b) \le 2\delta$.

Let $\varphi : \mathbb{R} \to \mathbb{R}$ be a bounded continuous function, $|\varphi| \le M$. Since φ is uniformly continuous in $[a, b]$, there exists $N = N_\delta$ and intervals $I_j = [a_j, a_{j+1}]$, $j = 1, \dots, N_\delta$ where $a = a_1 < a_2 < \cdots < a_{N+1} = b$ such that the oscillation of φ on every I_j is less than δ, $\max_{I_j} \varphi - \min_{I_j} \varphi \le \delta \forall j$. Morever, perturbating the extrema a_j if necessary, we can assume that all the points a_j are continuity points for F. Let $h(x) := \sum_{j=1}^N \varphi(a_j) \mathbb{1}_{I_j}(x)$. h is a simple function $h_{|I_j} = \varphi(a_j)$ and $h = 0$ in $\mathbb{R} \setminus [a, b]$. Moreover, $|\varphi(x) - h(x)| \le \delta$ on $]a, b]$. Since φ is bounded and $\mu_n(\mathbb{R}) = 1$,

$$
\begin{aligned}
\left| \int_{\mathbb{R}} (\varphi(x) - h(x))\, \mu_n(dx) \right| \\
&\le \int_{]-\infty,a]} |\varphi(x)|\, \mu_n(dx) + \int_{]b,+\infty[} |\varphi(x)|\, dx \\
&\quad + \int_{]a,b]} |\varphi(x) - h(x)|\, \mu_n(dx) \\
&\le 4M\delta + \delta = (4M + 1)\delta
\end{aligned}
$$

and, similarly,

$$
\left| \int_{\mathbb{R}} (\varphi(x) - h(x))\, \mu_n(dx) \right| \le (2M + 1)\delta,
$$

hence

$$
\begin{aligned}
&\left| \int_{\mathbb{R}} \varphi(x)\, \mu_n(dx) - \int_{\mathbb{R}} \varphi(x)\, \mu(dx) \right| \\
&\quad \le 2(3M + 1)\delta + \left| \int_{]a,b]} h(x)(\mu_n - \mu)(dx) \right| \\
&\quad \le 2(3M + 1)\delta + \sum_{j=1}^{N-1} |\varphi(a_j)| \Big(|F_n(a_{j+1}) - F(a_{j+1})| + |F_n(a_j) - F(a_j)| \Big).
\end{aligned}
$$

Since $F_n(a_j) \to F(a_j)$ for any $j = 1, \dots, N$, we get as $n \to \infty$

$$
\limsup_{n\to\infty} \left| \int_{\mathbb{R}} \varphi(x)\, \mu_n(dx) - \int_{\mathbb{R}} \varphi(x)\, \mu(dx) \right| \le 2(3M + 1)\delta.
$$

Since $\delta > 0$ is arbitrary, the claim is proven.

(ii) $\Rightarrow$ (i) Let $a \in \mathbb{R}$. It suffices to show that

$$F(a^-) \leq \liminf_{n\to\infty} F_n(a^-) \qquad \text{and} \qquad \limsup_{n\to\infty} F_n(a) \leq F(a). \tag{B.38}$$

In fact, from (B.38) one gets

$$F(a^-) \leq \liminf_{n\to\infty} F_n(a^-) \leq \liminf_{n\to\infty} F_n(a) \leq \limsup_{n\to\infty} F_n(a) \leq F(a),$$

i.e. $F(a) = \lim_{n\to\infty} F_n(a)$ if F is continuous at a.

For any $\delta > 0$, let $\varphi(t)$ be the piecewise linear function that is null for any $t \geq a$ and is identically 1 for $t \leq a - \delta$. Then

$$\begin{aligned} F(a-\delta) &= \int_{\mathbb{R}} \mathbb{1}_{]-\infty,a-\delta]}(t)\,\mu(dt) \\ &\leq \int_{\mathbb{R}} \varphi(t)\,\mu(dt) = \lim_{n\to\infty}\int_{\mathbb{R}} \varphi(t)\,\mu_n(dt) = \liminf_{n\to\infty}\int_{\mathbb{R}} \varphi(t)\,\mu_n(dt) \\ &\leq \liminf_{n\to\infty}\int_{\mathbb{R}} \mathbb{1}_{]-\infty,a[}(t)\,\mu_n(dt) \leq \liminf_{n\to\infty} F_n(a^-). \end{aligned}$$

As $\delta \to 0^+$, $F(a-\delta) \to F(a^-)$, so the first inequality in (B.38) is proven. Similarly, let $\varphi(t)$ be the piecewise linear function that is null for $t \geq a + \delta$ and is identically 1 for $t \leq a$. Then

$$\begin{aligned} F(a+\delta) &= \int_{\mathbb{R}} \mathbb{1}_{]-\infty,a+\delta]}(t)\,\mu(dt) \\ &\geq \int_{\mathbb{R}} \varphi(t)\,\mu(dt) = \lim_{n\to\infty}\int_{\mathbb{R}} \varphi(t)\,\mu_n(dt) = \limsup_{n\to\infty}\int_{\mathbb{R}} \varphi(t)\,\mu_n(dt) \\ &\geq \limsup_{n\to\infty}\int_{\mathbb{R}} \mathbb{1}_{]-\infty,a]}(t)\,\mu_n(dt) \geq \limsup_{n\to\infty} F_n(a). \end{aligned}$$

As $\delta \to 0^+$, $F(a+\delta) \to F(a)$ so that the second inequality in (B.38) holds. The proof of (B.38) is then complete.

Let $(\mathcal{E}, \mathbb{P})$ be a probability measure on a set Ω. We recall that the *law* F_X associated with an $\mathcal{E}$-measurable function $X : \Omega \to \mathbb{R}$ is the law of the image measure of $\mathbb{P}$ through X, i.e.

$$F_X(t) := \mathbb{P}\Big(\Big\{x \in \Omega \,\Big|\, X(x) \leq t\Big\}\Big).$$

Definition B.72 *Let $(\mathcal{E}, \mathbb{P})$ be a finite measure on a set Ω and let $\{X_n\}$, X be $\mathcal{E}$-measurable functions. We say that $\{X_n\}$* converges in law *to X if $F_{X_n}(t)$ converges to $F_X(t)$ pointwisely in all points of continuity of F_X.*

Proposition B.73 *Let $(\mathcal{E}, \mathbb{P})$ be a finite measure on a set Ω and let $\{X_n\}$, X be $\mathcal{E}$-measurable functions. If $X_n \to X$ in measure then X_n converges in law to X.*

Proof. If suffices to prove that

$$F_X(a^-) \le \liminf_{n\to\infty} F_{X_n}(a) \qquad \text{and} \qquad \limsup_{n\to\infty} F_{X_n}(a) \le F_X(a). \tag{B.39}$$

In fact, from (B.39) one gets

$$F_X(a^-) \le \liminf_{n\to\infty} F_{X_n}(a) \le \limsup_{n\to\infty} F_{X_n}(a) \le F_X(a),$$

hence $F_X(a) = \lim_{n\to\infty} F_{X_n}(a)$ if F_X is continuous at a.

Let us prove (B.39). Let $\delta > 0$. Since

$$\begin{aligned}\Big\{x \in \Omega \,\Big|\, X(x) \le a - \delta\Big\} &= \Big\{x \in \Omega \,\Big|\, X(x) \le a - \delta, X_n(x) \le a\Big\} \\ &\qquad \bigcup \Big\{x \in \Omega \,\Big|\, X(x) \le a - \delta,\ X_n(x) > a\Big\} \\ &\subset \Big\{x \in \Omega \,\Big|\, X_n(x) \le a\Big\} \bigcup \Big\{x \in \Omega \,\Big|\, |X_n - X| > \delta\Big\},\end{aligned}$$

we have

$$F_X(a - \delta) \le F_{X_n}(a) + \mathbb{P}(|X - X_n| > \delta).$$

Thus, passing to the limit with respect to n, we obtain

$$F_X(a - \delta) \le \liminf_{n\to\infty} F_{X_n}(a).$$

If we now let $\delta \to 0^+$, we get the first inequality (B.39). Similarly,

$$\begin{aligned}\Big\{x \in \Omega \,\Big|\, X_n(x) \le a\Big\} &= \Big\{x \in \Omega \,\Big|\, X_n(x) \le a,\ X(x) \le a + \delta\Big\} \\ &\qquad \bigcup \Big\{x \in \Omega \,\Big|\, X_n(x) \le a, X(x) > a + \delta\Big\} \\ &\subset \Big\{x \in \Omega \,\Big|\, X(x) \le a + \delta\Big\} \bigcup \Big\{x \in \Omega \,\Big|\, |X_n - X| > \delta\Big\}\end{aligned}$$

so that

$$F_{X_n}(a) \le F_X(a + \delta) + \mathbb{P}(|X - X_n| > \delta).$$

As $n \to \infty$ we get

$$\limsup_{n\to\infty} F_{X_n}(a) \le F_X(a + \delta),$$

so that, by letting $\delta \to 0^+$ we obtain the second inequality of (B.39).

B.4.8 Exercises

Exercise B.74 *Let $f : \mathbb{R}^n \to \mathbb{R}$ be $\mathcal{L}^n$-summable. Prove that the function $F : \mathbb{R}^n \times [0, \infty[\to \mathbb{R}$ defined by*

$$F(x, r) := \int_{B(x,r)} f(t)\, d\mathcal{L}^n(t)$$

is continuous.

Exercise B.75 *Let $f \in L^1(\mathbb{R})$. Prove that the following equalities hold for a.e. $x \in \mathbb{R}$:*

$$\lim_{r \to 0^+} \frac{1}{r} \int_0^r f(x+t)\, dt = f(x);$$

$$\lim_{r \to 0^+} \frac{1}{r} \int_{-r}^0 f(x+t)\, dt = f(x);$$

$$\lim_{r \to 0^+} \frac{1}{r} \int_r^{2r} f(x+t)\, dt = f(x).$$

Exercise B.76 *Let $\varphi : [a, b] \to [c, d]$ be continuous and piecewise differentiable. Let $h : [c, d] \to \mathbb{R}$ be absolutely continuous. Prove that $f \circ \varphi : [a, b] \to \mathbb{R}$ is absolutely continuous.*

Exercise B.77 *Let $\varphi : [a, b] \to [c, d]$ be continuous. Then φ is absolutely continuous if and only if the graph of φ has finite length.*

Exercise B.78 *Let $(\mathcal{E}, \mathbb{P})$ be a measure on a set Ω. Let $\{X_n\}$, X be $\mathcal{E}$-measurable functions and let $p \in [1, \infty[$. Assume that:*

(i) $X_n \to X$ *$\mathbb{P}$-a.e. $x \in \Omega$.*

(ii) $\int_\Omega |X_n|^p\, dx \to \int_\Omega |X|^p\, dx$.

Prove that $\int_\Omega |X_n - X|^p\, dx \to 0$.

Solution. For any positive n the function $Y_n := 2^{p-1}(|X|^p + |X_n|^p) - |X_n - X|^p$ is non-negative. Moreover, as $n \to \infty$, Y_n converges to $2^p|X|^p$ $\mathbb{P}$-a.e. Thus, by the Fatou lemma

$$2^p \int_\Omega |X|^p\, dx \le \liminf_{n\to\infty} \int_\Omega \left[2^{p-1}(|X|^p + |X_n|^p) - |X_n - X|^p\right]$$

$$= 2^p \int_\Omega |X|^p\, dx - \limsup_{n\to\infty} \int_\Omega |X_n - X|^p\, dx.$$

Appendix C

Systems of linear ordinary differential equations

We assume the reader is already familiar with first- and second-order linear ordinary differential equations (ODEs) with constant coefficients, either homogeneous or not. Here, we review results on the solutions to systems of N first-order linear ODEs.

C.1 Cauchy problem

We first consider the existence and uniqueness of a C^1 solution $X : [a, b] \to \mathbb{R}^N$ to the problem

$$\begin{cases} t_0 \in [a, b], \\ X(t_0) = X_0, \\ X'(t) = \mathbf{Q}X(t) + F(t) \qquad \forall t \in [a, b] \end{cases} \tag{C.1}$$

where $(t_0, X_0) \in [a, b] \times \mathbb{R}^N$ is the given *initial datum*, $\mathbf{Q}$ is a given real $N \times N$ matrix and $F : [a, b] \to \mathbb{R}^N$ is a given continuous function.

C.1.1 Uniqueness

Lemma C.1 (Grönwall) *Let $W \in C^1(]a, b[, \mathbb{R}^N)$ satisfy the inequality*

$$|W'(t)| \le \alpha + \beta\,|W(t)| \qquad \forall t \in]a, b[$$

for some $\alpha \ge 0$ and $\beta > 0$. Then

$$|W(t)| \le \Big(|W(t_0)| + \frac{\alpha}{\beta}\Big)e^{\beta|t-t_0|} \qquad \forall t,\ t_0 \in]a, b[.$$

A First Course in Probability and Markov Chains, First Edition. Giuseppe Modica and Laura Poggiolini.

Proof. Let $\epsilon > 0$. The function $z(t) := \sqrt{\epsilon^2 + |W(t)|^2}$ is strictly positive and of class $C^1(]a, b[)$, and

$$z'(t) = \frac{(W(t)|W'(t))}{\sqrt{\epsilon^2 + |W(t)|^2}} \leq |W'(t)| \frac{|W(t)|}{\sqrt{\epsilon^2 + |W(t)|^2}}$$
$$\leq |W'(t)| \leq \alpha + \beta|W(t)| \leq \alpha + \beta\, z(t).$$

Thus,

$$\frac{z'(t)}{\alpha + \beta z(t)} \leq 1 \qquad \forall t \in]a, b[$$

and integrating from t_0 to t we get

$$\frac{1}{\beta} \log \left| \frac{\alpha + \beta z(t)}{\alpha + \beta z(t_0)} \right| = \left| \int_{t_0}^{t} \frac{z'(s)}{\alpha + \beta z(s)}\, ds \right| \leq |t - t_0|,$$

i.e.

$$\left| \frac{\alpha + \beta z(t)}{\alpha + \beta z(t_0)} \right| \leq e^{\beta|t-t_0|} \qquad \forall t \in]a, b[.$$

Thus,

$$\beta|W(t)| \leq \alpha + \beta z(t) \leq (\alpha + \beta z(t_0))e^{\beta|t-t_0|}$$
$$= \left(\alpha + \beta\sqrt{\epsilon^2 + |W(t_0)|^2}\right) e^{\beta|t-t_0|}.$$

As $\epsilon \to 0$, we get the claim.

Theorem C.2 *Let* $\mathbf{Q} \in M_{N,N}$ *and* $F(t) \in C^0([a, b], \mathbb{R}^N)$*. Then the Cauchy problem (C.1) admits at most one solution.*

Proof. Let us introduce the maximum expansion norm of the matrix $\mathbf{Q} \in M_{N,N}$

$$||\mathbf{Q}|| := \sup_{|x|=1} |\mathbf{Q}x|. \tag{C.2}$$

so that

$$|\mathbf{Q}x| \leq ||\mathbf{Q}||\, |x| \qquad \forall x \in \mathbb{R}^N.$$

We point out that the sup in (C.2) is in fact a maximum since the map $x \mapsto |\mathbf{Q}x|$ is continuous.

Assume that $X_1(t)$ and $X_2(t)$ are two different C^1 solutions to (C.1). Then their difference $W(t) := X_1(t) - X_2(t)$ is a solution to the Cauchy problem:

$$\begin{cases} W'(t) = \mathbf{Q}W(t) & \forall t \in]a, b[, \\ W(t_0) = 0. \end{cases}$$

Thus, $|W'(t)| = |\mathbf{Q}W(t)| \leq ||\mathbf{Q}|| \, |W(t)|$ and the Grönwall lemma applied to $W(t)$ with $\alpha = 0$ and $\beta = ||\mathbf{Q}||$ yields

$$|W(t)| \leq |W(t_0)| \exp(||\mathbf{Q}|| \, |t - t_0|) \qquad \forall t \in]a, b[.$$

Since $W(t_0) = 0$, the claim is proven.

C.1.2 Existence

Let $\mathbf{Q} \in M_{N,N}(\mathbb{R})$. For each $z \in \mathbb{C}$, the power series $\sum_{k=0}^{\infty} \frac{\mathbf{Q}^k z^k}{k!}$ absolutely converges to a matrix denoted by $e^{\mathbf{Q}z}$, cf. (A.4),

$$e^{\mathbf{Q}z} := \sum_{k=0}^{\infty} \frac{\mathbf{Q}^k z^k}{k!}$$

Moreover, the series can be differentiated term by term, thus giving

$$(e^{\mathbf{Q}z})' = \sum_{k=1}^{\infty} \frac{\mathbf{Q}^k z^{k-1}}{(k-1)!} = \mathbf{Q} \sum_{k=1}^{\infty} \frac{\mathbf{Q}^{k-1} z^{k-1}}{(k-1)!} = \mathbf{Q}\, e^{\mathbf{Q}z}$$

and $e^{\mathbf{Q}(t+s)} = e^{\mathbf{Q}t} e^{\mathbf{Q}s}$ $\forall t, s \in \mathbb{R}$, see Proposition A.8. Since $e^{\mathbf{Q}0} = \mathrm{Id}$, we have the following.

Theorem C.3 *The only C^1 solution to the Cauchy problem (C.1) is the function $X : [a, b] \to \mathbb{R}^N$ defined by*

$$X(t) := e^{\mathbf{Q}(t-t_0)} X_0 + e^{\mathbf{Q}t} \int_{t_0}^{t} e^{-\mathbf{Q}s} F(s) \, ds.$$

Proof. It suffices to show that $X(t)$ satisfies all the equations in (C.1): we have

$$X(t_0) = e^{\mathbf{Q}0} X_0 = \mathrm{Id} X_0 = X_0,$$

$$X'(t) = \mathbf{Q}e^{\mathbf{Q}(t-t_0)} X_0 + \mathbf{Q}e^{\mathbf{Q}t} \int_{t_0}^{t} e^{-\mathbf{Q}s} F(s) \, ds + e^{\mathbf{Q}t} e^{-\mathbf{Q}t} F(t)$$

$$= \mathbf{Q}X(t) + F(t).$$

Corollary C.4 *Let $\mathbf{Q} \in M_{N,N}(\mathbb{R})$. Then $\mathbf{P}(t) := e^{\mathbf{Q}t}$ is the unique solution to*

$$\begin{cases} \mathbf{P}'(t) = \mathbf{Q}\mathbf{P}(t) & \forall t \in \mathbb{R}, \\ \mathbf{P}(0) = \mathrm{Id}. \end{cases} \qquad \text{(C.3)}$$

Proof. By Theorem C.3 the ith column of $\mathbf{P}(t)$ is the solution to the Cauchy problem

$$\begin{cases} X(0) = e_i, \\ X'(t) = \mathbf{Q}X(t) \end{cases} \tag{C.4}$$

where $e_i = (0, \dots, 1, \dots, 0)^T$ is the ith vector of the standard basis of $\mathbb{R}^N$. Thus $\mathbf{P}(t)$ is the only matrix solution of (C.3).

The exponential matrix $\mathbf{P}(t) = e^{\mathbf{Q}t}$ has several properties, see Proposition A.8. In particular, for any t, s the matrices $\mathbf{P}(t)$ and $\mathbf{P}(s)$ commute, since

$$\mathbf{P}(s)\mathbf{P}(t) = e^{\mathbf{Q}s}e^{\mathbf{Q}t} = e^{\mathbf{Q}(t+s)} = e^{\mathbf{Q}t}e^{\mathbf{Q}s} = \mathbf{P}(t)\mathbf{P}(s).$$

Moreover, $\mathbf{P}(t)$ commutes with $\mathbf{Q}$ and $\mathbf{P}(t)$ is invertible with inverse matrix

$$\mathbf{P}(t)^{-1} = (e^{\mathbf{Q}t})^{-1} = e^{\mathbf{Q}(-t)} = \mathbf{P}(-t) \qquad \forall t \in \mathbb{R}.$$

In particular, for any $t \in \mathbb{R}$ and $n \in \mathbb{Z}$

$$\mathbf{P}(t)^n = e^{\mathbf{Q}nt} = \mathbf{P}(nt).$$

The following proposition computes the determinant of $\mathbf{P}(t)$.

Proposition C.5 *Let* $\mathbf{Q} \in M_{N,N}(\mathbb{R})$, $\mathbf{P}(t) := e^{\mathbf{Q}t}$ *and let* $w(t) := \det \mathbf{P}(t)$. *Then* $w'(t) = \operatorname{tr}(\mathbf{Q})w(t)$ *so that*

$$w(t) = \exp(\operatorname{tr}(\mathbf{Q})t), \forall t \in \mathbb{R},$$

where $\operatorname{tr}(\mathbf{Q}) = \sum_{i=1}^N \mathbf{Q}_i^i$ *denotes the trace of the matrix* $\mathbf{Q} = (\mathbf{Q}_j^i)$.

Proof. The determinant of a matrix is a multilinear function of the columns. Let $\mathbf{P}(t) = [P_1(t) \mid P_2(t) \mid \dots \mid P_N(t)] \in M_{N,N}$. Then

$$\begin{aligned} \frac{d}{dt}\det \mathbf{P}(t) &= \det\,[P_1' \mid P_2 \mid \dots \mid P_N] \\ &\quad + \det\,[P_1 \mid P_2' \mid \dots \mid P_N] + \dots + \det\,[P_1 \mid P_2 \mid \dots \mid P_N']. \end{aligned}$$

Since $\mathbf{P}(0) = \mathrm{Id}$, we get

$$\begin{aligned} &\det\,[P_1'(0) \mid P_2(0) \mid \dots \mid P_N(0)] = (\mathbf{P}_1^1)'(0), \\ &\dots, \\ &\det\,[P_1(0) \mid P_2(0) \mid \dots \mid P_N'(0)] = (\mathbf{P}_N^N)'(0), \end{aligned}$$

so that

$$\frac{d\det \mathbf{P}(t)}{dt}(0) = \operatorname{tr}\mathbf{P}'(0) = \operatorname{tr}\mathbf{Q}. \tag{C.5}$$

Thus, recalling that $\mathbf{P}(t+s) = \mathbf{P}(s)\mathbf{P}(t)$,

$$w'(t) = \frac{d\det \mathbf{P}(t)}{dt}(t) = \frac{d\det \mathbf{P}(s+t)}{ds}(0)$$
$$= \frac{d\det (\mathbf{P}(s)\mathbf{P}(t))}{ds}(0) = \frac{d\det \mathbf{P}(s)}{ds}(0)\det \mathbf{P}(t) = \mathrm{tr}\,(\mathbf{Q})w(t).$$

Since $w(0) = 1$, we finally get $w(t) = \exp(\mathrm{tr}\,(\mathbf{Q})t)$.

C.2 Efficient computation of $e^{\mathbf{Q}t}$

The exponential matrix $\mathbf{P}(t) := e^{\mathbf{Q}t}$ can be computed via different techniques which have different degrees of numerical efficiency and accuracy also depending on the structure of $\mathbf{Q}$. This leads to a variety of methods and techniques to compute, or to approximately compute, $\mathbf{P}(t)$, see e.g. [21]. For instance, if $\mathbf{Q}$ is a Q-matrix, the *uniformization method* allows to easily approximate $\mathbf{P}(t)$ with bounded errors, see Proposition 6.23. Here we discuss two simple methods that apply to any matrix $\mathbf{Q}$.

C.2.1 Similarity methods

A first approach is by a *similarity transformation*. In fact, if $\mathbf{Q} = \mathbf{SJS}^{-1}$, i.e. if $\mathbf{Q}$ is $\mathbb{C}$-similar to a matrix $\mathbf{J}$ via an invertible matrix $\mathbf{S} \in M_{N,N}(\mathbb{C})$, then for any positive integer k, $\mathbf{Q}^k = \mathbf{SJ}^k\mathbf{S}^{-1}$, so that

$$e^{\mathbf{Q}t} = \sum_{k=0}^{\infty} \frac{\mathbf{SJ}^k\mathbf{S}^{-1}t^k}{k!} = \mathbf{S}\sum_{k=0}^{\infty} \frac{\mathbf{J}^k t^k}{k!}\mathbf{S}^{-1} = \mathbf{S}e^{\mathbf{J}t}\mathbf{S}^{-1}.$$

The computation thus boils down to the computation of $\mathbf{S}$, $\mathbf{S}^{-1}$ and the exponential of another matrix $\mathbf{J}$.

Example C.6 Assume there exists a basis $u_1, \dots, u_N$ of $\mathbb{C}^N$ of eigenvectors of $\mathbf{Q}$ and let $\lambda_1, \dots, \lambda_N$ be the corresponding eigenvalues. Then $\mathbf{Q} = \mathbf{SJS}^{-1}$ where $\mathbf{S} = [u_1|u_2|\cdots|u_n]$ and $\mathbf{J} = \mathrm{diag}(\lambda_1, \dots, \lambda_n)$. Since $\mathbf{J}$ is diagonal, one can easily show that

$$e^{\mathbf{J}t} = \mathrm{diag}(e^{\lambda_1 t}, e^{\lambda_2 t}, \dots, e^{\lambda_n t})$$

so that

$$e^{\mathbf{Q}t} = \mathbf{S}\,\mathrm{diag}(e^{\lambda_1 t}, e^{\lambda_2 t}, \dots, e^{\lambda_n t})\mathbf{S}^{-1} = \left[u_1 e^{\lambda_1 t}|u_2 e^{\lambda_2 t}|\dots|u_n e^{\lambda_n t}\right]\mathbf{S}^{-1}.$$

The same result can be proven by the following reasoning: let u be an eigenvector of $\mathbf{Q}$ and let λ be the associated eigenvalue. It is trivial to check that $x(t) := e^{\lambda t}u$ is the solution of the Cauchy problem $x' = \mathbf{Q}x$ $x(0) = u$. Thus, if there exists a basis $(u_1, \dots, u_N)$ of $\mathbb{C}^N$ where $u_1, \dots, u_N$ are eigenvectors of $\mathbf{Q}$, then the matrix

$$\mathbf{R}(t) := \Big[e^{\lambda_1 t}u_1 | e^{\lambda_2 t}u_2 | \dots | e^{\lambda_N t}u_N\Big],$$

where $\lambda_1, \dots, \lambda_N$ are the eigenvalues associated with $u_1, \dots, u_N$, is a solution to

$$\begin{cases} \mathbf{R}'(t) = \mathbf{Q}\mathbf{R}(t), \\ \mathbf{R}(0) = \Big[u_1|u_2|\dots|u_N\Big]. \end{cases}$$

Therefore, $\mathbf{S}(t) := \mathbf{R}(t)\mathbf{R}(0)^{-1}$ is a solution to

$$\begin{cases} \mathbf{S}'(t) = \mathbf{Q}\mathbf{S}(t), \\ \mathbf{S}(0) = \mathrm{Id}. \end{cases}$$

thus concluding that

$$e^{\mathbf{Q}t} = \mathbf{S}(t) = \mathbf{R}(t)\Big[u_1|u_2|\dots|u_N\Big]^{-1}.$$

A basis of $\mathbb{R}^N$ made by eigenvectors of $\mathbf{Q}$ may not exist, but in general, the *Jordan decomposition formula* holds, see e.g. [26]: one can find an invertible matrix $\mathbf{S}$ describing a change of basis in $\mathbb{C}^N$ such that $\mathbf{J} = \mathbf{S}^{-1}\mathbf{Q}\mathbf{S}$ is in the canonical Jordan form

$$\mathbf{J} = \begin{pmatrix} \boxed{\mathbf{J}_{1,1}} & 0 & 0 & \dots & 0 \\ 0 & \boxed{\mathbf{J}_{1,2}} & 0 & \dots & 0 \\ \vdots & \vdots & \vdots & \ddots & \vdots \\ 0 & 0 & 0 & \dots & \boxed{\mathbf{J}_{k,p_k}} \end{pmatrix}$$

where for any $i = 1, \dots, k$ and any $j = 1, \dots, p_i$

$$\mathbf{J}_{i,j} = \begin{cases} \lambda_i & \text{if } \mathbf{J}_{i,j} \text{ is a } 1 \times 1 \text{ matrix,} \\ \begin{pmatrix} \lambda_i & 1 & 0 & 0 & \dots & 0 \\ 0 & \lambda_i & 1 & 0 & \dots & 0 \\ 0 & 0 & \lambda_i & 1 & \dots & 0 \\ \vdots & \vdots & \vdots & \vdots & \ddots & \vdots \\ 0 & 0 & 0 & \dots & \lambda_i & 1 \\ 0 & 0 & 0 & \dots & 0 & \lambda_i \end{pmatrix} & \text{otherwise.} \end{cases}$$

Thus, one easily gets

$$e^{\mathbf{J}t} := \begin{pmatrix} \boxed{e^{\mathbf{J}_{1,1}t}} & 0 & \dots & 0 \\ 0 & \boxed{e^{\mathbf{J}_{1,2}t}} & \dots & 0 \\ \vdots & \vdots & \ddots & \vdots \\ 0 & 0 & \dots & \boxed{e^{J_{k,p_k}t}} \end{pmatrix};$$

where the matrices $e^{\mathbf{J}_{i,j}t}$ have to be computed.

If $\mathbf{J}' = \mathbf{J}_{i,j} = (\lambda)$ is 1×1, then obviously, $e^{t\mathbf{J}'} = e^{\lambda t}$. If $\mathbf{J}' = \mathbf{J}_{i,j}$ is a Jordan block of dimension $\ell \geq 2$, then

$$\mathbf{J}' = \lambda_i \mathrm{Id} + \mathbf{N}, \qquad \mathbf{N} = (\mathbf{N}^\alpha_\beta), \mathbf{N}^\alpha_\beta := \delta_{\alpha+1,\beta}.$$

Since $\mathbf{N}$ and Id commute and since $(\mathbf{N}^k)^\alpha_\beta = (\delta_{\alpha+k,\beta})$, so that $\mathbf{N}^k = 0$ for any $k \geq \ell$, we get

$$\begin{aligned} e^{\mathbf{J}'t} &= \sum_{n=0}^{\infty} \frac{(\lambda_i \mathrm{Id} + \mathbf{N})^n t^n}{n!} = \sum_{n=0}^{\infty}\sum_{k=0}^{n} \frac{t^n}{n!}\binom{n}{k}\lambda_i^{n-k}\mathbf{N}^k \\ &= \sum_{n=0}^{\infty}\sum_{k=0}^{\min(n,\ell-1)} \frac{t^n}{n!}\binom{n}{k}\lambda_i^{n-k}\mathbf{N}^k = \sum_{k=0}^{\ell-1}\sum_{n=k}^{\infty} \frac{t^n}{n!}\binom{n}{k}\lambda_i^{n-k}\mathbf{N}^k \\ &= \sum_{k=0}^{\ell-1}\left(\sum_{n=k}^{\infty}\frac{t^{n-k}\lambda_i^{n-k}}{(n-k)!}\right)\frac{t^k}{k!}\mathbf{N}^k = e^{\lambda_i t}\sum_{k=0}^{\ell-1}\frac{t^k}{k!}\mathbf{N}^k. \end{aligned}$$

Notice that $e^{\mathbf{J}t}$ depends only on the eigenvalues of $\mathbf{Q}$ and that each entry of $e^{\mathbf{J}t}$ is the product of an exponential function and a polynomial. Thus, $e^{\mathbf{Q}t} = \mathbf{S}e^{\mathbf{J}t}\mathbf{S}^{-1}$ depends only on the eigenvalues of $\mathbf{Q}$ and on its Jordan basis, which characterizes both $\mathbf{S}$ and $\mathbf{S}^{-1}$.

C.2.2 Putzer method

The Jordan decomposition of $\mathbf{Q}$ yields a satisfying analytical description of the exponential matrix $e^{\mathbf{Q}t}$. On the other hand, the computation of $\mathbf{S}$ and $\mathbf{S}^{-1}$, which depends on a basis of generalized eigenvectors of $\mathbf{Q}$, is often numerically unstable. Numerical algorithms that do not require an a priori knowledge of a basis of generalized eigenvectors of $\mathbf{Q}$ have been developed, see [21].

The following algorithm, the *Putzer method*, is often sufficiently efficient and precise. Let $\lambda_1, \dots, \lambda_N$ be the N eigenvalues of $\mathbf{Q}$. Consider the matrix

$$\mathbf{A} = \begin{pmatrix} \lambda_1 & 0 & 0 & \dots & 0 \\ 1 & \lambda_2 & 0 & \dots & 0 \\ \vdots & \vdots & \vdots & \ddots & \vdots \\ 0 & \dots & 1 & \lambda_{N-1} & 0 \\ 0 & 0 & \dots & 1 & \lambda_N \end{pmatrix}$$

and let $P(t)$ be the solution to the following Cauchy problem

$$\begin{cases} P'(t) = \mathbf{A}P(t), \\ P(0) = (1, 0, \dots, 0)^T, \end{cases}$$

i.e. $P(t)$ is the vector $P(t) = (p_1(t), p_2(t), \dots, p_N(t))^T$ whose first component is the solution to the problem

$$\begin{cases} p_1'(t) = \lambda_1 p_1(t), \\ p_1(0) = 1 \end{cases}$$

and, for $k = 1, 2, 3, \dots, N-1$

$$\begin{cases} p_{k+1}'(t) = \lambda_{k+1} p_{k+1}(t) + p_k(t), \\ p_{k+1}(0) = 0. \end{cases}$$

Notice that each component of $P(t)$ is obtained by solving a first-order linear nonhomogeneous ODE.

Let $\mathbf{M}_k$, $k = 0, 1, \dots N$, be the $N \times N$ matrices such that

$$\begin{cases} \mathbf{M}_0 := \mathrm{Id} & \text{if } k = 0, \\ \mathbf{M}_k := (\mathbf{Q} - \lambda_k \mathrm{Id})\mathbf{M}_{k-1} & \text{if } k \geq 1. \end{cases}$$

Notice that $\prod_{j=1}^N (z - \lambda_j) = \det(z\mathrm{Id} - \mathbf{Q}) =: p_{\mathbf{Q}}(z)$ is the characteristic polynomial of $\mathbf{Q}$, so that

$$\mathbf{M}_N = \prod_{j=1}^{N} (\mathbf{Q} - \lambda_j \mathrm{Id}) = p_{\mathbf{Q}}(\mathbf{Q}) = 0$$

by the Cayley–Hamilton theorem.

Proposition C.7 *With the notation above, the exponential matrix* $\mathbf{P}(t) := e^{\mathbf{Q}t}$ *is given by*

$$e^{\mathbf{Q}t} = \sum_{k=0}^{N-1} p_{k+1}(t)\mathbf{M}_k \qquad \forall t \in \mathbb{R}.$$

Proof. Let $\mathbf{Z}(t) := \sum_{k=0}^{N-1} p_{k+1}(t)\mathbf{M}_k$. It suffices to show, see Corollary C.4, that

$$\begin{cases} \mathbf{Z}'(t) = \mathbf{Q}\mathbf{Z}(t) \forall t, \\ \mathbf{Z}(0) = \mathrm{Id}. \end{cases}$$

Clearly, $\mathbf{Z}(0) = \sum_{k=0}^{N-1} p_{k+1}(0)\mathbf{M}_k = \mathbf{M}_0 = \mathrm{Id}$. Moreover,

$$\mathbf{Z}'(t) - \mathbf{Q}\mathbf{Z}(t) = \sum_{k=0}^{N-1} \left(p_{k+1}'(t)\mathbf{M}_k - p_{k+1}(t)\mathbf{Q}\mathbf{M}_k \right)$$

$$
\begin{aligned}
&= p_1'(t)\mathbf{M}_0 - p_1(t)\mathbf{Q}\mathbf{M}_0 \\
&\quad + \sum_{k=1}^{N-1}\Big(p_k(t)\mathbf{M}_k + \lambda_{k+1}p_{k+1}(t)\mathbf{M}_k - p_{k+1}(t)\mathbf{Q}\mathbf{M}_k\Big) \\
&= p_1(t)\Big(\lambda_1\mathrm{Id} - \mathbf{Q}\Big)\mathbf{M}_0 + \sum_{k=1}^{N-1} p_k(t)\mathbf{M}_k \\
&\quad + \sum_{k=1}^{N-1} p_{k+1}(t)\Big(\lambda_{k+1}\mathrm{Id} - \mathbf{Q}\Big)\mathbf{M}_k \\
&= -p_1(t)\mathbf{M}_1 + \sum_{k=1}^{N-1}\Big(p_k(t)\mathbf{M}_k - p_{k+1}(t)\mathbf{M}_{k+1}\Big) \\
&= -p_1(t)\mathbf{M}_1 + p_1(t)\mathbf{M}_1 - p_N(t)\mathbf{M}_N = 0.
\end{aligned}
$$

C.3 Continuous semigroups

Definition C.8 *A map* $\mathbf{P} : \mathbb{R}_+ \to M_{N,N}$ *is called a* continuous semigroup *on* $\mathbb{R}_+$ *if* $\mathbf{P}$ *is continuous at* 0^+ *and*

$$
\begin{cases}
\mathbf{P}(t+s) = \mathbf{P}(t)\mathbf{P}(s) & \forall t, s \geq 0, \\
\mathbf{P}(0) = \mathrm{Id}.
\end{cases}
\tag{C.6}
$$

As a consequence of Corollary C.4 and of the properties of the matrix exponential, Proposition A.8, one immediately sees the following.

Proposition C.9 *The exponential matrix* $\mathbf{P}(t) := e^{\mathbf{Q}t}$, $t \in \mathbb{R}$ *is a continuous semigroup on* $\mathbb{R}_+$.

Proof. Obviously, the map $t \mapsto \mathbf{P}(t) = e^{\mathbf{Q}t}$ is continuous. For any $s \in \mathbb{R}$, the function

$$
\mathbf{W}(t) := \mathbf{P}(t+s) - \mathbf{P}(t)\mathbf{P}(s) \qquad t \in \mathbb{R}.
$$

is a solution to the Cauchy problem

$$
\begin{cases}
\mathbf{W}'(t) = \mathbf{Q}\mathbf{W}(t) & \forall t \in \mathbb{R}, \\
\mathbf{W}(0) = 0,
\end{cases}
$$

Therefore, $\mathbf{W}(t) = 0$ $\forall t \in \mathbb{R}$, i.e. $\mathbf{P}(t+s) = \mathbf{P}(t)\mathbf{P}(s)$ $\forall t \in \mathbb{R}$. Since s is arbitrary, the claim follows.

The converse is also true and the following holds.

Theorem C.10 *A continuous semigroup* $\mathbf{P} : \mathbb{R}_+ \to M_{N,N}$ *on* $\mathbb{R}_+$ *is differentiable at every* $t \in \mathbb{R}_+$ *and, setting*

$$\mathbf{Q} := \mathbf{P}'_+(0) := \lim_{t\to 0^+} \frac{\mathbf{P}(t) - \mathrm{Id}}{t},$$

we have

$$\mathbf{P}(t) = e^{\mathbf{Q}t} \qquad \forall t \geq 0.$$

Proof. We first show that $t \to \mathbf{P}(t)$ is continuous at any $t > 0$. Since $\mathbf{P}(h) \to \mathrm{Id}$ as $h \to 0^+$, there exists a right-hand side neighbourhood of $h = 0$ such that $\mathbf{P}(h)$ is invertible for any h in such neighbourhood. Moreover, $\mathbf{P}(h)^{-1} \to \mathrm{Id}$ as $h \to 0^+$ and, by the semigroup property,

$$\mathbf{P}(t+h) = \mathbf{P}(t)\mathbf{P}(h) \to \mathbf{P}(t) \qquad \text{as } h \to 0^+,$$

$$\mathbf{P}(t-h) = \mathbf{P}(t)\mathbf{P}(h)^{-1} \to \mathbf{P}(t) \qquad \text{as } h \to 0^+.$$

Thus $t \mapsto \mathbf{P}(t)$ is continuous at any $t \geq 0$.

The differentiability follows by Theorem C.12, hence the matrix $\mathbf{Q} := (\mathbf{P})'_+(0)$ exists. Since

$$\frac{\mathbf{P}(t+s) - \mathbf{P}(t)}{s} = \frac{\mathbf{P}(s)\mathbf{P}(t) - \mathbf{P}(t)}{s} = \frac{\mathbf{P}(s) - \mathrm{Id}}{s}\mathbf{P}(t),$$

as $s \to 0^+$ we obtain that $\mathbf{P}(t)$ is a solution to the Cauchy problem

$$\begin{cases} \mathbf{P}'(t) = \mathbf{Q}\mathbf{P}(t) & \forall t \geq 0, \\ \mathbf{P}(0) = \mathrm{Id}. \end{cases}$$

Thus, by Proposition C.9,

$$e^{\mathbf{Q}z} := \sum_{k=0}^{\infty} \frac{\mathbf{Q}^k z^k}{k!}.$$

Lemma C.11 *Let* $f(t)$, $t \geq 0$, *be integrable, continuous at* 0 *and such that* $|f(t)| \leq C\,e^{kt}$ *for some non-negative constants* $C, k \geq 0$. *Then*

$$\int_0^\infty ne^{-ns} f(s)\,ds \to f(0) \qquad \text{as } n \to \infty.$$

Proof. Let $\epsilon > 0$. There exists $\delta > 0$ such that $|f(t) - f(0)| < \epsilon$ for any $t \in [0, \delta[$. Since $n\int_0^\infty e^{-ns}\,ds = 1$, we get

$$\begin{aligned} \left| \int_0^\infty ne^{-ns} f(s)\,ds - f(0) \right| &\leq \int_0^\infty ne^{-ns} |f(s) - f(0)|\,ds \\ &= \int_0^\delta ne^{-ns} |f(s) - f(0)|\,ds + \int_\delta^\infty ne^{-ns} |f(s) - f(0)|\,ds \\ &\leq \epsilon + Cne^{-n\delta}(1 + e^{k\delta}). \end{aligned}$$

As $n \to \infty$ we obtain

$$\left| \int_0^\infty n e^{-ns} f(s)\, ds - f(0) \right| \le \epsilon.$$

Since ϵ is arbitrary, the claim is proven.

Theorem C.12 *Let* $\mathbf{P}(t) : [0, +\infty[\to M_{N,N}(\mathbb{R})$ *be a semigroup continuous at every* $t \in \mathbb{R}_+$. *Then* $\mathbf{P}(t)$ *is differentiable.*

Proof. We first show that $||\mathbf{P}(t)||$ grows at most exponentially fast. We recall that we are considering the norm

$$||\mathbf{A}|| := \sup_{|x|=1} |\mathbf{A}x|$$

so that $||\mathbf{AB}|| \le ||\mathbf{A}||\,||\mathbf{B}||$ and $|\mathbf{A}^i_j| \le ||\mathbf{A}||\ \forall i, j$.

By the semigroup property, $||\mathbf{P}(t+s)|| \le ||\mathbf{P}(t)||\,||\mathbf{P}(s)||$, hence the function $w(t) := \sup_{s \le t} \log ||\mathbf{P}(s)||$ is nondecreasing and satisfies

$$\begin{cases} w(t+s) \le w(t) + w(s), \\ w(0) = 0. \end{cases}$$

In particular, $w(t) \le q w(1)$ for any positive rational number q. By approximation the same property holds for any non-negative t: $w(t) \le w(1)t\ \forall t$, so that $||\mathbf{P}(t)||_1 \le e^{k_0 t}$, $k_0 := w(1)$.

We now prove the claim of the theorem. For any $n > k_0$ consider the matrix

$$\mathbf{Z}_n := \int_0^\infty n e^{-ns} \mathbf{P}(s)\, ds.$$

From the above all the entries of $\mathbf{P}(t)$ grow at most exponentially fast. Thus, one applies Lemma C.11 to the sequence $\{\mathbf{Z}_n\}_{n > k_0}$ to get $\mathbf{Z}_n \to \mathbf{P}(0) = \mathrm{Id}$ as $n \to \infty$. In particular, if n is large enough, the matrix $\mathbf{Z}_n$ is invertible. Finally, consider the product

$$\begin{aligned} \mathbf{P}(t)\mathbf{Z}_n &= \mathbf{P}(t) \int_0^\infty n e^{-ns} \mathbf{P}(s)\, ds = \int_0^\infty n e^{-ns} \mathbf{P}(t)\mathbf{P}(s)\, ds \\ &= \int_0^\infty n e^{-ns} \mathbf{P}(t+s)\, ds = e^{nt} \int_t^\infty n e^{-nu} \mathbf{P}(u)\, du \\ &=: F_n(t). \end{aligned}$$

Since $F_n(t)$ is of class C^1 by the fundamental theorem of calculus, $\mathbf{P}(t) = F_n(t)\mathbf{Z}_n^{-1}$ is also of class C^1.

References

[1] Dall'Aglio, G. *Calcolo delle Probabilità*. Zanichelli, Bologna, 1987.

[2] Baclawski, K., Cerasoli, M., and Rota, G. *Introduzione alla Probabilità*. UMI, Bologna, 1984.

[3] Hajek, B. *ECE 313, Probability with Engineering Applications*. 2010. http://www.ifp.illinois.edu/~hajek/Papers/probability.html.

[4] Bañuelos, R. *Lecture Notes on Measure Theory and Probability*. 2003. http://www.math.purdue.edu/~banuelos/probability.pdf.

[5] Klenke, A. *Probability Theory*. Universitext. Springer-Verlag London Ltd, London, 2008. A comprehensive course, Translated from the 2006 German original. http://dx.doi.org/10.1007/978-1-84800-048-3.

[6] Varadhan, S. R. S. *Probability Theory*, vol. 7 of *Courant Lecture Notes in Mathematics*. New York University, Courant Institute of Mathematical Sciences, New York, 2001.

[7] Ambrosio, L., Da Prato, G., and Mennucci, A. *An Introduction to Measure Theory and Probability*. Lectures at Scuola Normale Superiore, Pisa, 2010.

[8] Feller, W. *An Introduction to Probability Theory and its Applications. Vol. I*. Third edition. John Wiley & Sons, Ltd, New York, 1968.

[9] Feller, W. *An Introduction to Probability Theory and its Applications. Vol. II*. Second edition. John Wiley & Sons, Ltd, New York, 1971.

[10] Mennucci, A., and Mitter, S. K. *Probabilità e Informazione*. Edizioni della Normale. Scuola Normale Superiore, Pisa, 2008.

[11] Hajek, B. *Communication Network Analysis*. 2006. http://www.ifp.illinois.edu/~hajek/Papers/networkanalysisDec06.pdf.

[12] Hajek, B. *An Exploration of Random Processes for Engineers*. 2009. http://www.ifp.illinois.edu/~hajek/Papers/randomprocJan09.pdf.

[13] Marsan, M., Balbo, G., Conte, G., Donatelli, S., and Franceschinis, G. *Modelling with Generalized Stochastic Petri Nets*. Wiley Series in Parallel Computing. John Wiley & Sons, Ltd, New York, 1994.

[14] Varadhan, S. R. S. *Stochastic Processes*, vol. 16 of *Courant Lecture Notes in Mathematics*. New York University, Courant Institute of Mathematical Sciences, New York, 2007.

A First Course in Probability and Markov Chains, First Edition. Giuseppe Modica and Laura Poggiolini.

[15] Vicario, E. *Metodi di Verifica e Testing*. Appunti del corso. Dipartimento di Sistemi e Informatica, University of Florence, 2008.

[16] Behrends, E. *Introduction to Markov Chains with Special Emphasis on Rapid Mixing*. Advanced Lectures in Mathematics. F. Vieweg and Son, Wiesbaden, 2000.

[17] Marinari, E., and Parisi, G. *Trattatello di Probabilità*. 2000. http://chimera.roma1.infn.it/ENZO/SISDIS/PROBABILITA/versione26.pdf.

[18] Liu, J. S. *Monte Carlo Strategies on Scientific Computing*. Springer Series in Statistics. Springer–Verlag, New York, 2001.

[19] Diaconis, P. The Markov chain Monte Carlo revolution. *Bull. AMS* 2009, 46, 179–205.

[20] Baldi, P. *Calcolo delle Probabilità e Statistica*. McGraw-Hill Libri Italia, Milan, 1992.

[21] Molery, C., and Van Loan, C. *Nineteen Dubious Ways to Compute the Exponential of a Matrix, Twenty Five Years Later*. SIAM Rev. 2003, 45. http://www.cs.cornell.edu/cv/researchpdf/19ways+.pdf.

[22] Norris, J. R. *Markov Chains*. Cambridge Series on Statistical and Probabilistic Mathematics Cambridge University Press, Cambridge, 1997.

[23] Kulkarni, V. G. *Modeling and Analysis of Stochastic Systems*. Chapman & Hall, London, 1995.

[24] Brémaud, P. *Markov Chains - Gibbs Fields, Monte Carlo Simulation and Queues*. Texts in Applied Mathematics, 31. Springer–Verlag, New York, 1999.

[25] Giaquinta, M., and Modica, G. *Mathematical Analysis, vol. 2. Approximation and Discrete Processes*. Birkhäuser, Boston, 2004.

[26] Giaquinta, M., and Modica, G. *Note di metodi matematici per Ingegneria Informatica, Edizione 2007*. Pitagora Ed., Bologna, 2007.

[27] Giaquinta, M., and Modica, G. *Mathematical Analysis, vol. 5. Foundations and Advanced Techniques for Functions of Several Variables*. Birkhäuser, Boston, 2012.

Index

A First Course in Probability and Markov Chains, First Edition. Giuseppe Modica and Laura Poggiolini.